MW00389854

Applied Univariate, Bivariate, and Multivariate Statistics Using Python

Applied Univariate, Bivariate, and Multivariate Statistics Using Python

A Beginner's Guide to Advanced Data Analysis

Daniel J. Denis

This edition first published 2021

Registered Office
John Wiley & Sons, Inc., 111 River Street, Hoboken, NJ 07030, USA

Editorial Office
111 River Street, Hoboken, NJ 07030, USA

For details of our global editorial offices, customer services, and more information about Wiley products visit us at www.wiley.com.

Wiley also publishes its books in a variety of electronic formats and by print-on-demand. Some content that appears in standard print versions of this book may not be available in other formats.

Library of Congress Cataloging-in-Publication Data
Names: Denis, Daniel J., 1974- author. | John Wiley & Sons, Inc., publisher.
Title: Applied univariate, bivariate, and multivariate statistics using Python Subtitle: A beginner's guide to advanced data analysis / Daniel J. Denis, University of Montana, Missoula, MT.
Description: Hoboken, NJ : John Wiley & Sons, Inc., 2021. | Includes bibliographical references and index.
Identifiers: LCCN 2020050202 (print) | LCCN 2020050203 (ebook) | ISBN 9781119578147 (hardback) | ISBN 9781119578178 (pdf) | ISBN 9781119578185 (epub) | ISBN 9781119578208 (ebook)
Subjects: LCSH: Statistics--Software. | Multivariate analysis. | Python (Computer program language).
Classification: LCC QA276.45.P98 D46 2021 (print) | LCC QA276.45.P98 (ebook) | DDC 519.5/302855133--dc23
LC record available at https://lccn.loc.gov/2020050202
LC ebook record available at https://lccn.loc.gov/2020050203

Cover image: © MR.Cole_Photographer/Getty Images
Cover design by Wiley

Set in 9.5/12.5 STIXTwoText by Integra Software Services, Pondicherry, India

10 9 8 7 6 5 4 3 2 1

To Kaiser

Contents

Preface

This book is an elementary beginner's introduction to **applied statistics using Python**. It for the most part assumes no prior knowledge of statistics or data analysis, though a prior introductory course is desirable. It can be appropriately used in a 16-week course in statistics or data analysis at the advanced **undergraduate** or beginning **graduate** level in fields such as **psychology**, **sociology**, **biology**, **forestry**, **education, nursing, chemistry**, **business**, **law**, and other areas where making sense of data is a priority rather than formal theoretical statistics as one may have in a more specialized program in a statistics department. Mathematics used in the book is minimal and where math is used, every effort has been made to unpack and explain it as clearly as possible. The goal of the book is to obtain **results using software rather quickly**, while at the same time not completely dismissing important conceptual and theoretical features. After all, if you do not understand what the computer is producing, then the output will be quite meaningless. For deeper theoretical accounts, the reader is encouraged to consult other sources, such as the author's more theoretical book, now in its second edition (Denis, 2021), or a number of other books on univariate and multivariate analysis (e.g., Izenman, 2008; Johnson and Wichern, 2007). The book you hold in your hands is merely meant to get your **foot in the door**, and so long as that is understood from the outset, it will be of great use to the newcomer or beginner in statistics and computing. It is hoped that you leave the book with a feeling of having better understood simple to relatively advanced statistics, while also experiencing a little bit of what Python is all about.

Python is used in performing and demonstrating data analyses throughout the book, but it should be emphasized that the **book is not a specialty on Python itself**. In this respect, the book does not contain a deep introduction to the software and nor does it go into the **language** that makes up Python computing to any significant degree. Rather, the book is much more "hands-on" in that code used is a starting point to generating useful results. That is, the code employed is that which worked for the problem under consideration and which the user can amend or adjust afterward when performing additional analyses. When it comes to coding with Python, there are usually several ways of accomplishing similar goals. In places, we also cite code used by others, assigning proper credit. There already exist a plethora of Python texts and user manuals that feature the software in much greater depth. Those users wishing to learn Python from scratch and become specialists in the software and aspire to become an efficient and general-purpose programmer should consult those sources

Applied Univariate, Bivariate, and Multivariate Statistics Using Python: A Beginner's Guide to Advanced Data Analysis, First Edition. Daniel J. Denis.

(e.g. see Guttag, 2013). For those who want some introductory exposure to Python on generating data-analytic results **and wish to understand what the software is producing**, it is hoped that the current book will be of great use.

In a book such as this, limited by a **fixed number of pages**, it is an exceedingly difficult and challenging endeavor to both instruct on statistics and software simultaneously. Attempting to cover univariate, bivariate, and multivariate techniques in a book of this size in any kind of respectable depth or completeness in coverage is, well, an impossibility. Combine this with including software options and the impossibility factor increases! However, such is the nature of books that attempt to survey a wide variety of techniques such as this one – one has to include only the **most essential of information to get the reader "going" on the techniques** and advise him or her to consult other sources for further details. Targeting the right mix of theory and software in a book like this is the most challenging part, but so long as the reader (and instructor) recognizes that this book is but a foot-in-the-door to get students "started," then I hope it will fall in the confidence band of a reasonable expectation. The reader wishing to better understand a given technique or principle will naturally find many narratives incomplete, while the reader hoping to find more details on Python will likewise find the book incomplete. On average, however, it is hoped that the current "mix" is of introductory use for the newcomer. It can be exceedingly difficult to enter the world of statistics and computing. This book will get you started. In many places, references are provided on where to go next.

Unfortunately, many available books on the market for Python are nothing more than slaps in the face to statistical theory while presenting a bunch of computer code that otherwise masks a true understanding of what the code actually accomplishes. Though **data science** is a welcome addition to the mathematical and applied scientific disciplines, and software advancements have made leaps and bounds in the area of quantitative analysis, it is also an unfortunate trend that understanding statistical theory and an actual understanding of statistical methods is sometimes taking a back seat to what we will otherwise call "generating output." **The goal of research and science is not to generate software output. The goal is, or at least should be, to understand in a deeper way whatever output that is generated.** Code can be looked up far easier than can statistical understanding. Hence, the goal of the book is to understand what the code represents (at least the important code on which techniques are run) and, to some extent at least, the underlying mathematical and philosophical mechanisms of one's analysis. We comment on this important distinction a bit later in this preface as it is very important. Each chapter of this book could easily be expanded and developed into a deeper book spanning more than 3–4 times the size of the book in entirety.

The objective of this book is to provide a pragmatic introduction to data analysis and statistics using Python, providing the reader with a starting point foot-in-the-door to understanding elementary to advanced statistical concepts while affording him or her the opportunity to apply some of these techniques using the Python language.

The book is the fourth in a series of books published by the author, all with Wiley. Readers wishing a deeper discussion of the topics treated in this book are encouraged to consult the author's first book, now in its second (and better) edition titled **Applied**

Univariate, Bivariate, and Multivariate Statistics: Understanding Statistics for Social and Natural Scientists, with Applications in SPSS and R (2021). The book encompasses a much more thorough overview of many of the techniques featured in the current book, featuring the use of both R and SPSS software. Readers wishing a book similar to this one, but instead focusing exclusively on R or SPSS, are encouraged to consult the author's other two books, **Univariate, Bivariate, and Multivariate Statistics Using R: Quantitative Tools for Data Analysis and Data Science** and **SPSS Data Analysis for Univariate, Bivariate, and Multivariate Statistics.** Each of these texts are far less theory-driven and are more similar to the current book in this regard, focusing on getting results quickly and interpreting findings for research reports, dissertations, or publication. Hence, depending on which software is preferred, readers (and instructors) can select the text best suited to their needs. Many of the data sets repeat themselves across texts. It should be emphasized, however, that all of these books are still at a relatively introductory level, even if surveying relatively advanced univariate and multivariate statistical techniques.

Features used in the book to help channel the reader's focus:

- **Bullet points** appear throughout the text. They are used primarily to detail and interpret output generated by Python. Understanding and interpreting output is a major focus of the book.
- **"Don't Forget!"**

Brief "don't forget" summaries serve to emphasize and reinforce that which is most pertinent to the discussion and to aid in learning these concepts. They also serve to highlight material that can be easily misunderstood or misapplied if care is not practiced. Scattered throughout the book, these boxes help the reader review and emphasize essential material discussed in the chapters.

- Each chapter concludes with a brief set of **exercises.** These include both conceptually-based problems that are targeted to help in mastering concepts introduced in the chapter, as well as computational problems using Python.
- Most concepts are **implicitly defined** throughout the book by introducing them in the **context** of how they are used in scientific and statistical practice. This is most appropriate for a short book such as this where time and space to unpack definitions in entirety is lacking. "Dictionary definitions" are usually grossly incomplete anyway and one could even argue that most definitions in even good textbooks often fail to capture the "essence" of the concept. It is only in seeing the term used in its proper context does one better appreciate how it is employed, and, in this sense, the reader is able to unpack the deeper intended meaning of the term. For example, defining a population as the set of objects of ultimate interest to the researcher is not enlightening. Using the word in the context of a **scientific example** is much more meaningful. Every effort in the book is made to accurately convey deeper conceptual understanding rather than rely on superficial definitions.
- Most of the book was written at the beginning of the COVID-19 pandemic of 2020 and hence it seemed appropriate to feature examples of COVID-19 in places throughout the book where possible, not so much in terms of data analysis, but rather in examples of how hypothesis-testing works and the like. In this way, it is hoped

examples and analogies "hit home" a bit more for readers and students, making the issues "come alive" somewhat rather than featuring abstract examples.

- Python code is "unpacked" and explained in many, though not all, places. Many existing books on the market contain explanations of statistical concepts (to varying degrees of precision) and then plop down a bunch of code the reader is expected to simply implement and understand. While we do not avoid this entirely, for the most part we guide the reader step-by-step through both concepts and Python code used. The goal of the book is in **understanding how statistical methods work**, not arming you with a bunch of code for which you do not understand what is behind it. Principal components code, for instance, is meaningless if you do not first understand and appreciate to some extent what components analysis is about.

Statistical Knowledge vs. Software Knowledge

Having now taught at both the undergraduate and graduate levels for the better part of fifteen years to applied students in the social and sometimes natural sciences, to the delight of my students (sarcasm), I have opened each course with a lecture of sorts on the differences between **statistical vs. software knowledge**. Very little of the warning is grasped I imagine, though the real-life experience of the warning usually surfaces later in their graduate careers (such as at thesis or dissertation defenses where they may fail to understand their own software output). I will repeat some of that sermon here. While this distinction, historically, has always been important, it is perhaps no more important than in the present day given the influx of computing power available to virtually every student in the sciences and related areas, and the relative ease with which such computing power can be implemented. Allowing a new teen driver to drive a Dodge Hellcat with upward of 700 horsepower would be unwise, yet newcomers to statistics and science, from their first day, have such access to the equivalent in computing power. The statistician is shaking his or her head in disapproval, for good reason. We live in an age where data analysis is available to virtually anybody with a laptop and a few lines of code. The code can often easily be dug up in a matter of seconds online, even with very little software knowledge. And of course, with many software programs coding is not even a requirement, as windows and GUIs (graphical user interfaces) have become very easy to use such that one can obtain an analysis in virtually seconds or even milliseconds. Though this has its advantages, it is not always and necessarily a good thing.

On the one hand, it does allow the student of applied science to "attempt" to conduct his or her data analyses. Yet on the other, as the adage goes, **a little knowledge can be a dangerous thing**. Being a student of the history of statistics, I can tell you that before computers were widely available, conducting statistical analyses were available only to those who could drudge through computations by hand in generating their "output" (which of course took the form of paper-and-pencil summaries, not the software output we have today). These computations took hours upon hours to perform, and hence, if one were going to do a statistical analysis, **one did not embark on such an endeavor lightly**. That does not mean the final solution would be valid necessarily, but rather folks may have been more likely to give **serious thought** to their analyses before conducting them. Today, a student can run a MANOVA in literally 5 minutes using software, but, unfortunately, this does not imply the student will understand what they have done or why they have done it. **Random assignment to conditions** may have never even been performed, yet in the haste to implement the software routine, the student failed to understand or appreciate how limiting their output would be. Concepts of **experimental design** get lost in the haste to produce computer output. However, the student of the "modern age" of computing somehow "missed" this step in his or her quickness to, as it were, perform "advanced statistics." Further, the result is "statistically significant," yet the student has no idea what Wilks's lambda is or how it is computed, nor is the difference between **statistical significance** and **effect size** understood. The limitations of what the student has produced are not appreciated and faulty substantive (and often philosophically illogical) conclusions follow. I kid you not, I have been told by a new student before that the only problem with the world is a lack of computing power. Once computing power increases, experimental design will be a thing of the past, or so the student believed. Some incoming students enter my class with such perceptions, failing to realize that discovering a cure for COVID-19, for instance, is not a computer issue. It is a scientific one. Computers help, but they do not on their own resolve scientific issues. Instructors faced with these initial misconceptions from their students have a tough road to hoe ahead, especially when forcing on their students fundamental linear algebra in the first two weeks of the course rather than computer code and statistical recipes.

The problem, succinctly put, is that in many sciences, and contrary to the opinion you might expect from someone writing a data analysis text, **students learn too much on how to obtain output at the expense of understanding what the output means or the process that is important in drawing proper scientific conclusions from said output**. Sadly, in many disciplines, a course in "Statistics" would be more appropriately, and unfortunately, called "**How to Obtain Software Output**," because that is pretty much all the course teaches students to do. **How did statistics education in applied fields become so watered down?** Since when did cultivating the art of analytical or quantitative thinking not matter? Faculty who teach such courses in such a superficial style should know better and instead teach courses with a lot more "statistical thinking" rather than simply generating software output. Among students (who should not necessarily know better – that is what makes them students), there often exists the illusion that simply because one can obtain output for a multiple regression, this somehow implies a multiple regression was performed correctly in line with the researcher's scientific aims. Do you know how to conduct a multiple regression? "Yes, I know how to do it in software." **This answer is not a correct answer to knowing how to conduct a multiple regression!** One need not even understand what multiple regression is to "compute one" in software. As a consultant, I have also had a client or two from very prestigious universities email me a bunch of software output and ask me "Did I do this right?" assuming I could evaluate their code and output without first knowledge of their scientific goals and aims. "Were the statistics done correctly?" Of course, without an understanding of what they intended to do or the goals of their research, such a question is not only figuratively, but also **literally impossible to answer** aside from ensuring them that the software has a strong reputation for accuracy in number-crunching.

This overemphasis on computation, software or otherwise, is not right, and is a real problem, and is responsible for many misuses and abuses of applied statistics in virtually every field of endeavor. However, it is especially poignant in fields in the social sciences because the objects on which the statistics are computed are often **statistical** or **psychometric entities** themselves, which makes understanding how statistical modeling works even more vital to understanding what can vs. what cannot be concluded from a given statistical analysis. Though these problems are also present in fields such as biology and others, they are less poignant, since the reality of the objects in these fields is usually more agreed upon. To be blunt, a *t*-test on whether a COVID-19 vaccine works or not is not too philosophically challenging. Finding the vaccine is difficult science to be sure, but analyzing the results statistically usually does not require advanced statistics. However, a regression analysis on whether social distancing is a contributing factor to depression rates during the COVID-19 pandemic is not quite as easy on a methodological level. One is so-called "hard science" on real objects, the other might just end up being a **statistical artifact**. This is why social science students, **especially those conducting non-experimental research**, need rather deep philosophical and methodological training so they do not read "too much" into a statistical result, things the physical scientist may never have had to confront due to the nature of his or her objects of study. Establishing scientific evidence and supporting a scientific claim in many social (and even natural) sciences is exceedingly difficult, despite the myriad of journals accepting for publication a wide variety of incorrect scientific claims presumably supported by bloated statistical analyses. Just look at the methodological debates that surrounded COVID-19, which is on an object that is relatively "easy" philosophically! Step away from concrete science, throw in advanced statistical technology and complexity, and you enter a world where establishing evidence is philosophical quicksand. Many students who use statistical methods fall into these pits without even knowing it and it is the instructor's responsibility to keep them grounded in what the statistical method can vs. cannot do. I have told students countless times, "No, the statistical method cannot tell you that; it can only tell you this."

Hence, for the student of empirical sciences, they need to be acutely aware and appreciative of the deeper issues of conducting their own science. This implies a heavier emphasis on not how to conduct a billion different statistical analyses, but on understanding the issues with conducting the "basic" analyses they are performing. It is a matter of fact that many students who fill their theses or dissertations with applied statistics may nonetheless fail to appreciate that very little of scientific usefulness has been achieved. **What has too often been achieved is a blatant abuse of statistics masquerading as scientific advancement**. The student "bootstrapped standard errors" (Wow! Impressive!), but in the midst of a dissertation that is scientifically unsound or at a minimum very weak on a methodological level.

A perfect example to illustrate how statistical analyses can be abused is when performing a so-called **"mediation"** analysis (you might infer by the quotation marks that I am generally not a fan, and for a very good reason I may add). In lightning speed, a student or researcher can regress Y on X, introduce Z as a mediator, and if statistically significant, draw the conclusion that "Z mediates the relationship between Y and X." That's fine, so long as it is clearly understood that what has been established is **statistical mediation** (Baron and Kenny, 1986), and not necessarily anything more. To say that Z mediates Y and X, in a real **substantive** sense, requires, of course, much more knowledge of the variables and/or of the research context or design. It first and foremost requires defining what one means by "mediation" in the first place. Simply because one computes statistical mediation does not, in any way whatsoever, justify somehow drawing the conclusion that "**X goes through Z on its way to Y,**" or anything even remotely similar. Crazy talk! Of course, understanding this limitation should be obvious, right? Not so for many who conduct such analyses. What would such a conclusion even mean? In most cases, with most variables, it simply does not even make sense, regardless of how much statistical mediation is established. Again, this should be blatantly obvious, however many students (and researchers) are unaware of this, failing to realize or appreciate that **a statistical model cannot, by itself, impart a "process" onto variables. All a statistical model can typically do, by itself, is partition variability and estimate parameters.** Fiedler et al. (2011) recently summarized the rather obvious fact that without the validity of prior assumptions, statistical mediation is simply, and merely, **variance partitioning**. Fisher, inventor of ANOVA (analysis of variance), already warned us of this when he said of his own novel (at the time) method that ANOVA was merely a way of "**arranging the arithmetic**." Whether or not that arrangement is meaningful or not has to come from the scientist and a deep consideration of the objects on which that arrangement is being performed. This idea, that the science matters more than the statistics on which it is applied, is at risk of being lost, especially in the social sciences where statistical models regularly "run the show" (at least in some fields) due to the difficulty in many cases of operationalizing or controlling the objects of study.

Returning to our mediation example, if the **context** of the research problem lends itself to a physical or substantive definition of mediation or any other physical process, such that there is good reason to believe Z is truly, substantively, "mediating," then the statistical model can be used as establishing support for this already-presumed relation, in the same way a statistical model can be used to quantify the generational transmission of physical qualities from parent to child in regression. The process itself, however, is not due to the fitting of a statistical model. Never in the history of science or statistics has a statistical model ever **generated** a process. It merely, and potentially, has only **described** one. Many students, however, excited to have bootstrapped those standard errors in their model and all the rest of it, are apt to draw substantive conclusions based on a statistical model that simply do not hold water. In such cases, one is better off not running a statistical model at all rather than using it to draw inane philosophically egregious conclusions that can usually be easily corrected in any introduction to a philosophy of science or research methodology course. **Abusing and overusing statistics does little to advance science. It simply provides a cloak of complexity.**

So, what is the conclusion and recommendation from what might appear to be a very cynical discussion in introducing this book? **Understanding the science and statistics must come first.** Understanding what can vs. cannot be concluded from a statistical result is the "hard part," not computing something in Python, at least not at our level of computation (at more advanced levels, of course, computing can be exceptionally difficult, as evidenced by the necessity of advanced computer science degrees). Python code can always be looked up for applied sciences purposes, but "statistical understanding" cannot. At least not so easily. Before embarking on either a statistics course or a computation course, students are strongly encouraged to take a rigorous **research design** course, as well as a **philosophy of science** course, so they might better appreciate the limitations of their "claims to evidence" in their projects. Otherwise, statistics, and the computers that compute them, can be just as easily misused and abused as used correctly, and sadly, often are. Instructors and supervisors need to also better educate students on the reckless fitting of statistical models and computing inordinate amounts of statistics without careful guidance on what can vs. cannot be interpreted from such numerical measures. **Design first, statistics second.**

Statistical knowledge is not equivalent to software knowledge. One can become a proficient expert at Python, for instance, yet still not possess the scientific expertise or experience to successfully interpret output from data analyses. The difficult part is not in generating analyses (that can always be looked up). The most important thing is to interpret analyses correctly in relation to the empirical objects under investigation, and in most cases, this involves recognizing the limitations of what can vs. cannot be concluded from the data analysis.

Mathematical vs. "Conceptual" Understanding

One important aspect of learning and understanding any craft is to know where and why making **distinctions** is important, and on the opposite end of the spectrum, where divisions simply blur what is really there. One area where this is especially true is in learning, or at least "using," a technical discipline such as mathematics and statistics to better understand another subject. Many instructors of applied statistics strive to teach statistics at a "conceptual" level, which, to them at least, means making the discipline less "mathematical." This is done presumably to attract students who may otherwise be fearful of mathematics with all of its formulas and symbolism. However, this distinction, I argue, does more harm than good, and completely misses the point. The truth of the matter is that **mathematics are concepts. Statistics are likewise concepts.** Attempting to draw a distinction between two things that are the same does little good and only provides more confusion for the student.

A linear function, for example, is a concept, just as a standard error is a concept. That they are symbolized does not take away the fact that there is a softer, more malleable "idea" underneath them, to which the symbolic definition has merely attempted to define. The sooner the student of applied statistics recognizes this, the sooner he or she will stop psychologically associating mathematics with "mathematics," and instead associate with it what it really is, a form of **conceptual development and refinement of intellectual ideas.** The mathematics is usually in many cases the "packaged form" of that conceptual development. Computing a *t*-test, for instance, is not mathematics. It is arithmetic. Understanding what **occurs** in the *t*-test as the mean difference in the numerator goes toward zero (for example) is not "conceptual understanding." Rather, it is mathematics, and the fact that the concepts of mathematics can be unpacked into a more verbal or descriptive discussion only serves to delineate the concept that already exists underneath the description. Many instructors of applied statistics are not aware of this and continually foster the idea to students that mathematics is somehow separate from the conceptual development they are trying to impart onto their students. Instructors who teach statistics as a series of recipes and formulas without any conceptual development at all do a serious (almost "malpractice") disservice to their students. Once students begin to appreciate that mathematics and statistics is, in a strong sense, a branch of philosophy "rigorized," replete with premises, justifications, and proofs and other analytical arguments, they begin to see it less as "mathematics" and adopt a deeper understanding of what they are engaging in. The student should always be critical of the **a priori associations they have made to any subject or discipline**. The student who "dislikes" mathematics is quite arrogant to think they understand the object enough to know they dislike it. It is a form of discrimination. Critical reflection and rebuilding of knowledge (i.e. or at least what one assumes to already be true) is always a productive endeavor. It's all "concepts," and mathematics and statistics have done a great job at rigorizing and symbolizing tools for the purpose of communication. Otherwise, "probability," for instance, remains an elusive concept and the phrase "the result is probably not due to chance" is not **measurable**. Mathematics and statistics give us a way to measure those ideas, those concepts. As Fisher again once told us, you may not be able to avoid chance and uncertainty, but if you can measure and quantify it, you are on to something. However, measuring uncertainty in a scientific (as opposed to an abstract) context can be exceedingly difficult.

Advice for Instructors

The book can be used at either the advanced undergraduate or graduate levels, or for self-study. The book is ideal for a 16-week course, for instance one in a Fall or Spring semester, and may prove especially useful for programs that only have space or desire to feature a single data-analytic course for students. Instructors can use the book as a **primary text** or as a **supplement** to a more theoretical book that unpacks the concepts featured in this book. Exercises at the end of each chapter can be assigned weekly and can be discussed in class or reviewed by a teaching assistant in lab. The goal of the exercises should be to get students **thinking critically and creatively**, not simply getting the "right answer."

It is hoped that you enjoy this book as a gentle introduction to the world of applied statistics using Python. Please feel free to contact me at daniel.denis@umontana.edu or email@datapsyc.com should you have any comments or corrections. For data files and errata, please visit www.datapsyc.com.

Daniel J. Denis
March, 2021

1

A Brief Introduction and Overview of Applied Statistics

CHAPTER OBJECTIVES

- How probability is the basis of statistical and scientific thinking.
- Examples of statistical inference and thinking in the COVID-19 pandemic.
- Overview of how null hypothesis significance testing (NHST) works.
- The relationship between statistical inference and decision-making.
- Error rates in statistical thinking and how to minimize them.
- The difference between a point estimator and an interval estimator.
- The difference between a continuous vs. discrete variable.
- Appreciating a few of the more salient philosophical underpinnings of applied statistics and science.
- Understanding scales of measurement, nominal, ordinal, interval, and ratio.
- Data analysis, data science, and "big data" distinctions.

The goal of this first chapter is to provide a global overview of the logic behind statistical inference and how it is the basis for analyzing data and addressing scientific problems. Statistical inference, in one form or another, has existed at least going back to the Greeks, even if it was only relatively recently formalized into a complete system. What unifies virtually all of statistical inference is that of **probability**. Without probability, statistical inference could not exist, and thus much of modern day statistics would not exist either (Stigler, 1986).

When we speak of the probability of an event occurring, we are seeking to know the **likelihood** of that event. Of course, that explanation is not useful, since all we have done is replace probability with the word likelihood. What we need is a more precise definition. Kolmogorov (1903–1987) established basic **axioms of probability** and was thus influential in the mathematics of modern-day probability theory. An **axiom** in mathematics is basically a statement that is assumed to be true without requiring any proof or justification. This is unlike a **theorem** in mathematics, which is only considered true if it can be rigorously justified, usually by other allied parallel mathematical results. Though the axioms help establish the mathematics of probability, they surprisingly do not help us define exactly what probability actually **is**. Some statisticians,

Applied Univariate, Bivariate, and Multivariate Statistics Using Python: A Beginner's Guide to Advanced Data Analysis, First Edition. Daniel J. Denis.

scientists and philosophers hold that probability is a **relative frequency**, while others find it more useful to consider probability as a **degree of belief**. An example of a relative frequency would be flipping a coin 100 times and observing the number of heads that result. If that number is 40, then we might estimate the probability of heads on the coin to be 0.40, that is, 40/100. However, this number can also reflect our degree of belief in the probability of heads, by which we based our belief on a relative frequency. There are cases, however, in which relative frequencies are not so easily obtained or virtually impossible to estimate, such as the probability that COVID-19 will become a seasonal disease. Often, experts in the area have to provide good guesstimates based on prior knowledge and their clinical opinion. These probabilities are best considered **subjective probabilities** as they reflect a degree of belief or disbelief in a theory rather than a strict relative frequency. Historically, scholars who espouse that probability can be nothing more than a relative frequency are often called **frequentists**, while those who believe it is a degree of belief are usually called **Bayesians**, due to Bayesian statistics regularly employing subjective probabilities in its development and operations. A discussion of Bayesian statistics is well beyond the scope of this chapter and book. For an excellent introduction, as well as a general introduction to the rudiments of statistical theory, see Savage (1972).

When you think about it for a moment, virtually all things in the world are probabilistic. As a recent example, consider the COVID-19 pandemic of 2020. Since the start of the outbreak, questions involving probability were front and center in virtually all media discussions. That is, the undertones of probability, science, and statistical inference were virtually everywhere where discussions of the pandemic were to be had. Concepts of probability could not be avoided. The following are just a few of the questions asked during the pandemic:

- What is the **probability of contracting the virus**, and does this probability vary as a function of factors such as pre-existing conditions or age? In this latter case, we might be interested in the **conditional probability** of contracting COVID-19 given a pre-existing condition or advanced age. For example, if someone suffers from heart disease, is that person at greatest risk of acquiring the infection? That is, what is the probability of COVID-19 infection being conditional on someone already suffering from heart disease or other ailments?
- What proportion of the general population has the virus? Ideally, researchers wanted to know how many people world-wide had contracted the virus. This constituted a case of **parameter estimation**, where the parameter of interest was the proportion of cases world-wide having the virus. Since this number was unknown, it was typically estimated based on sample data by computing a **statistic** (i.e. in this case, a proportion) and using that number to infer the true population proportion. It is important to understand that the statistic in this case was a proportion, but it could have also been a different **function of the data**. For example, a percentage increase or decrease in COVID-19 cases was also a parameter of interest to be estimated via sample data across a particular period of time. In all such cases, we wish to estimate a parameter based on a statistic.
- What proportion of those who contracted the virus will die of it? That is, what is the estimated total death count from the pandemic, from beginning to end? Statistics such as these involved **projections** of death counts over a specific period of time

and relied on already established model curves from similar pandemics. Scientists who study infectious diseases have historically documented the likely (i.e. read: "probabilistic") trajectories of death rates over a period of time, which incorporates estimates of how quickly and easily the virus spreads from one individual to the next. These estimates were all statistical in nature. Estimates often included **confidence limits** and bands around projected trajectories as a means of estimating the degree of **uncertainty in the prediction**. Hence, projected estimates were in the opinion of many media types "wrong," but this was usually due to not understanding or appreciating the limits of uncertainty provided in the original estimates. Of course, uncertainty limits were sometimes quite wide, because predicting death rates was very difficult to begin with. **When one models relatively wide margins of error, one is protected, in a sense, from getting the projection truly wrong.** But of course, one needs to understand what these limits represent, otherwise they can be easily misunderstood. Were the point estimates wrong? Of course they were! We knew far before the data came in that the point projections would be off. **Virtually all point predictions will always be wrong**. The issue is whether the data fell in line with the prediction bands that were modeled (e.g. see Figure 1.1). If a modeler sets them too wide, then the model is essentially quite useless. For instance, had we said the projected number of deaths would be between 1,000 and 5,000,000 in the USA, that does not really tell us much more than we could have guessed by our own estimates not using data at all! Be wary of "sophisticated models" that tell you about the same thing (or even less!) than you could have guessed on your own (e.g. a weather model that predicts cold temperatures in Montana in December, how insightful!).

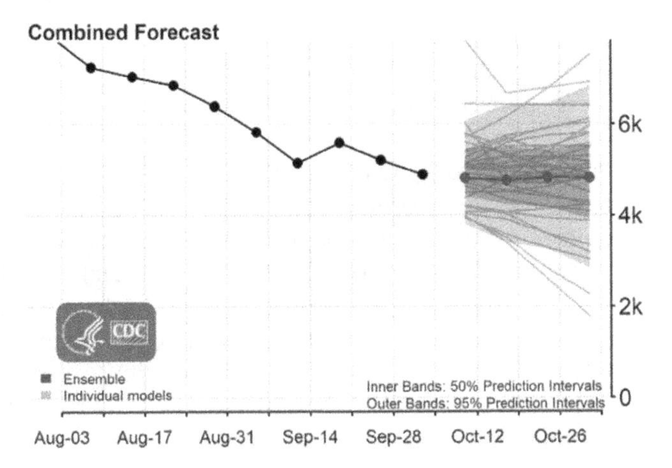

Figure 1.1 Sample death predictions in the United States during the COVID-19 pandemic in 2020. The connected dots toward the right of the plot (beyond the break in the line) represent a **point prediction** for the given period (the dots toward the left are actual deaths based on prior time periods), while the shaded area represents a **band of uncertainty**. From the current date in the period of October 2020 forward (the time in which the image was published), the shaded area increases in order to reflect greater uncertainty in the estimate. *Source: CDC (Centers for Disease Control and Prevention); Materials Developed by CDC. Used with Permission. Available at CDC* (www.cdc.gov) *free of charge.*

- **Measurement issues** were also at the heart of the pandemic (though rarely addressed by the media). What exactly constituted a COVID-19 case? Differentiating between individuals who died "of" COVID-19 vs. died "with" COVID-19 was paramount, yet was often ignored in early reports. However, the question was central to everything! "Another individual died of COVID-19" does not mean anything if we do not know the mechanism or etiology of the death. Quite possibly, COVID-19 was a **correlate** to death in many cases, not a **cause**. That is, within a typical COVID-19 death could lie a virtual infinite number of possibilities that "contributed" in a sense, to the death. Perhaps one person died primarily from the virus, whereas another person died because they already suffered from severe heart disease, and the addition of the virus simply complicated the overall health issue and overwhelmed them, which essentially caused the death.

To elaborate on the above point somewhat, **measurement issues** abound in scientific research and are extremely important, even when what is being measured is seemingly, at least at first glance, relatively simple and direct. If there are issues with how best to measure something like "COVID death," just imagine where they surface elsewhere. In psychological research, for instance, measurement is even more challenging, and in many cases adequate measurement is simply not possible. This is why some natural scientists do not give much psychological research its due (at least in particular subdivisions of psychology), because they are doubtful that the measurement of such characteristics as anxiety, intelligence, and many other things is even possible. **Self-reports** are also usually fraught with difficulty as well. Hence, assessing the degree of depression present may seem trivial to someone who believes that a self-report of such symptoms is meaningless. "But I did a complex statistical analysis using my self-report data." It doesn't matter if you haven't sold to the reader what you're analyzing was successfully measured. The most important component to a house is its **foundation**. Some scientists would require a more definite "marker" such as a biological gene or other more physical characteristic or behavioral observation before they take your ensuing statistical analysis seriously. Statistical complexity usually does not advance a science on its own. Resolution of measurement issues is more often the paramount problem to be solved.

The key point from the above discussion is that with any research, with any scientific investigation, scientists are typically interested in estimating population parameters based on information in samples. This occurs by way of probability, and hence one can say that virtually the entire edifice of statistical and scientific inference is based on the theory of probability. Even when probability is not explicitly invoked, for instance in the case of the easy result in an experiment (e.g. 100 rats live who received COVID-19 treatment and 100 control rats die who did not receive treatment), the elements of probability are still present, as we will now discuss in surveying at a very intuitive level how classical hypothesis testing works in the sciences.

1.1 How Statistical Inference Works

Armed with some examples of the COVID-19 pandemic, we can quite easily illustrate the process of statistical inference on a very practical level. The traditional and classical workhorse of statistical inference in most sciences is that of **null hypothesis**

significance testing (NHST), which originated with R.A. Fisher in the early 1920s. Fisher is largely regarded as the "father of modern statistics." Most of the classical techniques used today are due to the mathematical statistics developed in the early 1900s (and late 1800s). Fisher "packaged" the technique of NHST for research workers in agriculture, biology, and other fields, as a way to grapple with uncertainty in evaluating hypotheses and data. Fisher's contributions revolutionized how statistics are used to answer scientific questions (Denis, 2004).

Though NHST can be used in several different contexts, how it works is remarkably the same in each. A simple example will exemplify its logic. Suppose a treatment is discovered that purports to cure the COVID-19 virus and an experiment is set up to evaluate whether it does or not. Two groups of COVID-19 sufferers are recruited who agree to participate in the experiment. One group will be the **control group**, while the other group will receive the novel **treatment**. Of the subjects recruited, half will be randomly assigned to the control group, while the other half to the experimental group. This is an **experimental design** and constitutes the most rigorous means known to humankind for establishing the effectiveness of a treatment in science. Physicists, biologists, psychologists, and many others regularly use experimental designs in their work to evaluate potential treatment effects. You should too!

Carrying on with our example, we set up what is known as a **null hypothesis**, which in our case will state that the number of individuals surviving in the control group will be the same as that in the experimental group after 30 days from the start of the experiment. Key to this is understanding that the null hypothesis is about **population parameters**, not sample statistics. If the drug is not working, we would expect, under the most ideal of conditions, the same survival rates in each condition in the population under the null hypothesis. The null hypothesis in this case happens to specify a difference of zero; however, it should be noted that the null hypothesis does not always need to be about zero effect. The "null" in "null hypothesis" means it is the **hypothesis to be nullified by the statistical test**. Having set up our null, we then hypothesize a statement contrary to the null, known as the **alternative hypothesis**. The alternative hypothesis is generally of two types. The first is the **statistical alternative hypothesis**, which is essentially and quite simply a statement of the **complement** to the null hypothesis. That is, it is a statement of "not the null." Hence, if the null hypothesis is rejected, the statistical alternative hypothesis is automatically inferred. For our data, suppose after 30 days, the number of people surviving in the experimental group is equal to 50, while the number of people surviving in the control group is 20. Under the null hypothesis, we would have expected these survival rates to be equal. However, we have **observed a difference in our sample**. Since it is merely sample data, we are not really interested in this particular result specifically. Rather, we are interested in answering the following question:

What is the probability of observing a difference such as we have observed in our sample if the true difference in the population is equal to 0?

The above is the key question that repeats itself in one form or another in virtually every evaluation of a null hypothesis. That is, state a value for a parameter, then evaluate the probability of the sample result obtained in light of the null hypothesis. You might see where the argument goes from here. If the probability of the sample result is relatively high under the null, then we have no reason to reject the null hypothesis in favor of the statistical alternative. However, if the probability of the sample result is low under the null, then we take this as evidence that the null hypothesis may be false. We do not know if it is false, but we reject it because of the implausibility of the data in light of it. **A rejection of the null hypothesis does not necessarily mean the null is false.** What it does mean is that we will **act as though it is false** or potentially make scientific decisions based on its presumed falsity. Whether it is actually false or not usually remains an unknown in many cases.

For our example, if the number of people surviving in each group in our sample were equal to 50 spot on, then we definitely would not have evidence to reject the null hypothesis. Why not? Because a sample result of 50 and 50 lines up exactly with what we would expect under the null hypothesis. That is, it lines up perfectly with **expectation** under the null model. However, if the numbers turned up as they did earlier, 50 vs. 20, and we found the probability of this result to be rather small under the null, then it could be taken as evidence to possibly reject the null hypothesis and infer the alternative that the survival rates in each group are not the same. This is where the **substantive** or **research alternative hypothesis** comes in. Why were the survival rates found to be different? For our example, this is an easy one. If we did our experiment properly, it is hopefully due to the treatment. However, had we not performed a **rigorous experimental design**, then concluding the substantive or research hypothesis becomes much more difficult. That is, simply because you are able to reject a null hypothesis does not in itself lend credit to the substantive alternative hypothesis of your wishes and dreams. The substantive alternative hypothesis should naturally drop out or be a natural consequence of the rigorous approach and controls implemented for the experiment. If it does not, then drawing a substantive conclusion becomes very much more difficult if not impossible. **This is one reason why drawing conclusions from correlational research can be exceedingly difficult, if not impossible.** If you do not have a bullet-proof **experimental design**, then logically it becomes nearly impossible to know why the null was rejected. Even if you have a strong experimental design such conclusions are difficult under the best of circumstances, so if you do not have this level of rigor, you are in hot water when it comes to drawing strong conclusions. Many published research papers feature very little scientific support for purported scientific claims simply based on a rejection of a null hypothesis. This is due to many researchers not understanding or appreciating what a rejection of the null means (and what it does not mean). As we will discuss later in the book, **rejecting a null hypothesis is, usually, and by itself, no big deal at all.**

The goal of scientific research on a statistical level is generally to learn about population parameters. Since populations are usually quite large, scientists typically study statistics based on samples and make inferences toward the population based on these samples. Null hypothesis significance testing (NHST) involves putting forth a null hypothesis and then evaluating the probability of obtained sample evidence in light of that null. If the probability of such data occurring is relatively low under the null hypothesis, this provides evidence against the null and an inference toward the statistical alternative hypothesis. The substantive alternative hypothesis is the research reason for why the null was rejected and typically is known or hypothesized beforehand by the nature of the research design. If the research design is poor, it can prove exceedingly difficult or impossible to infer the correct research alternative. Experimental designs are usually preferred for this (and many other) reasons.

1.2 Statistics and Decision-Making

We have discussed thus far that a null hypothesis is typically rejected when the probability of observed data in the sample is relatively small under the posited null. For instance, with a simple example of 100 flips of a presumably fair coin, we would for certain reject the null hypothesis of fairness if we observed, for example, 98 heads. That is, the probability of observing 98 heads on 100 flips of a fair coin is very small. However, when we reject the null, we could be wrong. That is, rejecting fairness could be a mistake. Now, there is a very important distinction to make here. Rejecting the null hypothesis itself in this situation is likely to be a **good decision**. We have every reason to reject it based on the number of heads out of 100 flips. Obtaining 98 heads is more than enough statistical evidence in the sample to reject the null. However, as mentioned, a rejection of the null hypothesis does not necessarily mean the null hypothesis is false. All we have done is reject it. In other words, it is entirely possible that the coin is fair, but we simply observed an unlikely result. This is the problem with statistical inference, and that is, **there is always a chance of being wrong in our decision to reject a null hypothesis and infer an alternative**. That does not mean the rejection itself was wrong. It means simply that our decision may not turn out to be **in our favor**. In other words, we may not get a "lucky outcome." We have to live with that risk of being wrong if we are to make virtually any decisions (such as leaving the house and crossing the street or going shopping during a pandemic).

The above is an extremely important distinction and cannot be emphasized enough. Many times, researchers (and others, especially media) evaluate decisions based not on the **logic** that went into them, but rather on **outcomes**. This is a philosophically faulty way of assessing the goodness of a decision, however. The goodness of the decision should be based on whether it was made based on solid and efficient decision-making principles that a **rational agent** would make under similar circumstances, not whether the outcome happened to accord with what we hoped to see. Again, sometimes we experience lucky outcomes, sometimes we do not, even when our decision-making criteria is "spot on" in both cases. This is what the art of decision-making is all about. The following are some examples of popular decision-making events and the actual outcome of the given decision:

- The Iraq war beginning in 2003. Politics aside, a motivator for invasion was presumably whether or not Saddam Hussein possessed weapons of mass destruction. We know now that he apparently did not, and hence many have argued that the invasion and subsequent war was a mistake. However, without a proper **decision analysis** of the risks and probabilities beforehand, the quality of the decision should not be based on the lucky or unlucky outcome. For instance, if as assessed by experts in the area the probability of finding weapons of mass destruction (and that they would be used) were equal to 0.99, then the logic of the decision to go to war may have been a good one. The outcome of not finding such weapons, in the sense we are discussing, was simply an "unlucky" outcome. The decision, however, may have been correct. However, if the decision analysis revealed a low probability of having such weapons or whether they would be used, then regardless of the outcome, the actual decision would have been a poor one.

- The decision in 2020 to essentially shut down the US economy in the month of March due to the spread of COVID-19. Was it a good decision? The decision should not be evaluated based on the outcome of the spread or the degree to which it affected people's lives. The decision should be evaluated on the principles and logic that went into the decision beforehand. Whether a lucky outcome or not was achieved is a different process to the actual decision that was made. Likewise, the decision to purchase a stock then lose all of one's investment cannot be based on the outcome of oil dropping to negative numbers during the pandemic. It must be instead evaluated on the decision-making criteria that went into the decision. You may have purchased a great stock prior to the pandemic, but got an extremely unlucky and improbable outcome when the oil crash hit.
- The launch of **SpaceX** in May of 2020, returning Americans to space. On the day of the launch, there was a slight chance of lightning in the area, but the risk was low enough to go ahead with the launch. Had lightning occurred and it adversely affected the mission, it would not have somehow meant a poor decision was made. What it would have indicated above all else is that an unlucky outcome occurred. There is always a measure of **risk tolerance** in any event such as this. The goal in decision-making is generally to calibrate such risk and minimize it to an acceptable and sufficient degree.

1.3 Quantifying Error Rates in Decision-Making: Type I and Type II Errors

As discussed thus far, **decision-making is risky business**. Virtually all decisions are made with at least some degree of risk of being wrong. How that risk is distributed and calibrated, and the costs of making the wrong decision, are the components that must be considered before making the decision. For example, again with the coin, if we start out assuming the coin is fair (null hypothesis), then reject that hypothesis after obtaining a large number of heads out of 100 flips, though the decision is logical, reality itself may not agree with our decision. That is, the coin may, in reality, be fair. We simply observed a string of heads that may simply be due to chance fluctuation. Now, how are we ever to know if the coin is fair or not? That's a difficult question, since according to frequentist probabilists, we would literally need to flip the coin forever to get the true probability of heads. Since we cannot study an infinite population of coin flips, we are always restricted on betting based on the sample, and hoping our bet gets us a lucky outcome.

What may be most surprising to those unfamiliar with statistical inference, is that quite remarkably, statistical inference in science operates on the same philosophical principles as games of chance in Vegas! **Science is a gamble and all decisions have error rates.** Again, consider the idea of a potential treatment being advanced for COVID-19 in 2020, the year of the pandemic. Does the treatment work? We hope so, but if it does not, what are the risks of it not working? With every decision, there are error rates, and error rates also imply potential **opportunity costs**. Good decisions are made with an awareness of the benefits of being correct or the costs of being wrong. Beyond that, we roll the proverbial dice and see what happens.

If we set up a null hypothesis, then reject it, we risk a false rejection of the null. That is, maybe in truth the null hypothesis should not have been rejected. This type of error, **a false rejection of the null**, is what is known as a **type I error**. The probability of making a type I error is typically set at whatever the level of significance is for the statistical test. Scientists usually like to limit the type I error rate, keeping it at a nominal level such as 0.05. This is the infamous **$p < 0.05$ level**. However, this is an arbitrarily set level and there is absolutely no logic or reason to be setting it at 0.05 for every experiment you run. How the level is set should be governed by, you guessed it, your tolerance for risk of making a wrong decision. However, why is minimizing type I error rates usually preferred? Consider the COVID-19 treatment. If the null hypothesis is that it does not work, and we reject that null hypothesis, we probably want a relatively small chance of being wrong. That is, you probably do not want to be taking medication that is promised to work when it does not and nor does the scientific community want to fill their publication space with presumed treatments that in actuality are not effective. Hence, we usually wish to keep type I error rates quite low. It was R.A. Fisher, pioneer of modern-day NHST, who suggested 0.05 as a convenient level of significance. Scientists, afterward, adopted it as "gospel" without giving it further thought (Denis, 2004). As historians of statistics have argued, adopting "$p < 0.05$" was more of a social and historical phenomenon than a rational and scientific one.

However, error rates go both ways. Researchers often wish to minimize the risk of a type I error, often ignoring the **type II error** rate. A type II error is failing to reject a false null hypothesis. For our COVID-19 example, this would essentially mean failing to detect that a treatment is effective when in fact it is effective and could potentially save lives. If in reality the null hypothesis is false, yet through our statistical test we fail to detect its falsity, then we could potentially be missing out on a treatment that is effective. So-called "experimental treatments" for a disease (i.e. the "right to try") are often well-attuned to the risk of making type II errors. That is, the risk of not acting, even on something that has a relatively small probability of working out, may be high, because if it does work out, then the benefits could be substantial.

Virtually all decisions involve a certain degree of risk. A classical hypothesis test involves two error rates. The first is a type I error, which is a false rejection of the null hypothesis. The probability of making a type I error is equal to the significance level set for the test. The second is a type II error, which is failing to reject a false null hypothesis.

1.4 Estimation of Parameters

As has undoubtedly become clear at this point, statistical inference is about estimating parameters. If we regularly dealt with population data, then we would have no need for estimation. We would already know the parameters of the population and could simply describe features of the population using, aptly named, **descriptive statistics**. Referring again to our COVID-19 example, if researchers actually knew the true proportion of those suffering from the virus in the population, then we would know the parameter and could simply describe it via the population proportion. However, as discussed, we rarely if ever know the true parameters, for the reason that our populations are usually quite large and in some cases, as with the coin, may actually be

infinite in size. Hence, since we typically do not know the actual population parameters, we have to resort to using **inferential statistics** to estimate them. That is, we compute something on a sample and use that as an estimate of the population parameter. It should be remarked that the distinction between descriptive vs. inferential does not entail them to be mutually exclusive. When we compute a statistic on a sample, we can call that both a descriptive statistic as well as an inferential one, so long as we are using it for both purposes. Hence, we may compute the proportion of cases suffering from COVID-19 to be 0.01 in our sample, refer to that as a descriptive statistic because we are "describing the sample," yet when we use that statistic to infer the true proportion in the population, refer to it as an inferential statistic. Hence, it is best not to get too stuck on the meaning of "descriptive" in this case. An inferential statistic, however, typically always implies we are making an educated guess or inference toward the population.

Estimation in statistics usually operates by one of two types. **Point estimation** involves estimating a precise value of the parameter. In the case of our COVID-19 example, the precise parameter we are wishing to estimate would be the proportion of cases in the population suffering from the disease. If we obtain the value of 0.01 in the sample, for instance, we use this to estimate the true population proportion. You might think, at first glance, that point estimation is a great thing. However, we alluded to earlier why it can be problematic. How this is so is best exemplified by an example. Suppose you would like to catch a fish and resort to very primitive ways of doing so. You obtain a long stick, sharpen the end of it into a spear, and attempt to throw the spear at a fish wallowing in shallow waters. The spear is relatively sharp, so if it "hits" the fish, you are going to secure the catch. However, even intuitively, without any formality, you know there is a problem with this approach. The problem is that catching the fish will be extremely difficult, because even with a good throw, the probability of the spear hitting the fish is likely to be extremely small. It might even be close to zero. This is even if you are a skilled fisherperson!

So, what is the solution? **Build a net of course!** Instead of the spear, you choose instead to cast a net at the fish. Intuitively, you **widen your probability of catching the fish**. This idea of widening the net in this regard is referred to in statistics as **interval estimation**. Instead of estimating with merely a point (sharp spear), you widen the net in order to increase your chances of catching the fish. Though interval estimation is a fairly broad concept, in practice, one of the most common interval estimators is that of a **confidence interval**. Hence, when we compute a confidence interval, we are estimating the value of the parameter, but with a wider margin than with a point estimator. Theoretically at least, the margin of error for a point estimator is equal to zero because it allows for no "wiggle room" regarding the location of the parameter. So, what is a good margin of error? Just as the significance level of 0.05 is often used as a default significance level, 95% confidence intervals are often used. A 95% confidence interval has a 0.95 probability of capturing (or "covering") the true parameter. That is, if you took a bunch of samples and on each computed a confidence interval, 95% of them would capture the parameter. If you computed a 99% interval instead, then 99% of them would capture the parameter.

The following is a **95% confidence interval for the mean** for a z-distribution,

$$\bar{y} - 1.96\sigma_M < \mu < \bar{y} + 1.96\sigma_M$$

where $\bar{y}$ is the sample mean and σ_M is the standard error of the mean, and, when unpacked, is equal to $\sigma/\sqrt{n}$ (we will discuss this later). Notice that it is the population mean, μ, that is at the center of the interval. However, μ is not the random variable here. Rather, the sample on which $\bar{y}$ was computed is the random sample. The population parameter μ in this case is assumed to be **fixed**. What the above confidence interval is saying, probabilistically, is the following:

Over all possible samples, the probability is 0.95 that the range between

$\bar{y} - 1.96\sigma_M$ **and** $\bar{y} + 1.96\sigma_M$ **will include the true mean,** μ.

Now, it may appear at first glance that increasing the confidence level will lead to a better estimate of the parameter. That is, it might seem that increasing the confidence interval from 95% to 99%, for instance, might provide a more precise estimate. A 99% interval looks as follows:

$$\bar{y} - 2.58\sigma_M < \mu < \bar{y} + 2.58\sigma_M$$

Notice that the critical values for z are more extreme (i.e. they are larger in absolute value) for the 99% interval than for the 95% one. But, shouldn't increasing the confidence from 95% to 99% help us "narrow" in on the parameter more sharply? At first, it seems like it should. However, this interpretation is misguided and is a prime example of how intuition can sometimes lead us astray. Increasing the level of confidence, all else equal, actually **widens** the interval, not narrows it. What if we wanted full confidence, 100%? The interval, in theory, would look as follows:

$$\bar{y} - \infty\sigma_M < \mu < \bar{y} + \infty\sigma_M$$

That is, we are quite sure the true mean will fall between negative and positive infinity! A truly meaningless statement. The morale of the story is this – **if you want more confidence, you are going to have to pay for it with a wider interval**. Scientists usually like to use 95% intervals in most of their work, but there is definitely no mathematical principle that says this is the level one should use. For some problems, even a 90% interval might be appropriate. This again highlights the importance of understanding research principles, so that you can appreciate why the research paper features this or that level of confidence. **Be critical (in a good way). Ask questions of the research paper.**

1.5 Essential Philosophical Principles for Applied Statistics

As already briefly discussed, one reason why many students (especially those outside of the mathematical sciences) have an initial disdain for learning statistics is that they have the impression that statistics is mathematics, and since they dislike mathematics, they naturally believe they will dislike statistics. However, it is important to realize that **statistics is not mathematics**. Even a deeply mathematical version of statistics (the so-called field of **mathematical statistics**) is still, by itself, not mathematics! It is statistics, first and foremost. The range of mathematics used in the communication of statistical concepts varies greatly from source to source, just as it does from course to course in college. An analogy can be drawn to physics. Physics itself is not mathematics. Mathematics, however, is the most useful **vehicle** for communicating ideas in physics. For instance, the idea of the **instantaneous rate of change** is not, by itself, mathematics. However, it is very well defined in mathematics as the **derivative**. Historically it began as a concept, and evolved eventually over thousands of years into a mathematical expression so we could actually work with the philosophical concept. If this challenges your view on what mathematics actually is, that is a good thing! Most people who do not know better associate mathematics with numbers and equations. Yes, it's that too, but it's so much more. It is not number-crunching, it is a discipline of concepts rigorized for the sake of communication and manipulation, and agrees a lot (but not always) with the physical world.

Even more important than essential mathematics is probably the essential **philosophical principles** that underlie scientific and statistical analyses. You may ask what on earth a discussion of philosophy has to do with statistics and science? Everything! Now, I am not talking about the kind of philosophy where we question who we are and the meaning of life, lay in bed as Descartes did, and eventually conclude "I think, therefore I am," wear a robe, smoke a pipe, and contemplate our own existences. Most empirically-trained scientists are practicing an **empirical philosophy**. Philosophy is at the heart of all scientific and non-scientific disciplines. "**Philosophy of science**" is a branch of philosophy that, loosely put, surveys the assumptions of science and what it means to establish evidence in the sciences. Translated, it means "**thinking about what you are doing, rather than just doing it.**" Philosophy of science asks questions such as the following:

- What does it mean to establish **scientific evidence** and why is scientific evidence so valued vs. other forms of evidence?
- Why is the **experiment** usually considered the gold-standard of scientific evidence, and why is correlational research often not nearly as prized?
- Why is science **inductive** rather than **deductive**? What are some of the problems associated with induction that make drawing scientific conclusions so difficult?
- Is establishing **causation** possible, and if so, what does it mean to establish it?
- Why has **science** adopted a statistical approach to establishing evidence in most of its fields? What is so special about the statistical approach? If statistics can be so misleading, would we be better off not using them at all? Does the use of statistics advance science or hinder it?
- Do **multivariate statistics** help clarify or otherwise confuse and impede the search for scientific evidence? Do procedures such as factor analysis actually advance the cause of science or delay it due to its problematic issues? Does "finding" a cluster in cluster analysis provide evidential support for real, substantive clusters? Or, does it simply provide evidence of clusters on an abstract mathematical level?

For the conscientious reader, it does not take long to realize that the above questions are important ones to answer if one is to make any sense of the scientific evidence they have obtained from their own study of nature. Now, for relatively simple experiments with non-controversial variables, philosophical issues do not arise as much, if at all. For instance, correlating heart rate with blood pressure does not typically require extensive philosophical examination of underlying methodology. It is an easy correlation on non-controversial variables. This is why biological sciences are often considered "harder" than the softer sciences, not in their level of difficulty necessarily, but because it is generally much easier to establish convincing evidence in those sciences. A cure for COVID-19 is difficult to come by, but once it does arrive, we can visually observe people living longer who once had the disease. In some areas of social science, however, including fields such as economics, psychology, etc., establishing evidence is much harder, simply because the matter under investigation does not lend itself to such neat and nice definition and experimentation. That is, establishing convincing evidence in such fields is often quite complex, especially if non-experimental methods are used on variables for which measurement and even "existence" can be controversial. You can correlate depression with anxiety all you like, but you should also first be able to defend the idea that asking people about their depression symptoms on a questionnaire is actually a reasonable or valid way to measure something called "depression." Some would say it is not, and that self-report is a very weak way to establish evidence of anything other than, well, the self-report of what people say! Hence, finding that depression is linked to overall well-being (or lack thereof), for example, means little if we first do not agree on how these constructs are defined and measured.

Other issues that at first glance might appear purely "mathematical" have groundings in philosophical principles. For example, whether a research variable should be considered continuous or discrete in a scientific sense cannot be answered via mathematics; it must be answered through thoughtful consideration in a philosophical or scientific sense. We survey this issue now.

1.6 Continuous vs. Discrete Variables

A **variable** in mathematics is usually represented by a symbol such as x or y, etc. It is usually indexed by a subscript such as "i" to indicate that it represents the complete set of possible values that the symbol can take on. For example, for a variable x_i, the "i" implies that the variable in question can take on any of the ith values in the given set of possibilities. For instance, for a sample of 10 individuals in a room, weight is a variable. It is a variable because not everyone in the room has the same weight. Hence, x_i in this case implies that there are $i = 1$ to $i = 10$ values for weight, not all necessarily distinct from one another. Now, if everyone in the room had the same weight, such that it was a **constant** instead of a variable, then the "sub-i" index would not be required, at least not for describing this particular set of individuals.

There are two types of variables in mathematics that are requisite knowledge for understanding applied statistics, especially when it comes to generating statistical models and conducting moderate to advanced techniques. A variable is generally considered to be either discrete or continuous. A **discrete** variable, crudely defined, can take on values for which in between those values none are possible. For example, a variable with numbers 0, 1, 2 is discrete if there is no possibility of values between 0 and 1 or between 1 and 2. If, on the other hand, any values between 0 and 1 and 1 and 2 are theoretically possible, then the variable is no longer discrete. Rather, it is continuous. For a **continuous** variable then, any values are possible, even if only theoretically, as depicted in the following (below) graphical and more formal definition of continuity.

In this plot, continuity is said to exist at the given point $f(x_0)$ on the y-axis if for small changes on the x-axis, either above or below x_0 (i.e. $x_{0\,+}\,\delta_2$ or $x_{0\,-}\,\delta_2$), we have an equally

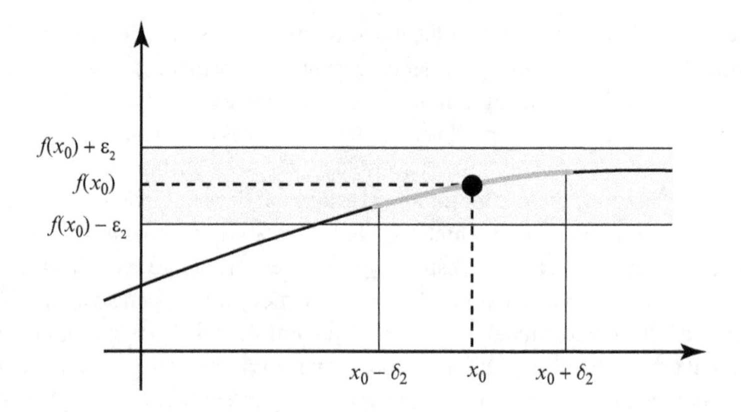

allowable small change on the y-axis (i.e. $f(x_0) + \varepsilon_2$ or $f(x_0) - \varepsilon_2$). These changes can be made extremely small, and actually as small as we wish them to be on a theoretical level. That is, changes in delta δ_2 and epsilon ε_2 can be made infinitesimally small right up to the point $f(x_0)$. **Informally, continuity implies a sense of narrowing infinitely on a given point, such that we can make smaller and smaller, in fact infinitesimally smaller, divisions.** Though we have only skimmed the formal definition of continuity here, this general idea is the most formal definition that currently exists for what continuity is mathematically. The point is to emphasize that true continuity is something that exists in theory only, and is, or at least can be, rigorously defined. For further details on this precise definition of continuity, see Bartle and Sherbert (2011), who discuss the foundations of calculus in much more detail. The foundations of calculus come generally under the topic of the branch of mathematics called **real analysis.** The essence of continuity does not lie in the mathematical definition of it, however, and has likely existed since early thought. Mathematics provided a precise definition for it so it could be used by, and communicated between, other mathematicians and scientists. Continuity, however, is first and foremost an **idea**. If you understand what we are getting at here, you will start to see mathematics in many cases as a bit more of a surface layer to deeper concepts, rather than simply the "mathematics" you may have associated earlier in your studies.

Now, in practice, one cannot measure to an infinite number of decimal places in practical research. One can never keep narrowing in on a given value by an infinite number of refined slices. Hence, while researchers may sometimes like to believe their variables have an underlying continuity to them, they are always far from being truly continuous. Only philosophically speaking (read: theoretical mathematics and its underlying philosophy) can a variable be continuous. So why is this brief discussion of continuity vs. discreteness important to the scientist? It is important since the starting point to any statistical analysis is in determining whether one's variables are best considered discrete or continuous. Though there is much more flexibility in statistical models than the following may suggest, it is nonetheless useful to give a broad overview of where traditional statistical models use continuous vs. discrete variables. When using ***z*-tests** and ***t*-tests** for means, as well as most **ANOVA-type models**, it is understood that the dependent or response variable is continuous in nature, or at minimum, has sufficient distribution such that we may consider it to be of a general continuous nature. The independent variables in these models are discrete or categorical, indicating the different populations on the response.

As an example, suppose we wanted to measure the **VO2-max** of participants treated with a new COVID-19 medication vs. those not treated. VO2-max is essentially a measure of oxygen uptake during exercise of greater intensity (e.g. a Tour de France cyclist has better VO2-max than you and I). The VO2-max variable is the response, which is considered continuous, as a function of the independent variable treatment vs. control. For this, we are in the realm of ***z*-tests** or ***t*-tests for means**, or we could also perform an ANOVA on these variables. A regression analysis is also an option since we can operationalize the independent variable as a binary dummy-coded predictor. When we flip things around, such that the grouping variable is now the response and VO2-max is the predictor, we are in the realm of **discriminant analysis** or **logistic regression** on two groups. Here, we would like to predict group membership based on the continuous predictor. Notice that these models are answering different research questions, but at their core, it stands that they must have great technical similarity. As we will see as we progress, indeed they do. **Within a *t*-test, for example, can be considered, at least on a conceptual level, to house a very primitive discriminant function!** When doing a *t*-test, we don't "see" the idea of a discriminant function simply because it is not a question we are asking. Nonetheless, it is there in concept underlying the technique. Once you understand the commonality of what underlies virtually all of these models, they will quickly lose their mystery. **You will be less inclined to survey a decision-tree using statistical methods and see different procedures. What you will rather see is one larger model with special cases and peculiarities in each method.**

Most statistical models, even if used for different research purposes and to answer different research questions, are quite technically similar at their core. One of the goals of learning and understanding statistical modeling is to grasp as quickly as possible this similarity so that you realize that differences in approach often have more to do with differences in research questions rather than differences in underlying technical details.

1.6.1 Continuity Is Not Always Clear-Cut

Having explained the distinction between continuity and discreteness at a mathematical level, at times it can be quite difficult to turn these distinctions into practice. Since, as mentioned, there are no measurably truly continuous variables in a practical sense, the question then becomes when to consider a variable as continuous or not. After all, the number of coins in my pocket can hardly be considered a continuous variable. However, for the number of coins in the entire United States, we might get away with treating the variable as continuous, even if it is not. There are so many coins that computing such things as average number of coins, a measure that assumes continuity, is not that farfetched. Even census data often reports continuous measures on otherwise discrete variables. "The average number of members per household is 3.4" the census may report. Obviously, this is nonsensical since fractions of household members cannot exist! However, since it is convenient to use an arithmetic mean to describe such things, we are implicitly treating the variable as somewhat continuous. The key take-away point is to always inquire about the data that you are computing measures of central tendency or variation on. Do not assume that because the variable is being treated as somewhat continuous in statistical computation that it is in fact continuous in its true nature. It is best to start with the premise that continuity is a **theoretical entity** and then see how far from that the presumably "continuous" measured research variable veers from it.

1.7 Using Abstract Systems to Describe Physical Phenomena: Understanding Numerical vs. Physical Differences

One of the key starting points to using and applying statistics to real phenomena is to understand and appreciate the difference between the tool you are using and the "stuff" you are applying it to. They are often not one-to-one. Simply because we represent a difference numerically does not imply that the difference exists on a physical level. Making this distinction is extremely important, especially in today's age where everything is about "data" and hence it is simply taken for granted that what we choose to measure is "real" and our measuring tool and system can capture such differences. In some cases, it can, but in others, automatically equating numerical differences with actual substantive differences is foolish.

As an example, suppose I developed a questionnaire to assess your degree of pizza preference. Suppose I scaled the questionnaire from 0 to 10, where "0" indicates a dislike for pizza and "10" indicates a strong preference. Suppose you circle "7" as your choice and your friend circles "5." Does that mean you prefer pizza more than your friend? Not necessarily. Simply because you have selected a higher number may not mean you enjoy pizza more. It may simply mean you selected a higher number. **The measured distance between 5 and 7 may not equate to an actual difference in pizza preference.**

Scales of measurement (Stevens, 1946) have been developed to try to highlight these and other issues, but, as we will see, they are far from adequate in solving the measurement problem. Everything we measure is based on a scale. We attempt to capture the phenomena and assign a numerical measurement to it. A **nominal scale** is one in which labels are simply given to values of the variable. For example, "short" vs. "tall" when measuring height would represent a variable measurable on a nominal scale. However, we can do better. Since "tall" presumably contains more height than "short," we can say **tall > short** (i.e. tall is greater than short) and assign the variable measurable on an **ordinal scale**. The next level of measurement is that of an **interval scale** in which distances between values on the scale are presumed to be equal. For example, the difference in the number of coins in my pocket from 0 to 5 is the same distance between the number of coins from 5 to 10. If the scale has an absolute zero point, meaning that a measurement of "0" actually means "zero coins," then the scale takes on the extra property of being a **ratio scale**.

A lot has been made of scales of measurement historically. Their importance is probably overstated in the literature. Where they are especially useful is in helping the researcher understand and better appreciate that simply because they obtain a number to represent something, or a difference in numbers to represent a difference in that "something," **it does not necessarily mean a precise correspondence between numbers and reality has occurred**. In many social sciences especially, assuming that such a correspondence exists is a very unrealistic idea. True, that a difference in weight from 100 pounds to 150 pounds represents the same distance as between 150 to 200 pounds, both numerically and physically, for many social variables this correspondence is likely to simply not exist, or, at a minimum, be tremendously difficult to justify. What is more, associating change on an x-axis with change on a y-axis can be done quite easily numerically, but whether it means something physically is an entirely different question.

As an example, in chemistry and nutrition, the **oxidative stability** of an oil is a measure of how quickly the oil starts to degrade when heated and exposed to light. Presumably, consumers would prefer, on this basis, an oil with more oxidative stability than less (frying at very high temperatures can apparently degrade the oil). In a recent study (Guillaume and Ravetti, 2018), it was found that the oxidative stability for olive oil was higher than the oxidative ability of, say, sunflower oil. Hence, one might be tempted to select olive oil instead of sunflower oil on this basis. However, does the difference in oxidative values translate into anything meaningful, or is it simply a measure of numerical difference that for all purposes is somewhat academic? Olive oil may be more stable, but is that "more" amount really worth not using sunflower oil if you indeed prefer sunflower? It's very easy when analyzing and interpreting data to fall into the **ranking trap**, where simply because one element ranks higher than another falsely implies a pragmatic or even meaningful increase on a physical level. The headline may be that "Olive oil is #1," but is #10 practically pretty much the same anyway, or is the **utility** of the difference in oils enough to influence one's **decision**? The ranking differences may be inconsequential to the decision. For example, if I told you your primary doctor ranked 100th out of 100 individuals graduating out of his or her graduating class, you may at first assume your doctor is not very good. However, the differences between ranking quantities may be extremely slight or so small when translated on a practical level to not matter at all or, at minimum, be negligible. Differences may even be due to measurement error and hence not exist beyond chance. Likewise, the pilot of your aircraft may be virtually as competent as the best pilot out there, but still ranks lower on an imperfect measure. **Do not simply assume that the numerical change in what is being assessed represents a meaningful difference when applied to change on a scientific (as opposed to numerical) level. Numerical differences do not necessarily equate to equivalent physical changes. Instead of being eager to include a bunch of measures into your thesis, dissertation or publication, a good idea might be to work on, and deeply validate, what is being measured in the first place.** Can something like self-esteem be measured? That is not a small or inconsequential question. You can pick up an existing questionnaire that purports to measure it or you can first critically evaluate whether it is something measurable at all. Regardless of whether we can correlate it with an existing measure does not provide fundamental validity. It only provides **statistical validity**. The ultimate **psychometric issue** may still remain. For instance, how will you convince your committee that what you have measured is actually a good measure of self-esteem?

1.8 Data Analysis, Data Science, Machine Learning, Big Data

In recent years, the "data explosion" has gripped much of science, business, and almost every field of inquiry. Thanks to computers and advanced data warehouse capacities that could have only been dreamt of in years past (and will seem trivial in years to come), the "data deluge" is officially upon us. The facility by which statistical and software analyses can be conducted has increased dramatically. New designations for quantitative analyses often fall under the names of **data science** and **machine learning**, and because data is so cheap to store, many corporations, both academic and otherwise, can collect and store massive amounts of data – so much so, that analysis of such data sometimes falls under the title of "**big data**." For example, world population data regarding COVID-19 were analyzed in an attempt to spot trends in the virus across age groups, extent of comorbidity with other illnesses, among other things. Such analyses are usually done on very large and evolving databases. The mechanisms for storing and accessing such data are, rightly so, not truly areas of "statistics" per say, and have more to do with **data engineering** and the like. The field of **machine learning**, an area primarily in **computer science**, is an emerging area that emphasizes modern software technology in analyzing data, deciphering trends, and visually depicting results via advanced and sophisticated graphics. As you venture further into data analysis in general, some of the algorithms you may use may come from this field.

Though the fields of data science, machine learning, and other allied fields are relatively new and exciting, it is nonetheless important for the reader to not simply and automatically associate new words with necessarily new "things." Human beings are creatures of **psychological association**, and so when we hear of a new term, we often create a new **category** for that term in our minds, and we assume that since there is a new word, there must be an equivalent new category. However, that does not necessarily imply the new association we have created is one-to-one with the reality of the object. The new vehicle promoted by a car company may be an older design "updated" rather than an entirely new vehicle. Hence, when you hear new terminology in quantitative areas, it is imperative that you never stop with the word alone, but instead delve in deeper to see what is actually "there" in terms of new substance. Why is this approach to understanding important? It is important because otherwise, especially as a newcomer to these areas, you may come to believe that what you are studying is entirely novel. Indeed, it may be "new," but it may not be as novel or categorically different from the "old" as you may at first think. Likewise, humanistic psychology of the 1950s was not entirely new. The Greeks had very similar ideas. The marketing was new, but the ideas were generally not.

As an example, suppose you are fitting a model to data in machine learning and are concerned about **overfitting** the model to your data, which, in general, means you are fitting a functional form that too closely matches up to the obtained data you have, potentially allowing for poor replication and generalizability if attempted. You may read about overfitting in a machine learning book and believe the concept applies to machine learning. That is, you may believe overfitting is a property of machine learning models only! How false! While it is a term often used in machine learning, it definitely is not a term specific to the field. Historically, not only has the term and concept of overfitting been used in statistics, but prior to the separation of statistics from mathematics, examples of scientists being concerned about overfitting are scattered throughout history! Hence, "overfitting" is not a concept unique to the field from which you are learning, no more than **algorithms** are unique to computer science. Historically, algorithms have existed forever, and even the Babylonians were using primitive algorithms (Knuth, 1972). If you are not at least somewhat aware of history, you may come to believe new words and terms necessarily imply new "things." The concept is usually old news, however. That does not imply the new use of the word is not at least somewhat unique and that it is not being applied to a new algorithm (for example). However, it is likely that the concept has existed well before the word was paired with the thing it is describing in a given field. Likewise, if you believe that **support vector machines** have anything to do with **machines** (and I have had students assume there must be a "machine" component within its mathematics!), you need to remember that words are imperfect descriptors for what is actually there. Indeed, much of **language** in general is nothing more than approximations to what we truly wish to communicate. As any linguist will tell you, language is far from a precise method of communication, but it is often the best we can do. Likewise, with music, a series of notes played on the piano with the goal of communicating a sentiment or emotional quality will necessarily not do so perfectly. It is an **approximation**. But how awkward it would be for the musician to follow up his or her performance with "What I meant to say was ..." or "What is really behind those notes is ..." Notions of **machine learning**, **data science**, **statistics**, **mathematics**, all conjure up associations, but you need to unpack and unravel those associations if you are to understand what is really there. In other words, just as an abstract numerical system may not perfectly coincide with the representation of physical phenomena, so it is true that an abstract linguistic system (of which you might say numerical systems might be a special case) rarely coincides perfectly with the objects it seeks to describe.

This discussion is not meant to start a "turf war" over the priority of human intellectual invention. Far from it. If we were to do that, then we would have to also acknowledge that though Newton and Leibniz put the final touches on the calculus, the idea that they "invented" it, in the truest sense of the word, is a bit of a far cry. **Priority disputes** in the history of human discovery usually prove futile and virtually impossible to resolve, even among those historians who study the most ancient of roots of intellectual invention on a full-time basis. That is, even assigning priority to ancient discoveries of intellectual concepts is exceedingly difficult (especially without lawyers!), which further provides evidence that "modern" concepts are often not modern at all. As another example, the concept of a **computer** may not have been a modern invention. Historians have shown that its primitive origin may possibly go back to Charles Babbage and the "Analytical Engine," and its concept probably goes far beyond that in years as well (Green, 2005). As the saying goes, **the only things we do not know is the history we are unaware of** or, as Mark Twain once remarked, few if any ideas are original, and can usually be traced back to earlier ones.

1.9 "Training" and "Testing" Models: What "Statistical Learning" Means in the Age of Machine Learning and Data Science

One aspect of the "data revolution" with data science and machine learning leading the way has been the emphasis on the concept of **statistical learning**. As mentioned, simply because we assign a new word or phrase to something does not necessarily mean that it represents the equivalent of something entirely new. The phrase "statistical learning" is testimony to this. In its simplest and most direct form, statistical learning simply means fitting a **model** or **algorithm** to data. The model then "learns" from the data. But what does it learn exactly? It essentially learns what its estimators (in terms of selecting optimal values for them) should be in order to maximize or minimize a function. For example, in the case of a simple linear regression, if we take a model of the type $y_i = \alpha + \beta x_i + \varepsilon_i$ and fit it to some data, the model "learns" from the data what are the best values for a and b, which are estimators for α and β, respectively. Given that we are using ordinary least-squares as our method of estimation, the regression model "learns" what a and b should be such that the sum of squared errors is kept to a minimum value (if you don't know what all this means, no worries, you'll learn it in Chapter 7). The point here for this discussion is that the "learning" or "training" consists simply of selecting scalars for a and b such that the function of minimizing the sum of squared errors is satisfied. Once that occurs, **the model is said to have learned or been "trained" from the data**. This is, at its most essential and rudimentary level, what statistical learning actually means in many (not all) contexts. If we subject that model to new data after that, thus "sharpening" its scalars, the model "updates" what its estimators should be in order to continue optimizing a function. Note that this more or less parallels the idea of human learning, in that the model (or "you") is "learning from experience" as a new experience is incorporated into knowledge. For example, a worker learns how to maximize his or her potential in a job through trial and error, otherwise known as "experience." If one day his or her boss corrects him or her, that new "data" is incorporated into the learning mechanism. If on another day the individual is reinforced for doing something right, that is also incorporated into the learning mechanism. Of course, we cannot see the scalars or estimators (they are largely metaphorical in this case), but you get the idea. **Learning "optimizes" some function though exposure to new experience**. In classical learning theory in psychology, for instance, the rat in a Skinner box learns that if he presses the lever, he will receive a pellet of food. If he doesn't press the lever, he doesn't receive food. The rat is optimizing the function (its in his little brain, and its metaphorical, we can't see it) that will allow him to distinguish which response gets the food. This is learning! When the rat is "trained" enough, he starts making predictions nearly perfectly with very few errors. So it also is with the statistical model; it does an increasingly good job at "getting it right" as it is trained on increasingly more data (i.e. more "experience"). It also "learns" from what it did wrong, just as the rat learns that if he doesn't press the lever, he doesn't eat.

Is any of this "new?" Of course not! In a very real way, pioneers of regression in the 1890s, with the likes of Karl Pearson and George Udny Yule (see Denis and Docherty, 2007), were computing these same regression coefficients on their own data, though not with the use of computers. However, back then it was not referred to as a model learning from data; it was simply seen as a novel statistical method that could help address a social problem of the day. Even earlier than that, Legendre and Gauss in the early nineteenth century (1800s) were developing the method of least-squares that would eventually be used in applying regression methods later that century. Again, they were not called statistical learning methods back then. The idea of calling them learning methods seems to have arisen mostly in statistics, but is now center stage in data science and machine learning. However, a lot of this is due to the zeitgeist of the times, where "zeitgeist" means the "spirit of the times" we are in, which is one of computers, artificial intelligence, and the idea that if we supply a model with enough data, it can eventually "fly itself" so to speak. Hence, the idea here is of "training" as well. This idea is very popular in digit recognition, in that the model is supplied with enough data that it "learns" to discriminate between whether a number is a "2" for instance, or a "4" by learning its edges and most of the rest of what makes these numbers distinct from one another. Of course, the training of every model is not always done via ordinary least-squares regression. Other models are used, and the process can get quite complex and will not always follow this simple regression idea. Sometimes an algorithm is designed to search for patterns in data, which in this case the statistical method is considered to be **unsupervised** because it has no a priori group structure to guide it as in so-called **supervised** learning. **Principal components**, **exploratory factor analysis**, and **cluster analysis** are examples of this. However, even in these cases, optimization criteria have been applied. For example, in principal components analysis, we are still maximizing values for scalars, but instead of minimizing the sum of squared (vertical) errors, we are instead maximizing the variance in the original variables subjected to the procedure (this will all become clear how this is done when we survey PCA later in the book).

Now, in the spirit of statistical learning and "training," **validating** a model has become equally emphasized, in the sense that after a model is trained on one set of data, it should be applied to a similar set of data to estimate the error rate on that new set. But what does this mean? How can we understand this idea? Easily! Here are some easy examples of where this occurs:

- The pilot learns in the simulator or test flights and then his or her knowledge is "validated" on a new flight. The pilot was "trained" in landing in a thunderstorm yesterday and now that knowledge (model) will be evaluated in a new flight on a new storm.
- Rafael Nadal, tennis player, learns from his previous match how to not make errors when returning the ball. That learning is evaluated on new data, which is a new tennis match.
- A student in a statistics class learns from the first test how to adjust his or her study strategies. That knowledge is validated on test 2 to see how much was learned.

Of course, we can go on and on. The point is that the idea of statistical learning, including concepts of machine learning, are meant to exemplify the zeitgeist we find ourselves in, which is one of increased **automation**, **computers**, **artificial intelligence**, and the idea of machines becoming more and more self-sufficient and learning themselves how to make "optimal" decisions (e.g. self-driving cars). However, what is really going on "behind the scenes" is essential mathematics and usually a problem of optimization of scalars to satisfy particular constraints imposed on the problem.

In this book, while it can be said that we do "train" models by fitting them, we do not cross-validate them on new data. Since it is essentially an introduction and primer, we do not take that additional step. However, you should know that such a step is often a good one to take if you have such data at your disposal to make cross-validation doable. In many cases, scientists may not have such cross-validation data available to them, at least not yet. Hence, "splitting the sample" into a training and test set may not be do-able due to the size of the data. However, that does not necessarily mean testing cannot be done. It can be, on a new data set that is assumed to be drawn from the same population as the original test set. Techniques for cross-validation do exist that minimize having to collect very large validation samples (e.g. see James et al., 2013). Further, to use one of our previous metaphors, validating the pilot's skill may be delayed until a new storm is available; it does not necessarily have to be done today. Hence, and in general, when you fit a model, you should always have it in mind to validate that model on new data, data that was not used in the training of the model. Why is this last point important? Quite simply because if the pilot is testing his or her skills on the same storm in which he or she was trained, it's hardly a test at all, because he or she already knows that particular storm and knows the intricacies and details of that storm, so it is not really a test of new skills; it is more akin to a test of how well he or she remembers how to deal with that specific storm and (returning to our statistical discussion) capitalizes on chance factors. This is why if you are to cross-validate a model, it should be done on new "test" data, never the original training data. If you do not cross-validate the model, you can generally expect your model fit on the training data in most cases to be more **optimistic** than not, such that it will appear that the model fits "better" than it actually would on new data. This is the primary reason why cross-validation of a model is strongly encouraged. Either way, clear communication of your results is the goal, in that if you fit a model to training data and do not cross-validate it on test data, inform your audience of this so they can know what you have done. If you do cross-validate it, likewise inform them. Hence, in this respect, it is not "essential" that you cross-validate immediately, but what is essential is that you are honest and open about what you have done with your data and clearly communicate to your readers why the current estimates of model fit are likely to be a bit inflated due to not immediately testing out the model on new data. In your next study, if you are able to collect a sample from the same population, evaluate your model on new data to see how well it fits. That will give you a more honest assessment of how good your model really is. For further details on cross-validation, see James et al. (2013), and for a more thorough and deeper theoretical treatment, see Hastie et al. (2009).

1.10 Where We Are Going From Here: How to Use This Book

This introductory chapter has surveyed a few of the more salient concepts related to applied statistics. We have surveyed and reviewed the logic of statistical inference, why inference is necessary even in the age of "big data," as well as discussed some of the fundamental principles, both mathematical and philosophical, on which applied statistics is based. In the following chapter, we begin our discussion of Python software, the software used to demonstrate many of the methods surveyed in the book. The software, however, **any software**, is not in any way a panacea to understanding what underlies its use. As we will discuss in the following chapter, what is most essential is to first understand the statistical procedures and concepts that underlie the code used to communicate and run these procedures.

Review Exercises

1. Discuss what is meant by an "**axiom**" in mathematics, and why such axioms are important to building the structure of theoretical and applied statistics.
2. Explore whether you consider **probability** to be a **relative frequency** or **degree of belief**. How would you define probability? Why would you define it in this way? Explain.
3. Summarize the overall purpose of **inferential statistics**. What is the "big picture" behind why statistical inference is necessary, even in the age of "big data?"
4. Why and how are **measurement issues** so important in science? How does this importance differ, if at all, between sciences such as physics and biology vs. psychology and economics? What are the issues at play here?
5. Explore whether **self-reports** actually tell you anything about what you are seeking to measure from an individual. Can you think up a situation where you can have full confidence that it is a valid measure? Can you think up a different situation where a self-report may not be measuring the information you seek to know from a research participant? Explain and explore.
6. Brainstorm and highlight a few of the measurement issues involved in the **COVID-19 pandemic**. What are some ways in which statistics could be misleading in reporting the status of the pandemic? Discuss as many of these as you can come up with.
7. Give a description and summary of how **null hypothesis significance testing** works and use a COVID-19 example in your discussion as an example to highlight its concepts and logic.
8. Distinguish between a **type I** and **type II error** and explain why virtually all **decisions** made in science have **error rates**. Which error rate would you consider the most important to minimize? Why?
9. Distinguish between **point estimation** and a **confidence interval**. Theoretically at least, when does a confidence interval become a point estimate?
10. Why is a basic understanding of the **philosophy of science** mandatory knowledge for the student or researcher in the applied sciences? Why is simply learning statistics and research design not enough?
11. Distinguish between **deductive** logic and **inductive** logic. Why is science necessarily inductive?
12. Explore and discuss why **causation** can be such a "sticky" topic in the sciences. Why is making causal statements so difficult if not impossible? Use an example or two in your exploration of this important issue.

13. Distinguish between a **continuous** vs. a **discrete** variable. Why is continuity in a research variable controversial? Explain.
14. How is it the case that **numerical** differences do not necessarily translate to equivalent differences on the **physical** variable being measured? Explore this issue with reference to **scales of measurement**.
15. Explore differences that may exist between **data analysis**, **data science**, **machine learning**, and **big data**. What might be the differences between these areas of investigation?

2

Introduction to Python and the Field of Computational Statistics

CHAPTER OBJECTIVES

- Brief introduction and overview of Python software and how to install it on your machine.
- Why your focus should be on science and statistics first and foremost, not software.
- How to import packages into Python.
- How to compute *z*-scores in Python.
- Compute basic statistics such as means, medians, and standard deviations.
- How to build a dataframe or load external datasets into Python.
- Survey of essential linear and matrix algebra for statistics.

This chapter provides a brief introduction and overview to Python software with an emphasis, of course, on applications in statistics and data analysis. Unlike R and SPSS, two very popular statistical software packages used by statisticians, researchers, data scientists, and others, Python is a more general **computing language,** which, though useful for statistics, was not designed specifically for statistical applications such as R and SPSS (and SAS, STATA, and others). Python is used for **general computing purposes** by those generally in computer science and programming, and hence can be employed to generate a variety of computer programs and coding, the least of which is for statistical analyses. For many data-analytic tasks, software such as SPSS or Minitab or others provides more than sufficient capacity for analyzing data arising from the social, behavioral, medical, biological, and other sciences. Excel is also used by many for applications in finance, economics, and related areas. Having said all this, **why use Python at all then?**

The rationale for learning a programming language such as Python is similar to that of learning R. Aside from enjoyment value (some people like learning software for its own sake, the phrase "computer geek" did not evolve from nowhere!), it provides one with more flexibility and possibilities than "canned" software packages such as SPSS, SAS, etc. Furthermore, visualization capacities in R and Python far outweigh capacities in more traditional packages such as SPSS and Excel. When you learn a programming

Applied Univariate, Bivariate, and Multivariate Statistics Using Python: A Beginner's Guide to Advanced Data Analysis, First Edition. Daniel J. Denis.

language such as Python, your creative capacity and range is greatly expanded and enriched, because you are essentially engaging in the same computational "basics" as even advanced programmers use to carry out extraordinary things, from producing sophisticated video computer games to launching a **SpaceX** vehicle. Learning a language such as Python is the starting point to doing such things, whereas the same cannot be said exactly for learning more canned packages. These latter programs already come "pre-packaged" to carry out statistical operations and functions and are based on an underlying language that simplifies things to get to the statistics more quickly. Python, on the other hand, is a more **primitive computing language** (by "primitive," we do not mean more "basic" or "elementary"), and hence when you are computing in Python, you are beginning more with the actual "sticks and stones" of computing (though Python, like R, also has packages for statistics). It gives you the opportunity to start with its most basic elements and building blocks. It is what computer scientists need to do regularly to program successfully, not simply use window GUI unsure of what the computer is actually doing behind the scenes. It is an exciting venture to embark upon because learning Python even at a very elementary level, as in this book, you enter the field of computer programming in general, a primer of sorts on your way to working in **computational statistics**. Of course, you will likely be spending most of your career computing statistics, and not programming per se, but the point is that with Python, you could theoretically do a lot more if you ever so choose. Learning Python to some degree also makes it easier to learn other languages should you choose to down the line. Analogous to learning one instrument such as the guitar, learning a second instrument such as the piano, though still difficult and challenging, becomes a bit easier because the "unifier" called "music" links both of them. If you learn Python, you can learn other languages as well (though some will be much more challenging).

Contrary to SPSS or similar packages, when you learn Python, you are learning a very general-purpose computing language that can be used far beyond statistical applications. Though in this book our focus is on statistical modeling using Python, the sky is the limit in using Python for many programming needs you may come across in your career. Learning Python also allows you to generalize to other languages as well, such as Java. Though the languages are different, the learning curve to mastering other languages is greatly accelerated once you learn one of them.

2.1 The Importance of Specializing in Statistics and Research, Not Python: Advice for Prioritizing Your Hierarchy

Having introduced this chapter and book featuring Python, I am now going to tell you why, as an aspiring scientist, you should probably not "specialize" in Python or any other software specifically. What I mean by this is that you should not claim Python as "your software" and close your doors to other programs, nor should you aim to necessarily master Python if your goal is that of being a scientist. Why not? Allow me to explain.

As a scientist, your primary obligation and specialization should be to your **field of study**. For example, if you are a biologist, then your focus should be on, well, biology. Perhaps you are passionate about the study of wolves. You thus busy yourself with the study of such animals, focusing perhaps on their habitat, socialization patterns, and other fascinating things about these amazing animals. You do field trips to Yellowstone and follow a pack of wolves (not too closely, of course). In studying for your doctorate, however, you will likely need to conduct a study on wolves and analyze results quantitatively. You may even need to program an apparatus to study them in greater detail. The results you analyze will undoubtedly require the use and interpretation of statistics, which will thus require the use of software to conduct such analyses.

Hence, we already have a **hierarchy of specialization**, from biologist, to statistician, to computer programmer. Now, you are primarily a biologist, since, as mentioned previously, being a biologist is a full-time job. You can spend many lifetimes studying biology and only scratch the surface regarding what there is to know. However, without understanding statistics, you will not be able to interpret your scientific findings, and hence being part "applied statistician" to some degree is imperative. Otherwise excellent scientists who do not understand statistics are not excellent scientists at all, since they are unable to interpret research findings at a level of any critical depth, and are thus restricted to "trusting" to a great degree what is reported in published papers. The ideal combination is to be both an **applied scientist** and **applied statistician** (or data scientist, etc.), because it is only in having knowledge of both fields that you can confidently interpret research findings and all of its technicalities. For example, if one is unsure of what constitutes a **rigorous** vs. **non-rigorous** research design, one is simply unable to properly interpret empirical results and assign any meaning to them. Poor interpretations of experimental or non-experimental results may follow, leading to misinformed policy decisions or strategies you invoke in your clinical practice.

So what about computer programming? As is well known, computer science is a field unto itself. Computer scientists, among other things, busy themselves with writing and inventing new algorithms and developing new software. **They are specialists.** Our society is greatly influenced by such professionals, and as is true of the biologist, being a computer scientist is a full-time job. However, as a scientist and applied statistician, computer programming should correctly and most appropriately be ranked as your third priority in specialization. This is because beyond the specialized knowledge possessed by computer scientists, your goal, again as a biologist for example, is to use a programming language to address a biological problem, which, as mentioned, will often be analyzed using statistics. Hence, it is imperative to first understand the statistics you are using before applying them to the computer. This cannot be emphasized enough.

If you do not understand the logic behind the statistics you are using, then it will matter little if you can plot down a bunch of computer code; and in fact, it can be dangerous to have a lot of computer knowledge but not understand the statistics you are computing. The old adage, "knowing just enough to be dangerous," is true in this regard.

Hence, as a scientist, you are encouraged to prioritize science and statistics, and then dig for the computer code as you require it to address your scientific questions. In this respect, the computer code is the **machinery** you need to carry out what you have first intellectualized as part of your scientific reasoning process. Now, as you learn more statistics, the computational needs will likely also increase, but it is so important not to put computer programming before statistics, and statistics before your science, otherwise you run the risk of making very serious errors in interpretation of your substantive findings. Soon, the computer code and software fascination become a priority over the ignored weaknesses of your research design, which leads to faulty interpretation of data. It becomes a "gong show" of demonstrating how much computing, not science, you can do. The result? A bunch of statistical analyses on what is otherwise very weak science. Social scientists, in particular, are especially prone to this, to use statistics as a cover-up (often unknowingly) to an otherwise poor research design.

It is imperative from the outset that you do not associate software knowledge with statistical knowledge. Learning software is a separate endeavor from understanding the statistics that underlie the software, or learning how to think statistically in general. As an applied scientist, you are encouraged to first specialize in your chosen field first and foremost, then focus on understanding how statistical thinking is used in your field, and then finally learn how to implement statistical computation in software.

2.2 How to Obtain Python

With the earlier caveats in mind, we now proceed to introduce the software we will use for our analyses. **Python** is free software and is quite easily loaded onto your desktop or laptop computer. Most people who use Python access it through **Anaconda**, which consists of a number of Python computing environments. For instance, one may choose to use **IPython** or **Spyder** or other ways of accessing the software. You can access Anaconda via www.anaconda.com. From there, select "Get Started," and then "Install Anaconda Individual Edition" and download for your given operating system. In our case, we will download **Python 3.8 for Windows, 64-bit** (circled in the figure):

Once you have successfully downloaded the software, you can choose the **development environment** on which you wish to run Python. Many researchers and data scientists prefer to use the **Spyder** interface (shown in the image with the arrow pointing toward it), as it allows you to enter code and see output next to it in an adjacent window, almost analogous to using R in this regard. Others enjoy using **IPython**, while still others choose to use **Jupyter**. Whichever one you choose for our purpose and level of introduction is immaterial. They all run Python and the code presented in this book will work in any of them. Our goal for this book is not to survey computational environment possibilities, but instead to get on with learning how to use the software to perform statistical operations in analyzing one's data.

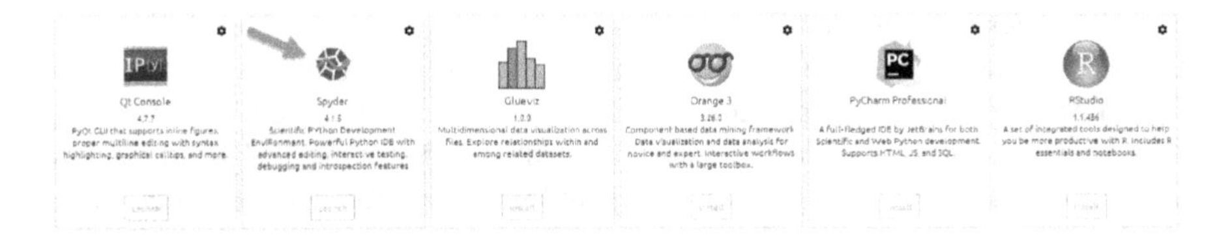

2.3 Python Packages

Similar to R software in which users can load packages to carry out specialized operations for the task at hand (e.g. one may require a particular package to carry out survival analyses, for instance), Python contains a series of packages that users can load into Python as needed. Common packages that you will see over and over include **Numpy** ("Numerical Python"), **Pandas**, **Matplotlib**, and many others. The good news is that the Anaconda distribution comes "pre-packaged" with many of these packages. Packages can be easily "imported" for use by the simple command **import**. For example, let's import **numpy**:

```
In [1]: import numpy
```

While the above will successfully import **numpy**, when we use **numpy**, we typically need to append the **numpy** name to the object we are working with. For example, suppose we wished to use **numpy** to sum a set of values for a variable. We will first give Python some data to work with and construct the variable (or vector) `x`:

```
x = numpy.array([0, 1, 2, 3, 4, 5])
numpy.sum(x)
Out[3]: 15
```

We see that Python has successfully added the numbers in the vector using `numpy.sum()`. However, instead of having to write out "numpy" each time, wouldn't it be easier to simply abbreviate it to a shorter label? This is exactly what most programmers do. That is, instead of importing **numpy** as itself, it is instead imported as "`np`":

```
import numpy as np
x = np.array([0, 1, 2, 3, 4, 5])
np.sum(x)
Out[6]: 15
```

We can see that other than identifying **numpy** as `np`, everything else in the earlier computation is the same. Hence, when you import packages, it is often useful to import them with abbreviations to make them easier to use and refer to later when you need them. Had we really wanted to, we could have imported **numpy** as:

```
import numpy as numpywhichisquitefascinating
numpywhichisquitefascinating.sum(x)
Out[8]: 15
```

Of course, doing the above would be a gigantic waste of our time. It is much easier to simply call it `np` and get on with more important things. As for **numpy**, as is true for all Python packages, there are far too many functions to summarize here. Doing good data analysis means having a book or library (or Internet connection) available to you at all times so you can dig up code and functions as you require them. **Memorizing code, or attempting to learn all code, is a waste of your time and will be a goal forever unrealized. Learning how to problem-solve given novel data situations is the skill you actually need to learn.** Have you ever noticed that your mechanic does not always have the immediate answers, but knows where to look? Likewise, your veterinarian may not immediately know the problem with your dog, but a good veterinarian will know the path to pursue to finding the answer. Your approach to data analysis and computing should be that you may not have an immediate and complete grasp of the problem, but you know where you should look to find out, and you know the difference between being successful in the search vs. not. Recall from our previous discussion, however, that this should not be your goal regarding your science or knowledge of statistics. Understanding science and associated statistics is much more difficult to "dig up" as your knowledge requirements increase. Start with the solid scientific and statistical base, then dig up code as you require it.

For example, suppose you have created an array of data, which can be easily produced by `np.array()`:

```
x = np.array([1, 5, 8, 2, 7, 4])
Out[12]: array([1, 5, 8, 2, 7, 4])
```

Suppose you then want to know the value of a given element of the array. At first glance, you may not know how to retrieve this value. It is a relatively easy task in Python, but you may still not know how to do it. That's okay! That is why you have your Python books on your shelf or are equally able to access sources online. With even a bit of digging, you will find that you can easily access the fourth element by (counting "1" as zero):

```
x = np.array([1, 5, 8, 2, 7, 4])
x[4]
Out[16]: 7
```

Notice how easy that was! You may have not known beforehand how to index that number, but in seconds, you got what you needed from the manual or Internet, and now you are all set. You are not going to "memorize" how to index, but in time, you will simply remember instead of having to look it up. In addition, if you are coming to Python from R or similar software, you may already have an idea of how to index (or perform other functions), but need a bit more detail or specifics on how to get it done in Python. That is what the Internet or a source book is for. The process is one of "trial and error." Unless you are in a computer programming class, your instructor should not be testing you on how to index or to perform any other computer functions (to use this example). What they should be testing you on is whether you are able to access resources that allow you **independence** in using this or any other software. Therefore, in this book, while we could go on listing a catalogue of ways of doing almost an infinite number of things in Python, it is much easier to refer you to books already written that contain all the functions you may need as you progress in your career. Since this book is not a book on Python per se and our focus is on **understanding statistics and scientific principles**, we only survey a few of the most useful and in-demand functions so you can get on with analyzing data quickly and efficiently. The rest will be up to you and your "digging skills" in accessing Python resources as you require them, remembering that each and every data analysis is unique and may require different functions.

As another example, suppose we needed to join or concatenate arrays in **numpy**. Suppose in addition to our `x` array, we also have a `y` array:

```
y = np.array([4, 6, 8, 2, 4, 1])
y
Out[20]: array([4, 6, 8, 2, 4, 1])
```

Now, let us concatenate the arrays using `np.concatenate()`:

```
np.concatenate([x, y])
Out[21]: array([1, 5, 8, 2, 7, 4, 4, 6, 8, 2, 4, 1])
```

Notice that the `y` array now follows the `x` array.

2.4 Installing a New Package in Python

As mentioned, Anaconda comes already installed with most of the packages you will need for this book. However, at times you may need to install a new package. You can do this quite easily through **PIP**, which is a package-management system for installing new packages. In Anaconda, go to your base root directory:

Click on the arrow, and then select "Open terminal." Once in the new terminal, you can install a new package using **PIP**. For example, suppose we wanted to install the package **pandas**, we would enter in the prompt `pip install pandas`:

```
Anaconda Prompt (anaconda3)
(base) C:\Users\sewav>pip install pandas
Requirement already satisfied: pandas in c:\users\sewav\anaconda3\lib\site-packages (1.1.3)
Requirement already satisfied: python-dateutil>=2.7.3 in c:\users\sewav\anaconda3\lib\site-packages (from pandas) (2.8.1)
Requirement already satisfied: numpy>=1.15.4 in c:\users\sewav\anaconda3\lib\site-packages (from pandas) (1.19.2)
Requirement already satisfied: pytz>=2017.2 in c:\users\sewav\anaconda3\lib\site-packages (from pandas) (2020.1)
Requirement already satisfied: six>=1.5 in c:\users\sewav\anaconda3\lib\site-packages (from python-dateutil>=2.7.3->pandas) (1.15.0)
```

Of course, since pandas comes with Anaconda already, the system reports it as "already satisfied." However, later in the book (or in your future ventures using Python), you will need to install a package that does not already come with the Anaconda distribution. Once installed (as earlier), you can then return to your favorite Python environment and when needing the package, import it. Once you are back in Spyder (or the environment you prefer), you would code "`import pandas`" (or whatever package it is) and then it will be ready for use. We will demonstrate how this works as we proceed with computational examples in the book. If the earlier method of installation via PIP is not working for you, it may be because you are not accessing the correct folder on your computer and will need to do a bit of online digging and problem solving to rectify the issue (try Googling phrases such as "installing Python packages using PIP" to learn what may be going wrong).

2.5 Computing *z*-Scores in Python

Standardizing a distribution to *z*-**scores** means converting it from a raw distribution with a given mean and variance to a distribution with a mean of zero and variance of one. To obtain a *z*-score, we first subtract the mean from every observation (thereby "centering" the data). This produces a deviation of the form $x - \mu$. To get the *z*-score, we divide this deviation by the standard deviation, as follows:

$$z = \frac{x - \mu}{\sigma}$$

Notice that the *z*-score is, in effect, generating a **ratio of one deviation to another**. In the numerator is the deviation of the given score from the mean, $x - \mu$, while in the denominator is the "average" or "standard" deviation, σ. The question then becomes, **how big is the deviation for the given score relative to the average deviation**? This is essentially what a *z*-score is telling us. When you see the *z*-score as the ratio that it is, it begins to make more sense, and you do not simply associate with it being a "standardized" score and be satisfied with that "definition." Yes, the new mean will be 0 and the standard deviation 1, but the construction of the z-score has very much meaning in it. In general, whenever you look at formulas, always break them down into components and figure out what each is communicating. What is the formula really saying? Make no mistake, it is communicating a **concept.** It is much more than a bunch of symbols.

As an easy example, suppose one has a distribution of IQ scores measured on a sample or population of individuals. Suppose the mean is equal to 110 with standard deviation of 2. Suppose also that a given individual in the data scored 114. What is that person's *z*-score? To obtain the *z*-score, we calculate:

$$z = \frac{x - \mu}{\sigma} = \frac{114 - 110}{2} = \frac{4}{2} = 2$$

We can see for a score of 114 that the individual lies 2 standard deviations above the mean. If the distribution is normal, then, as shown, we can conclude that the given individual scores higher than approximately 97.7% of the given IQ distribution (i.e. $+2\sigma$):

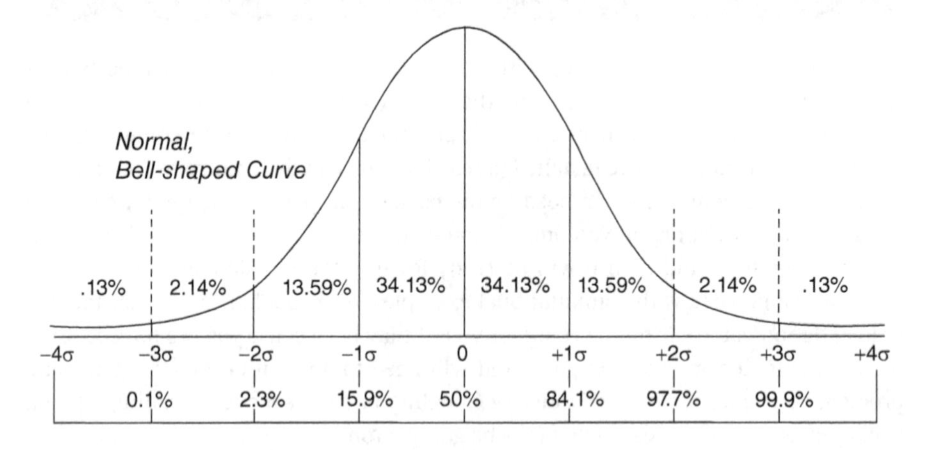

We can easily compute z-scores in Python. To demonstrate, we will first load a data set that we will refer to often throughout the book. On this fictitious data achievement scores in mathematics were recorded as a function of which teacher and textbook students were paired with (we will describe the details of how we build the dataframe a bit later and print only a few cases of the output here):

```
data = {'ac' : [70, 67, 65, 75, 76, 73, 69, 68, 70, 76, 77, 75, 85,
86, 85, 76, 75, 73, 95, 94, 89, 94, 93, 91],
'teach' : [1, 1, 1, 1, 1, 1, 2, 2, 2, 2, 2, 2, 3, 3, 3, 3, 3, 3, 4,
4, 4, 4, 4, 4],
'text' : [1, 1, 1, 2, 2, 2, 1, 1, 1, 2, 2, 2, 1, 1, 1, 2, 2, 2, 1,
1, 1, 2, 2, 2]}
import pandas as pd
df = pd.DataFrame(data)
df
Out[82]:
    ac  teach  text
0   70      1     1
1   67      1     1
2   65      1     1
3   75      1     2
4   76      1     2
5   73      1     2
6   69      2     1
7   68      2     1
8   70      2     1
9   76      2     2
10  77      2     2
```

Suppose we would like to transform our `ac` variable from the achievement data into one of z-scores. Let us begin with a computation that will compute them "manually," meaning that we are going to tell Python exactly what to do to get the scores. That is, we are going to compute a function that will compute z-scores for us:

```
import statistics
ac = df['ac']
z_numerator = (ac - statistics.mean(ac))
z_denominator = statistics.stdev(ac)
z = z_numerator/z_denominator

z
Out[33]:
0     -0.937148
1     -1.248091
2     -1.455387
3     -0.418910
4     -0.315262
5     -0.626205
```

Note in our computation we first took `ac` scores then subtracted the mean to give us `z_numerator`, then computed `z_denominator` using the **statistics** module `statistics.stdev` to get the standard deviation. The *z*-score is then computed as the numerator divided by the denominator, `z_numerator/z_denominator`. Throughout the book we present computations for various measures, but will not always describe the computation in detail as we did here. With practice, you will begin to see the code and be able to decipher what is being computed. This is true of more advanced computing as well; always investigate the code for ideas and hints as to what is being computed.

Now, as you may imagine, computing statistics the long way is inefficient and very time-consuming. We can often do much better by using **packages** directly. We now generate the *z*-scores using **scipy** by first importing `stats`:

```
from scipy import stats
ac = df['ac']
stats.zscore(ac, axis=0, ddof=1)
Out[28]:
array([-0.93714826, -1.24809146, -1.45538693, -0.41890959, -0.31526186,
       -0.62620506, -1.04079599, -1.14444373, -0.93714826, -0.31526186,
       -0.21161412, -0.41890959,  0.61756775,  0.72121548,  0.61756775,
       -0.31526186, -0.41890959, -0.62620506,  1.65404508,  1.55039735,
        1.03215868,  1.55039735,  1.44674962,  1.23945415])
```

Note how the *z*-score is requested. The function `zscore` requests `zscores` from the variable `ac` in the dataframe `df`. Had we wanted to compute *z*-scores for a different variable from the dataframe `df`, then instead of `['ac']` we would have specified whatever variable we wished. Next, let us generate a histogram of `ac` raw scores and then one on our newly computed *z*-score distribution. For this, we use `matplotlib.pyplot`:

```
import matplotlib.pyplot as plt
ac = df['ac']
z = stats.zscore(ac, axis=0, ddof=1)
ac_hist = plt.hist(ac)
ac.z_hist = plt.hist(z)
```

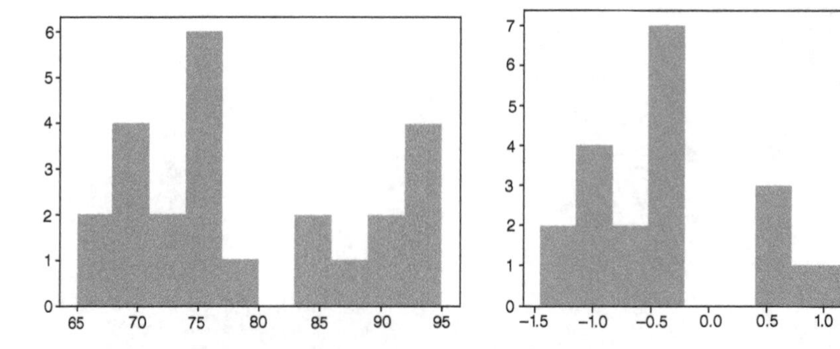

Notice that the plots, whether in raw `ac` scores or *z*-scores, have the same distribution (the scaling and grouping of intervals along the *x*-axis makes them look slightly different). **Converting to z-scores does not change the shape of the original distribution.** That is, the fact that we linearly transformed them to *z*-scores does not change its shape. The computation $z = (x - \mu) / \sigma$ simply generates a new mean (of zero) and standard deviation of one. This is an example of a linear transformation. Had we transformed this into something that is **non-linear**, then the distribution would have changed. Often students will assume that transforming to *z*-scores will "normalize" an otherwise skewed or abnormal distribution, but this is not the case. If you have a skewed distribution and wish it to be more normal in shape, then a non-linear transformation may be called for, such as taking **squares** or **square roots**. A logarithmic transformation on data may also prove useful at times.

We can also easily obtain measures of **skewness** and **kurtosis**, these coming from the package **scipy**:

```
import scipy as sp
sp.stats.skew(ac)
Out[190]: 0.3882748054677566

sp.stats.kurtosis(ac)
Out[189]: -1.2190774612249433
```

The **skewness** value of 0.388 indicates a slight **positive skew** (greater than 0), while the **kurtosis** value of −1.219 (less than 0) indicates that the distribution's tails are not so pronounced, or, roughly equivalent, that the distribution is a bit flatter than one would have with no kurtosis (the so-called **mesokurtic** distribution). This latter definition should be used with caution (see DeCarlo, 1997), but is often a useful crude and "working rule" to define kurtosis (one can use scipy's `kurtosistest()` to evaluate kurtosis inferentially). While measures of skewness and kurtosis are sometimes useful and meaningful in some work, in data analysis, graphical representations of data usually suffice. For the most part, you need not worry too much about numerical measures of skewness and kurtosis. Visualization is much more powerful and useful in this regard to detect deviations from ideal shapes.

A z-score is a comparison of one deviation to another. The deviation in the numerator is the observed deviation for the given observation from the mean of the data. The deviation in the denominator is the average deviation in the data, the so-called "standard" deviation. While converting a distribution of raw scores to standard scores generates a new distribution with mean 0 and variance 1, it does not "normalize" the distribution. Hence, the distribution of z-scores will have the same shape as the original distribution. It will not suddenly become normal due to the transformation to z-scores.

2.6 Building a Dataframe in Python: And Computing Some Statistical Functions

There are many ways of building and accessing data in Python. We will survey a few of these as we progress through the book. For now, we wish to demonstrate how a dataframe can be built from scratch by first entering the data right into our work area and then obtaining a few statistics on the dataframe to get us going. In fact, we have already done this just above when we surveyed the achievement data. Recall that we had laid out of our data on each variable and then joined it into a dataframe:

```
data = {'ac' : [70, 67, 65, 75, 76, 73, 69, 68, 70, 76, 77, 75, 85,
86, 85, 76, 75, 73, 95, 94, 89, 94, 93, 91],
'teach' : [1, 1, 1, 1, 1, 1, 2, 2, 2, 2, 2, 2, 3, 3, 3, 3, 3, 3, 4,
4, 4, 4, 4, 4],
'text' : [1, 1, 1, 2, 2, 2, 1, 1, 1, 2, 2, 2, 1, 1, 1, 2, 2, 2, 1,
1, 1, 2, 2, 2]}
import pandas as pd
df = pd.DataFrame(data)
df
Out[82]:
    ac  teach  text
0   70      1     1
1   67      1     1
2   65      1     1
```

To recap what we did earlier, the line `df = pd.DataFrame(data)` is what is creating the dataframe from the set of numbers contained in `data`. Our new dataframe is thus `df`. As mentioned, we will revisit this data later in the book when we use it for statistical models. For now, we simply wish to demonstrate a few of the computations we can perform on the data file to help us learn more about it and to demonstrate the power of Python. Let us first demonstrate how we can compute a few functions from the "ground-up" so to speak. For example, suppose we wish to compute the mean of `ac`. We have already seen how we can extract the variable `ac`. We print only a few cases here:

```
df['ac']
Out[107]:
0     70
1     67
2     65
3     75
```

Now let us compute the arithmetic mean of `ac`. We can do this easily by giving the object a new name:

```
ac = df['ac']
mean = sum(ac)/len(ac)
mean
Out[111]: 79.04166666666667
```

In the above, we summed all observations in `ac` by `sum(ac)` and then divided this sum by the length of `ac`, denoted `len(ac)`. The length is the number of values that went into the sum. Hence, what the above is computing is simply

$$\bar{y} = \frac{\sum_{i=1}^{n} y_i}{n}$$

The mean is 79.04. Note that this is the **arithmetic mean**. There are other means, such as the **harmonic** and **geometric** means that are beyond the scope of this book but are sometimes useful in data analysis in particular contexts. The point for now is that when you hear "mean" do not necessarily assume it is the arithmetic mean. It usually will be, but not always. Of course, we could have found the mean using one of Python's built-in functions. By importing **statistics**, we could have also computed the mean by using `statistics.mean()`:

```
import statistics
mean = statistics.mean(ac)
mean

Out[115]: 79.04166666666667
```

We could have also just as easily used **numpy** to compute the mean:

```
import numpy as np
np.mean(ac)

Out[116]: 79.04166666666667
```

Obtaining a **standard deviation** is just as easy. A standard deviation, by definition, is the square root of the variance, computed as follows:

$$s = \sqrt{\frac{\sum_{i=1}^{n}(x_i - \bar{x})^2}{n-1}}$$

Notice we are taking the sum of squared deviations from the mean for each score and then dividing by the number of observations that went into the sum. For the current computation, since we wish to use the sample standard deviation as an **unbiased estimator** of the population standard deviation, we lose a **degree of freedom** in the denominator. If we wished to only compute the standard deviation on the sample and not use it as an estimator of the population parameter, then we could use the following:

$$s = \sqrt{\frac{\sum_{i=1}^{n}(x_i - \bar{x})^2}{n}}$$

Usually, however, since we typically do not have the population standard deviation at our disposal and wish to use the sample standard deviation as an estimator, $n-1$ is used. For a proof as to why $n-1$ generates an unbiased estimator, see Hays (1994) or virtually any book on mathematical statistics. For a less technical explanation and one that is a bit more intuitive, see Denis (2021).

The variance of the data can be computed via `np.var(ac) = 89.20`. Now let us get the standard deviation. We will use the **math** module for this, specifically the `math.sqrt()` function:

```
import math
math.sqrt(89.20)

Out[147]: 9.444575162494075
```

To get the population standard deviation, we compute:

```
sd = statistics.pstdev(ac)
sd
```

```
Out[162]: 9.444924415908378
```

Hence, we can say that, on average, scores deviate from the mean by approximately 9.44 scores in this distribution. If we wanted the **median** of our variable, the value that divides the data into two equal parts, we could also easily compute this:

```
statistics.median(ac)
Out[157]: 76.0
```

Note that the median is a bit less than the mean of 79, indicating that a few of the higher-end scores are pulling the distribution up a bit toward the right side of the distribution. This is so because the mean is **sensitive** to extreme scores whereas the median is not. In cases where you have a highly skewed distribution especially, computing the median rather than the mean provides a more **resistant** (i.e. resistant to extreme scores) measure of central tendency. We could also try the mode for `ac`, but this is what we might get:

```
statistics.mode(ac)
StatisticsError: no unique mode; found 2 equally common values
```

This is hardly an "error," as it simply means there is no **unique mode** in the data. However, one could just as easily, if not somewhat informally, say that there are more than two modes. So long as you are communicating this clearly to your reader, it does not really matter how you say it, so long as the message gets through.

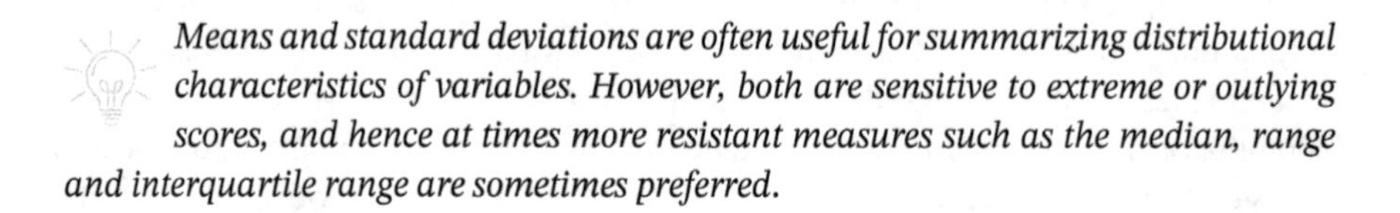

Means and standard deviations are often useful for summarizing distributional characteristics of variables. However, both are sensitive to extreme or outlying scores, and hence at times more resistant measures such as the median, range and interquartile range are sometimes preferred.

2.7 Importing a .txt or .csv File

Usually, importing and reading a text file is relatively straightforward. However, sometimes it may require some adjustment. Here is a problem you may run into. Suppose you would like to load a text or .csv file, but that it appears as follows when you load it (only a few observations of the entire data set are shown). We are correctly using `pd.read_csv` to attempt to read it, but we get the following, when we should be getting simply numbers under the headers `verbal`, `quant`, `analytic` and `group`:

```
import pandas as pd
iq_data = pd.read_csv('iq.data.txt')


iq_data
```

```
Out[295]:
    Unnamed: 0 verbal quant analytic group
0            0     56.00\t56.00\t59.00\t.00
1            1     59.00\t42.00\t54.00\t.00
2            2     62.00\t43.00\t52.00\t.00
```

The file is coming out this way because Python is not recognizing the spaces between numbers. This can be fixed easily by the following, in which we specify a `delimiter`:

```
dataset=pd.read_csv("iq.data.txt",delimiter="\t")


dataset
Out[316]:
verbal quant analytic group
56.0 56.0 59.0 0.0
59.0 42.0 54.0 0.0
62.0 43.0 52.0 0.0
```

We see that now the file is read correctly. In general, if a file is not being imported properly, it may require a bit of digging online or through Python sources to figure out why and then fix things. Never underestimate how long it can take just to get a file working correctly. Data tasks like this can easily consume a lot of work. In data science lingo, ensuring that your data is organized properly is often referred to as **"tidying"** a data set (see Wickham and Grolemund, 2017, for more details).

2.8 Loading Data into Python

Entering data directly into Python is fine as we did earlier on the achievement data, and for small projects, may be a sufficient way to get your data into Python. Sometimes, entering data by hand is a nice idea, especially if the data set is small, because it allows you to screen the data as you enter it. For example, if you are entering ages of people, and you come across an age of 200, you will immediately recognize that something is amiss. Often, however, data come in very large sizes and we are best loading it from **external files**. Sometimes, we would like to load it from R software, since R has many **prepackaged** data sets. One nice data feature of Python is the ability to access data sets in R via `get_rdataset()`. Below we access the **iris** data using **seaborn**, which is a historic data set containing the lengths and widths of sepals and petals. As shown, we request the data from `sns` and then request a few cases:

```
import seaborn as sns
iris = sns.load_dataset('iris')


iris.head()
Out[133]:
   sepal_length  sepal_width  petal_length  petal_width species
0           5.1          3.5           1.4          0.2  setosa
1           4.9          3.0           1.4          0.2  setosa
```

```
2            4.7           3.2           1.3           0.2   setosa
3            4.6           3.1           1.5           0.2   setosa
4            5.0           3.6           1.4           0.2   setosa
```

2.9 Creating Random Data in Python

Instead of our own data, sometimes we would like to create random data, especially for the sake of demonstrating statistical principles. It provides a powerful tool for teaching. What is random data generated by a computer? It is not what you may think. **No computer can actually generate truly random data**. It is best called **"pseudo-random"** since it is still based on an algorithm, which is a deterministic way of generating data no matter how much we would like to believe it to be probabilistic. Hence, when we ask any computer to generate random data, it is a bit different from flipping coins, where one might say a true random process is present (although one could argue a higher power knows the deterministic algorithm at work in the coin flips, but we as mortal beings certainly do not).

We can use `np.random.randn()` to generate the random numbers we desire. We will do so for 100 random digits (printing only a few cases):

```
import numpy as np
np.random.randn(100)
Out[54]:
array([ 0.02269355, -0.07788592,  0.45105672, -1.19752485,  0.87797989,
        0.46153428, -0.43629477,  1.28920105, -0.74609082, -0.02370795,
       -0.13875894,  0.16783418,  0.68196274, -0.07277099,  0.81438243,
        1.34040424, -1.84640245, -0.30100139, -0.21200374,  0.99749304,
```

2.10 Exploring Mathematics in Python

Not surprisingly, Python is able to compute just about any mathematical function we might desire. Since statistics is built on the foundations of mathematics and philosophy, surveying some of these mathematical capabilities is well worth our time. In some cases (such as in coursework for good practice), you may wish to build up the programs yourself so that you become completely aware of what the function is actually doing. In other cases, simply obtaining the function may be enough. For example, we can easily obtain the constants for e and π from the **math** module:

```
import math
math.e
Out[154]: 2.718281828459045

math.pi
Out[155]: 3.141592653589793
```

The constant e is **Euler's number**, found virtually everywhere in mathematics, statistics, and science more generally, whereas π is another constant found in innumerable places, well-known for being the ratio of the circumference to diameter of a circle.

These are **constants** that find themselves in many key formulas and functions in even so-called elementary statistics. For example, the equation for the normal distribution features both of them:

$$f(x_i, \mu, \sigma^2) = \frac{1}{\sqrt{2\pi\sigma^2}} e^{-(x_i - \mu)^2 / 2\sigma^2}$$

where, in the equation, μ is the population mean for the given density, σ^2 is the population variance, π is the "pi" constant we just mentioned, equal to approximately 3.14, and e is the other constant, equal to approximately 2.71. The value for x_i is a given value of the independent variable, assumed to be a real number. The values for π and e are regarded as constants because they remain the same regardless of the values of the other "inputs" to the function rule. That is, we can essentially imagine an infinite number of normal distributions as defined by the above rule, but all will have π and e in them, and these numbers will not change. They are "constants" across variations of the function rule. This is in contrast to x_i,, which is a **variable** in the expression, as its value will change depending on what value we input for it.

The normal distribution is a family of distributions defined by a function rule. There are literally an infinite number of normal distributions possible, each defined by parameters μ and σ^2. Exact normal distributions will never appear in nature. Since they are defined by a function rule first and foremost, the best empirical data will do is approximate a normal density, but it will never exactly mirror one.

2.11 Linear and Matrix Algebra in Python: Mechanics of Statistical Analyses

As we began discussing in the previous chapter, underlying every statistical analysis you or I perform with software is a wealth of **mathematical** and **philosophical** foundations. For example, when we conduct a **principal components analysis**, which is a **multivariate** statistical procedure, what we are seeing in software output is but the surface to a whole underground of mathematical and philosophical structures. This, again, is one of the dangers of conducting analyses one may not understand at a deeper level. It becomes much too easy to draw research and scientific conclusions from software output that are not defensible since the underlying mathematics or philosophy may simply not allow for such conclusions to be drawn. The software does not "know" how a logical process functions toward drawing a scientific conclusion. The computer simply computes what you tell it to, and it does so extremely reliably and very fast. To a computer, a complex statistical analysis is a series of computations and nothing more. It is not aware of how you are using such computations or what conclusions you hope to draw from them. For example, the computer knows how to compute an arithmetic mean, but it does not "understand" the advantages or limitations of using the mean as a measure of central tendency. It is up to you to know that and to use the computations generated by software in an intelligent manner. **As a scientist, you should be very familiar with the underpinnings of the tools you work with.** You may not appreciate why this is so important at first glance, but as you become more experienced, you will soon realize that not understanding the mathematics and essential philosophy "behind the scenes" of your software output can have rather serious implications. **Always strive to understand the method you are using at the deepest level you can before using it in Python or any other software. This includes understanding both the mathematical and philosophical issues underlying the method you are using.**

Linear and matrix algebra is based on vectors and matrices. A **vector** in mathematics is usually defined as a **line segment having magnitude and direction.** This is the classic **physics** definition of the term and it is not difficult to see why when we consider physical applications. For example, vectors help aircraft determine desired direction and arrival points. Hence, when air traffic controllers are communicating navigation information to pilots, precise vector information provides reliable coordinates. In **computer science**, a vector may also simply be considered as a listing of numbers. For example, the numbers (5, 8, 9, 10) can be considered a vector. Hence, in this computer science sense of the word, vectors do not necessarily have to have magnitude or direction as in the sense of an implied physical force in physics. Via this simple example, one immediately gets the correct sense that many terms in mathematics, even though seemingly rigorously defined, can still have slightly different meanings depending on the given area of application. This is again why attempting to understand definitions outside of their context can be challenging, such that if you attempted to memorize a definition for vectors, you may not be aware when it is used in a slightly different context than the definition you originally learned. Definitions, especially in statistics, do their best to capture the essence of the given concept or construct, but only experience with the concept will encourage and foster a deeper understanding.

As an example of two vectors that have been added, consider the following:

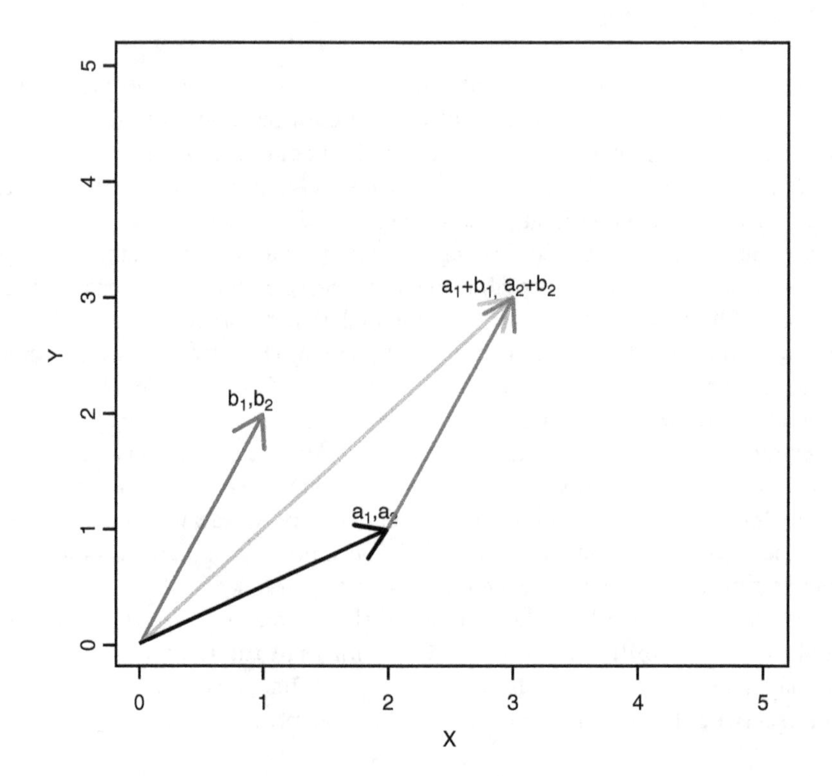

Each tip of a vector represents a given point in the plane. For instance, the point a_1, a_2 corresponds to the point $x = 2$ and $y = 1$. Though the addition of the two vectors need not really concern us, it is enough to know for now that two vectors can be added, multiplied, and subject to many other mathematical operations. The vectors shown are in two-dimensional space, but mathematically there is no limit to how much we can generalize the dimensions of vectors. Mathematically, we could be working in 100 dimensions if we chose and the principles behind vector operations will remain the same. Mathematicians can work in virtually as many dimensions as they choose, since their abstract work does not necessarily need to correlate with the empirical world or "real" objects. Pure (not applied) mathematics is (or can be considered to be) an **abstract** system.

A **matrix** in linear algebra can be defined as many vectors "glued" together, such that the matrix now has more than a single column or single row. An example of a matrix is the following:

$$\mathbf{A} = \begin{bmatrix} a_{11} & a_{12} & \cdots & a_{1n} \\ a_{21} & a_{22} & \cdots & a_{2n} \\ \vdots & \vdots & \ddots & \vdots \\ a_{m1} & a_{m2} & \cdots & a_{mn} \end{bmatrix} = \begin{pmatrix} a_{11} & a_{12} & \cdots & a_{1n} \\ a_{21} & a_{22} & \cdots & a_{2n} \\ \vdots & \vdots & \ddots & \vdots \\ a_{m1} & a_{m2} & \cdots & a_{mn} \end{pmatrix} = (a_{ij}) \in \mathbb{R}^{m \times n}$$

Notice the matrix has m rows and n columns, and is typically written with square or round brackets as above. Many advanced statistical procedures operate directly on vectors and matrices. For instance, **principal components analysis** as well as **factor analysis** typically use a **covariance** or **correlation matrix** as input to the analysis. For example, the following is a covariance matrix:

$$K_{XX} = \begin{bmatrix} E[(X_1 - E[X_1])(X_1 - E[X_1])] & E[(X_1 - E[X_1])(X_2 - E[X_2])] & \cdots & E[(X_1 - E[X_1])(X_n - E[X_n])] \\ E[(X_2 - E[X_2])(X_1 - E[X_1])] & E[(X_2 - E[X_2])(X_2 - E[X_2])] & \cdots & E[(X_2 - E[X_2])(X_n - E[X_n])] \\ \vdots & \vdots & \ddots & \vdots \\ E[(X_n - E[X_n])(X_1 - E[X_1])] & E[(X_n - E[X_n])(X_2 - E[X_2])] & \cdots & E[(X_n - E[X_n])(X_n - E[X_n])] \end{bmatrix}$$

Along the **main diagonal** (top left to bottom right) of the matrix are variances, whereas along the **off-diagonal** (i.e. above or below the main diagonal) are covariances. The notation "E" denotes **expectation**, which is a long-run weighted average. Expectations are used heavily in theoretical statistics and probability to give us an idea of what is being estimated in the long run by a sample quantity. If we standardize the covariance matrix, we obtain the correlation matrix:

$$\text{corr}(\mathbf{X}) = \begin{bmatrix} 1 & \frac{E[(X_1 - \mu_1)(X_2 - \mu_2)]}{\sigma(X_1)\sigma(X_2)} & \cdots & \frac{E[(X_1 - \mu_1)(X_n - \mu_n)]}{\sigma(X_1)\sigma(X_n)} \\ \frac{E[(X_2 - \mu_2)(X_1 - \mu_1)]}{\sigma(X_2)\sigma(X_1)} & 1 & \cdots & \frac{E[(X_2 - \mu_2)(X_n - \mu_n)]}{\sigma(X_2)\sigma(X_n)} \\ \vdots & \vdots & \ddots & \vdots \\ \frac{E[(X_n - \mu_n)(X_1 - \mu_1)]}{\sigma(X_n)\sigma(X_1)} & \frac{E[(X_n - \mu_n)(X_2 - \mu_2)]}{\sigma(X_n)\sigma(X_2)} & \cdots & 1 \end{bmatrix}$$

Notice that to get the correlation matrix from the covariance matrix, we divide each entry by the product of standard deviations for X_1 and X_2, where σ denotes the standard deviation and $\sigma(X_1)\sigma(X_2)$ is the product of standard deviations for X_1 and X_2.

We can build a covariance matrix in Python quite easily. Let us first build up some fictitious data on two variables by generating a dataframe using `pd.Dataframe()`:

```
cov_matrix = pd.DataFrame([(10, 20), (5, 15), (20, 12), (8, 17)],
columns=['var1','var2'])

cov_matrix
Out[189]:
   var1  var2
0    10    20
1     5    15
2    20    12
3     8    17
```

The above denotes the scores on `var1` to be 10, 5, 20 and 8, and so on for `var2`. We have named the object `cov_matrix`, even though it is not a matrix yet, but simply a listing of two variables. To get the matrix, we attach `.cov()` to it:

```
cov_matrix.cov()
Out[190]:
           var1       var2
var1  42.250000 -12.333333
var2 -12.333333  11.333333
```

For reasons we will discuss in the following chapter, we often prefer the **correlation matrix** to the covariance matrix for many analyses. As mentioned, the correlation matrix is the standardized version of the covariance matrix. We can easily standardize the covariance matrix by using `.corr()` instead of `.cov()`:

```
cov_matrix.corr()
Out[196]:
          var1      var2
var1  1.000000 -0.563622
var2 -0.563622  1.000000
```

We can see from the matrix that the correlation between `var1` and `var2` is equal to –0.563622, meaning that the correlation between the two variables is moderately negative. We also see in the main diagonal from top left to bottom right that there are values of 1.000000. This is as a result of the correlation of a variable with itself being perfect and positive. Again, we delay our discussion of covariance and correlation to the following chapter, where we unpack the nature of correlation and explain why the correlation coefficient is bounded by values of –1.0 and 1.0.

2.11.1 Operations on Matrices

Earlier we constructed a covariance and correlation matrix, but the sky is the limit regarding the types of matrices one can build in Python or other software. For instance, the following is a matrix made up entirely of zeroes having four rows and four columns:

```
import numpy as np
np.zeros((4, 4))
Out[287]:
array([[0., 0., 0., 0.],
       [0., 0., 0., 0.],
       [0., 0., 0., 0.],
       [0., 0., 0., 0.]])
```

We could have built the same matrix using the following as well by specifically designating the matrix as `numpy.empty()`. An **empty matrix** is one with zeros within the entire matrix:

```
import numpy as np
np.empty((4, 4))
Out[289]:
array([[0., 0., 0., 0.],
       [0., 0., 0., 0.],
       [0., 0., 0., 0.],
       [0., 0., 0., 0.]])
```

We could also easily build a matrix having values of 1.0 everywhere. For example, the following matrix has four rows and two columns with values of 1.0. Notice that instead of `numpy.zeroes()` (which would generate zeros) we are using `numpy.ones()`:

```
import numpy as np
np.ones((4, 2))
Out[290]:
array([[1., 1.],
       [1., 1.],
       [1., 1.],
       [1., 1.]])
```

When we have different values in our matrix, we can build it by specifying each of them. The following is a 2 × 2 matrix having values of 1, 2, 3, 4:

```
import numpy as np
A = np.matrix('1 2; 3 4')

A
Out[160]:
matrix([[1, 2],
        [3, 4]])
```

We could have also used the `np.array()` function to build the same matrix. We name this matrix B to distinguish it from the above, even though it contains the same elements:

```
import numpy as np
B = np.array([[1, 2],
[3, 4]])
B
```

```
Out[165]:
array([[1, 2],
       [3, 4]])
```

The **transpose** of a matrix is defined by exchanging rows for columns and columns for rows. That is, if our original matrix is

$$\mathbf{B} = \begin{bmatrix} b_{11} & b_{12} \\ b_{21} & b_{22} \end{bmatrix}$$

then the transpose is defined by $\mathbf{B}'$:

$$\mathbf{B}' = \begin{bmatrix} b_{11} & b_{21} \\ b_{12} & b_{22} \end{bmatrix}$$

Notice that the first row, b_{11} b_{12}, is now the first column, and likewise for the second row. In Python, we can easily compute the transpose by again a simple extension to `np`, this time `np.transpose()`:

```
import numpy as np
np.transpose(B)
Out[178]:
array([[1, 3],
       [2, 4]])
```

Notice that the rows of the old matrix are now the columns of the new matrix.

Another important measure as it concerns matrices is the **trace of a matrix**, which consists of summing the values along the main diagonal of the matrix. For example, the trace of the matrix B transpose that we just computed is equal to 5, since the values along the main diagonal (top left to bottom right) consist of values 1 and 4 for a total of $1 + 4 = 5$. More generally, the trace for a 3×3 matrix

$$\mathbf{A} = \begin{bmatrix} a_{11} & a_{12} & a_{13} \\ a_{21} & a_{22} & a_{23} \\ a_{31} & a_{32} & a_{33} \end{bmatrix}$$

is equal to $a_{11} + a_{22} + a_{33}$. The trace of **non-square matrices** is not defined since for non-square matrices there exists no **principal diagonal** (i.e. no "top left to bottom right" idea). We can easily compute the trace in Python by this time adding the extension of `matrix.trace` on `np`. We compute this for our original matrix B:

```
import numpy as np
np.matrix.trace(B)
Out[198]: 5
```

We can also request the elements along the main diagonal specifically:

```
import numpy as np
np.matrix.diagonal(B)
Out[201]: array([1, 4])
```

So why bother computing something called the trace? Computing the trace is especially important when computing a variety of multivariate test statistics, as we will see later in the book when we consider multivariate procedures. Its relevancy is also especially apparent when working with principal components analysis, as **the sum of variances of all variables is equal to the trace of the covariance matrix**. As we will see, this sum will also equal the **sum of eigenvalues** across the matrix. Hence, traces of matrices are used quite regularly in the communication of multivariate results. In the context of matrix theory, we now briefly survey the very important concepts of eigenvalues and eigenvectors.

2.11.2 Eigenvalues and Eigenvectors

At first glance, multivariate procedures can seem formidable. Though they are rather complex, there is a unifying component to them that remarkably encompasses much of their complexity, and that is in the mathematical objects of **eigenvalues** and **eigenvectors**. In most univariate and multivariate procedures, the essential computation that is occurring "behind the scenes" is a computation of eigenvalues and eigenvectors. Though a deep investigation into these concepts is well beyond the scope of this book, we can nonetheless get the main ideas across by a simple survey of them. Be forewarned that they will make much more sense when we survey **principal components analysis** and other dimension-reduction techniques later in the book. Or at minimum, you will see how they are used in applied analysis and situate them in some context other than abstract mathematics.

Suppose we have a square matrix of dimension $n \cdot n$. We will call this matrix **A**. Now, it is a result of mathematics that the following equality is true with regards to matrix **A**:

$$\mathbf{Ax} = \lambda \mathbf{x}$$

In this equation, **x** is a vector and λ is a scalar, that is, an ordinary real number. In English, what the above equality says is this: Multiplying vector **x** by the matrix **A** generates the same result as multiplying vector **x** by the scalar λ. The scalar λ is called an **eigenvalue** of the matrix **A** and the vector **x** is called an **eigenvector** associated with λ. The equivalency $\mathbf{Ax} = \lambda \mathbf{x}$ when worked on slightly algebraically can be written as follows:

$$\mathbf{Ax} - \lambda \mathbf{x} = \mathbf{0}$$
$$(\mathbf{A} - \lambda \mathbf{I})\mathbf{x} = \mathbf{0}$$

If the determinant of this expression does not equal 0, that is, if $|\mathbf{A} - \lambda \mathbf{I}| \neq \mathbf{0}$, then this also implies that $(\mathbf{A} - \lambda \mathbf{I})$ has an inverse, which further implies that the only vector **x** that will provide a solution is that of $\mathbf{x} = \mathbf{0}$. The solution $\mathbf{x} = \mathbf{0}$ is known as the trivial solution. To obtain a solution that is not equal to 0, $|\mathbf{A} - \lambda \mathbf{I}|$ is set to 0 and then values for λ are found that, when substituted back into $(\mathbf{A} - \lambda \mathbf{I})\mathbf{x} = \mathbf{0}$, give a solution for **x**. The equation $|\mathbf{A} - \lambda \mathbf{I}| = \mathbf{0}$ is known as the **characteristic equation**, and for a square matrix, that is, one that has n rows and n columns, the characteristic equation will have a total of n roots. But what is a root? A **root** is where the function intersects the x-axis, a value that once inserted into the equation will make the value of the equation equal to 0. Recall that to compute the roots of a quadratic equation, we can use the quadratic formula:

$$x = \frac{-b \pm \sqrt{b^2 - 4ac}}{2a}$$

where the values of a, b, and c are from the general quadratic equation, $ax^2 + bx + c = 0$. These roots in the context of eigenanalysis are referred to as **eigenvalues**. For the $n \cdot n$ matrix, the characteristic equation will have $\lambda_1, \lambda_2, \ldots, \lambda_n$ eigenvalues, where some values may be the same.

We can obtain the eigenvalues and eigenvectors of a matrix in Python quite easily. For example, the eigenvalues and eigenvectors of the matrix B can be obtained from **linalg** in **numpy** using the extension `.eig()`:

```
B = np.array([[1, 2], [3, 4]])
B
Out[7]:
array([[1, 2],
       [3, 4]])

results = np.linalg.eig(B)

results
Out[46]:
(array([-0.37228132,  5.37228132]), array([[-0.82456484, -0.41597356],
       [ 0.56576746, -0.90937671]]))
```

The first array reports the eigenvalues for the 2 × 2 matrix. The eigenvalues are thus –0.37 and 5.37, while the corresponding eigenvectors are –0.82, 0.56 and –0.41, –0.90, respectively. Again, such computations will have much more meaning when we consider them later in the book in the context of actual statistical procedures such as **discriminant analysis** and **principal components analysis**.

Review Exercises

1. Explore and discuss why specializing in your field of study first, then statistics second, and then software computations third, should be your priority. What are the dangers inherent in specializing in software first at the expense of the other two?
2. Discuss the nature of a **z-score** in statistics. Specifically, what does the ratio $x - \mu$ to σ tell us? Second, will standardizing a raw distribution make the distribution normal? Why or why not? Explain.
3. The **arithmetic mean** is a common measure of **central tendency**. But what is it exactly? Describe in words what the arithmetic mean actually accomplishes. Look at the formula and "unpack" its meaning as much as you can. How would you describe it to someone?
4. Describe in words what a **standard deviation** is by unpacking the formula for it in some detail. Explain also why it is necessary to square deviations. What would be

the consequence of leaving them **unsquared**? What is the square of the standard deviation called and what does it accomplish?

5. For an increasingly **large sample size**, what is the consequence of dividing by n rather than $n-1$ in the standard deviation? Play with some numbers and allow n to get extremely large. What happens as n goes toward infinity?
6. Take a good look at the function rule for the **normal density**. What does the formula reduce to for a standardized z-distribution? Perform the relevant substitutions to obtain the simplified form of the distribution.
7. Distinguish between a **vector** and a **matrix** in linear algebra.
8. Distinguish between a **covariance** vs. a **correlation** matrix. How is one the standardized counterpart of the other? Explain.
9. Why is the **upper triangular** the same as the **lower triangular** in a covariance or correlation matrix? Draft a sample matrix with numbers to demonstrate.
10. An **identity matrix** is one in which there are values of 1.0 along the main diagonal of the matrix from top left to bottom right, and values of 0.0 everywhere else. Construct an identity matrix in Python of dimension 3×3.
11. Discuss the nature of the equation $\mathbf{Ax} = \lambda\mathbf{x}$. In words, what is it communicating? Draw a figure depicting its main features.
12. Recall the **quadratic equation** from high school and featured in the chapter. What does the equation actually accomplish? In other words, what does it mean to calculate the "**roots of an equation**" and when is the quadratic equation necessary for such computation?

3

Visualization in Python

Introduction to Graphs and Plots

CHAPTER OBJECTIVES

- Gain an appreciation for the importance of visualization in data analysis and data science.
- Appreciate how graphs can communicate essential information that tables could never.
- Understand how perception of a data set or research finding can be even more important than the data itself.
- Why when it comes to scientific tables and graphs, simple is usually better than complex.
- Why percentage increases or percentage decreases can be misleading.
- How to produce scatterplots, histograms, and other simple but effective tools of visualization.
- A survey of bubble graphs, pie charts, heatmaps, correlograms, and other graphical tools.

In this chapter, we briefly explore some of the numerous visualization possibilities in Python and feature a discussion of graphics and data visualization in general. In reality, we only scratch the surface, as the capacities for visualization in Python greatly exceed what we can accomplish here. The importance of visualization in data analysis and data science cannot be overstated. **Graphs are essential to the effective communication of data.** As pointed out by many (e.g. see Tufte, 2011), however, graphs can just as well **mislead** as they can elucidate, and hence it is imperative that we develop an acuity and aptitude for being able to decipher exactly what a graphic is or is not communicating. Especially in these days of online visualizations literally bombarding us every day, the average consumer of a graph is likely to only spend a **few seconds** interpreting an image before redirecting his or her mouse to another application. Hence, the **perception** yielded by a graph can be long-lasting, even if that perception is faulty. For example, you may see a graph on the effects of COVID-19, and without realizing that you have misinterpreted it, carry that perception with you possibly forever! You never once critically evaluate or revise your perception. Then, when you are conversing with someone about COVID-19, you implement that

Applied Univariate, Bivariate, and Multivariate Statistics Using Python: A Beginner's Guide to Advanced Data Analysis, First Edition. Daniel J. Denis.

perception into a discussion as if it were reality! It is a faulty perception, yet you are carrying on with it as though it were **real**. As with the communication of any and all information, if the perception of what is being communicated is faulty, then regardless of the truth behind the information, it is the perception that will unfortunately dominate. In this sense, **perception becomes reality**. Therefore, how you choose to communicate information, whether scientific or otherwise, can have tremendous consequences. Statistics and visualization are especially good at misleading, if not used judiciously.

As an example of how powerful graphical perception can be, consider the following image comparing the magnitude of nuclear weapons in 1945 compared to 2020. How much energy and firepower existed in nuclear weapons at the time relative to today? Reporting a numerical measurement would likely not be that meaningful because it lacks context. On the other hand, a graphic representing a comparison of nuclear energy in the form of a basic visualization is much more enlightening (see Figure 3.1).

Regarding this plot, had the author simply cited a number in kilotons or similar units, for instance, unless the reader is a scientist familiar with the area, he/she does not leave the graph with a strong perception. **Visualizations have the power to convey a sense of urgency**. This visualization makes one go "Wow!," something a number could never do. As we will see, Python is more than capable of generating quite effective and beautiful graphics. We survey some of these throughout the chapter, while highlighting which **packages** are most essential in generating quality scientific graphics and also where to go to explore graphics beyond what we cover in the chapter.

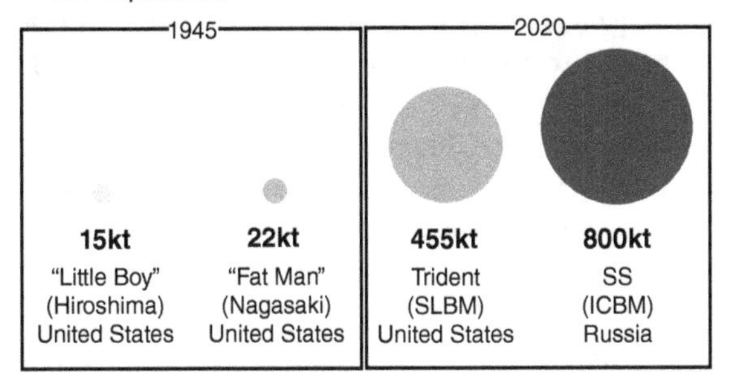

Figure 3.1 A Comparison of Nuclear Power in 1945 and 2020. Reproduced with permission from Statista (https://www.statista.com/chart/3714/nuclear-weapons-in-comparison).

3.1 Aim for Simplicity and Clarity in Tables and Graphs: Complexity is for Fools!

As Dr. Malcolm in the film Jurassic Park once said, "Your scientists were so preoccupied with whether or not they **could**, they didn't stop to think if they **should**!" The same is true for graphs. Just because software allows us to "dress up" graphics in the most interesting and complicated ways does not mean we should, and in most cases adding **aesthetics** to a graphic for the purpose of making it look **complex** is the sign of an amateur researcher rather than a professional scientist. Good graphs should be built with the clearest, ethical, and most effective honest communication motives possible in mind, not with the intention of making a fashion statement and manipulating the perceiver by way of complexity. Next time you see a graphic, and it is such a beautiful picture, ask yourself whether the author could have communicated the finding in a more efficient and parsimonious way. Too often, you will find the graph was constructed to "impress" rather than to communicate. This is truly unfortunate. Fortunately, good scientific journals will usually balk at such graphics and request instead graphs that prioritize a clear and ethical transfer of information from the author to the reader. What is perhaps most concerning about poorly constructed graphics is that often they may contain correct information, but it is simply that the information that they are communicating is unclear or purposely misleading. Again, it is all about what is perceived, not necessarily what the table or graph actually contains (that "perception becomes reality" idea we earlier alluded to).

For example, consider Table 3.1 describing "hot spots" for COVID-19 as of June 22, 2020. At first glance, it would appear that Montana especially is a true hot spot and cases must be surging due to the 316.7% increase. Though the percentage increase figure is correct, the perception one may be gleaning from the table, if not careful, can be completely misguided unless one simultaneously considers competing statistics. It turns out that the seven-day rolling average (yet another COVID-19 statistic) of daily cases in Montana was only equal to 17.9 for the period under review. Compare this with Texas, for instance, where the seven-day average was 3,939.9. However, Texas only had a 153.7% increase over 14 days, much less than that of Montana's. So what is going on? What is going on is that getting a large **percentage increase** in Montana is not that difficult, since there are relatively few people who live in Montana to begin with (winter storms in late September (2019) and a high of freezing at Halloween in the same year apparently does not always encourage migration). However, remember that from 1 to 2 is a 100% increase. That is, 2 is 100% more than 1. Likewise, increasing from 1 to 3 is a 200% increase. The **absolute increase** in both cases is still exceedingly small. On the other hand, 1,000 to 1,500 is only a 50% increase, even though we are adding 500 more units. **Percentage increases or decreases can be extremely misleading, and hence must be interpreted with caution, preferably with sideline knowledge of the corresponding absolute increases**. While the percentage increase is informative, it does not tell the full story and can lead to a faulty perception of reality. A look at the total number of overall cases (Figure 3.2) reveals that Montana was one of the states with the fewest numbers, as shown (note that Montana is only lightly shaded, indicating total cases in the range of 0 to 1,000).

Table 3.1 Percentage increases in COVID-19 in 14 days as of June 22, 2020.

State	14-Day Percentage Change in Cases
Alabama	82.2%
Arizona	135.2%
California	52.2%
Florida	183.5%
Georgia	63.5%
Hawaii	233.3%
Idaho	144.9%
Kansas	77.7%
Mississippi	25.5%
Missouri	26.4%
Montana	316.7%
Nevada	98.7%
Ohio	38.8%
Oklahoma	266.6%
Oregon	103.7%
South Carolina	137.1%
Texas	153.7%
Utah	42.0%
West Virginia	72.9%

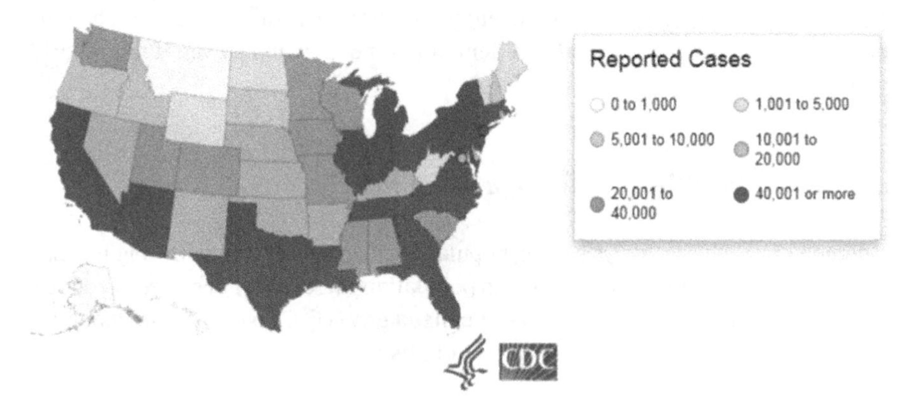

Figure 3.2 Total reported COVID-19 cases. *Source: CDC (Centers for Disease Control and Prevention). Used with Permission. Available at CDC (www.cdc.gov) free of charge.*

Percentage increases in general (and not specifically related to COVID-19) are also often stated statistics for general **population growth** by state. Over the past few years (as of the time of this writing), much has been made about the population growth of southern Idaho. Apparently, Californians are moving to Idaho in droves. It has often been reported as the fastest growing state in the country, growing by approximately 2.5% over the past few years as of the time of this writing. However, a 2.5%

increase is still quite small in absolute numbers if the base from which the growth is taking place is quite small. "Idaho is growing like crazy!" media will report and yet the total population of the entire state as of 2020 is approximately one-fifth of the total population of New York City and not even one-half of the total population of Los Angeles! Yes, growing at a fast rate, but that is easy to do when the base is not very much to begin with.

Likewise, Bozeman, Montana, is apparently, again as of the time of this writing, one of the fastest growing towns in the USA. Total population as of 2020? Approximately 52,000! In 2010 the city had about 37,000 residents. Hence, in a 10-year period the city grew by approximately 40%, since 15,000 (i.e. 52,000–37,000) is approximately 40% of 37,000. A 15,000 increase in residents is really not that much at all over a 10-year period when you think about it. The **rate of growth** is large, but the absolute increase is still rather small. Therefore, the Internet headline of "Bozeman is fastest growing city and has grown by 40% in the last 10 years" is 100% accurate, but entirely misleading if you are taking something away (your perception) from the data that is not actually there. To a naïve consumer, the headline report of it being the fastest growing town may be enough to carry with them the perception that the town is being overrun by flocks of people. Hardly the case. Bozeman is growing, but the number of cows in the city limits (approximately 3 to 1 apparently) still far outweighs the number of residents! **Beware of percentage changes and interpret them with caution**. Or, at minimum, pair them up with absolute changes in the numbers to get a better feel for the data. Always be mindful of results you are interpreting in data analysis and be sure to put them into **context** of what is actually being communicated. Always ask for more context when interpreting any data or scientific result. It will usually reveal a new "base" (i.e. a new "denominator-type" number) upon which you can then more intelligently interpret the result before you. The number you are interpreting always has a denominator somewhere, even if not reported by the research.

3.2 State Population Change Data

To demonstrate a basic analysis using population change data, the following is census data on a number of indicators related to population increases or decreases by state in the United States. The data is available at **census.gov** (US Census Bureau, Population Division). We show only the first five cases of this data:

```
import pandas as pd
data = pd.read_csv("population.csv")
data.head()
Out[17]:
  SUMLEV REGION DIVISION ... NRANK_PPCHG2017 NRANK_PPCHG2018  NRANK_PPCHG2019
0     10      0        0 ...               X               X                X
1     20      1        0 ...               4               4                4
2     20      2        0 ...               3               3                3
3     20      3        0 ...               1               1                1
4     20      4        0 ...               2               2                2

[5 rows x 67 columns]
```

We will define and extract the **state** variable through the following, listing only a few of the cases at the top and then at the bottom of the file:

```
state = data['NAME']
state

Out[30]:
0            United States
1         Northeast Region
2           Midwest Region
55                 Wyoming
56             Puerto Rico
Name: NAME, dtype: object
```

We will also extract the population change for 2019. This variable is named `PPOPCHG_2019` in the data file and we call it `pop_change_2019`, again listing only a few of the cases. Notice that Python allows us to simply name a variable by pulling it out of a dataframe. We could have named `pop_change_2019` something else had we wanted to, so long as we are still pulling `PPOPCHG_2019` out of the file:

```
pop_change_2019 = data['PPOPCHG_2019']
pop_change_2019

Out[32]:
0      0.475078
1     -0.113864
2      0.135376
3      0.811608
```

We can see that, overall, the United States has grown by 0.475% (i.e. the first number in the list), while the Northeast region has decreased by 0.113% (the second number), and so on. While the above is the percentage increase or decrease, below are the actual **raw numbers** corresponding to this increase or decrease. The raw numbers are given by the variable `NPOPCHG_2019`:

```
N_pop_change_2019 = data['NPOPCHG_2019']
N_pop_change_2019
Out[47]:
0      1552022
1       -63817
2        92376
3      1011015
```

That is, the total increase in population in the United States is 1,552,022, for the Northeast region a decrease of 63,817, and so on. Notice that the percentage increase or decrease, other than the positive or negative signs, does not tell us the full story. The raw numbers give us the true sense of the **magnitude of change**. Indeed, if we produce a scatterplot of the two variables to visualize how percentage change is associated with actual numerical change, we obtain the following:

```
import matplotlib.pyplot as plt
import numpy as np
```

```
plt.plot(data["NPOPCHG_2019"], data["PPOPCHG_2019"], "o")
plt.xlabel("NPOPCHG_2019")
plt.ylabel("PPOPCHG_2019")
Out[26]: Text(0, 0.5, 'PPOPCHG_2019')
```

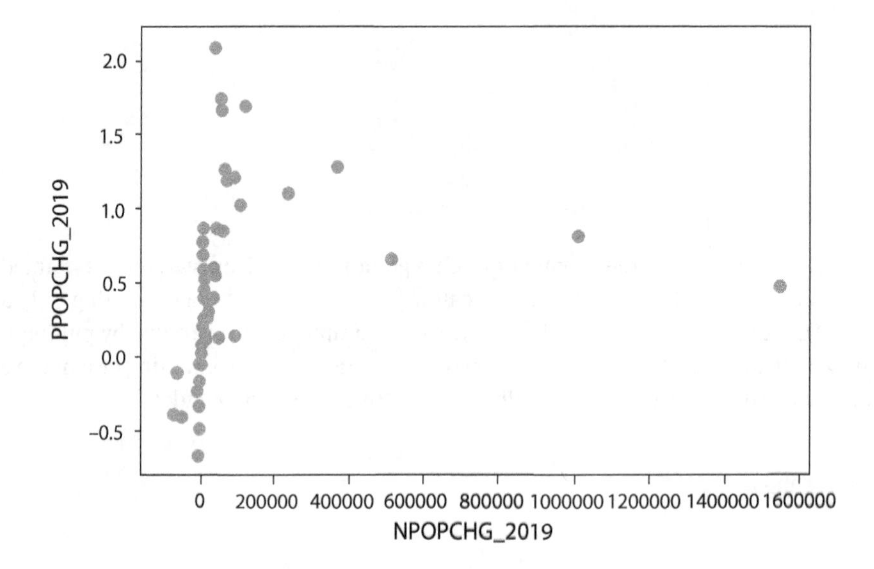

We see that a few points associated with large raw numbers are not necessarily the ones associated with huge percentage increases or decreases, otherwise the plot would reveal that as the raw numbers increase, the percentage change would also increase. As we can see, this is not the case. A percentage increase or decrease is not necessarily related to an absolute increase or decrease, again highlighting the point just discussed. Hence, the news report that gun sales have increased by 200% in a small town, for instance, may simply mean a few more people bought guns, but still a very inconsequential increase in absolute number. On the other hand, an increase of 20% in gun sales in New York City would imply a lot more guns on the streets, simply because the base from which we are expressing the percentage increase is much larger. For this increase, the absolute number will also be greater in number.

3.3 What Do the Numbers Tell Us? Clues to Substantive Theory

Returning again to the COVID-19 data, recall that the total number of cases may be equally or even more informative than the percentage change, as the following graph depicts for the period ending October 9, 2020. The shading represents the number of cases of COVID-19 per 100,000 people by county in the state of California, where a darker shade represents more cases than a lighter shade (Figure 3.3).

At first glance, it may be argued that what is most vital in COVID-19 case information is the number of cases per square mile or some other measure of **density** information. Although this information might be useful, it is less likely to be of value for people

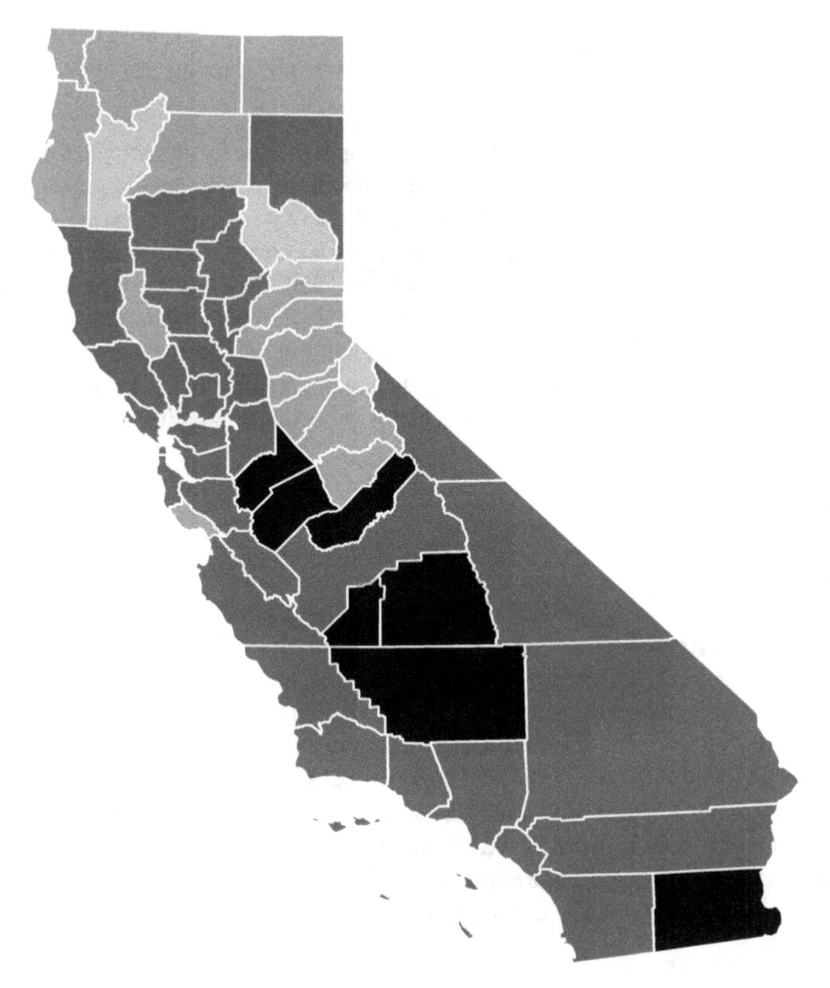

Figure 3.3 Map of COVID-19 Outbreak in the State of California as of October 9, 2020, where a darker shade represents more cases than a lighter shade. *(Source: California Department of Public Health. Used with Permission.)*

who want to know where they may have a greater chance of contracting the illness, since it is highly contagious. **Reasons** or **causes** as to why some counties have higher numbers and potentially higher densities than other counties is, of course, a complex question. At first glance, it may appear that potentially **higher temperatures** in the southern part of the state (assuming those temperatures are higher) are not subduing the virus, as was originally thought they might. Recall it was originally hypothesized early in 2020 that with increasing temperature, the virus may not survive or spread and would be more limited than in the winter months when the virus presumably began. However, the density information may encourage us to ask further questions of the data, such as the following:

- Are counties in the south more populated than counties in the north? Hence if temperature is diminishing spread, perhaps it does not have as much opportunity to do so in the north as much as in the south.

- What were the social distancing and mask requirement regulations in the different counties? Did the southern counties have more stringent or less stringent requirements than the north?

We can see through examples such as this that graphics rarely answer questions, but may help us understand further **which new questions to ask**. That is, graphics and visualizations may help us conjure up new **hypotheses** that we can subject to data analysis to seek out answers. **Data itself, however, never equals theory.** In reference to our example, all we see is a greater number of individuals with COVID-19 in the southern parts of the state. The "how and why" behind those numbers is where theory comes in. Hence, you can perform data analyses all you like, but informing theory and obtaining scientific explanation can still be very challenging, especially if you are unable to perform a rigorous experiment to rule out many competing alternative hypotheses.

3.4 The Scatterplot

One of the most common and useful graphs in the history of statistics is that of the **scatterplot**. The scatterplot depicts the **bivariate relationship** between two variables or the **multivariate** relationship among many more variables (e.g. three-dimensional scatterplots), and hence it is appropriate for bivariate or multivariate data. However, it becomes unwieldy for visualizations in higher than three dimensions. The classic two-dimensional scatterplot can be easily generated using **matplotlib** in Python. **Matplotlib** is one of the more popular packages in Python for generating a variety of graphics, which include static, animated, and interactive visualizations. You can read more on **matplotlib** by referring to its website at **matplotlib.org**.

As an example of a simple scatterplot in Python, we build a plot from scratch on two generic variables *x* and *y*. The data for this example are fictitious:

```
x = [10, 15, 16, 23, 27, 38, 43, 56, 57, 60]
y = [5, 8, 9, 13, 16, 20, 40, 45, 67, 75]
import matplotlib.pyplot as plt
import numpy as np
plt.hist2d(x, y, bins=(50, 50), cmap=plt.cm.Reds)
Out[6]:
```

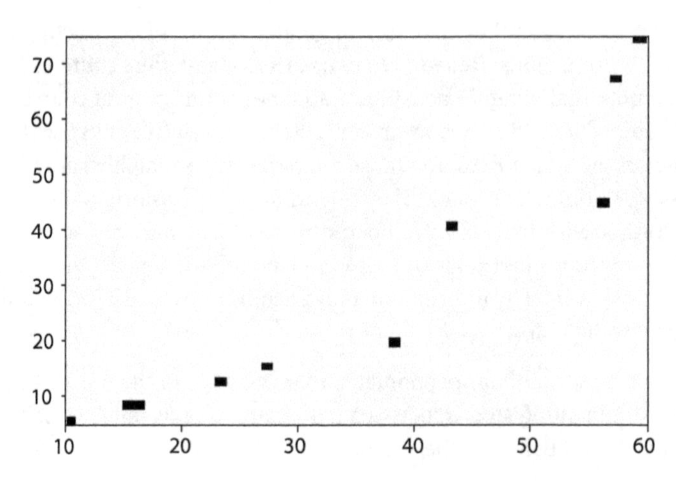

We can easily add a title to the plot quite simply by applying `plt.title("Scatterplot")`. We could also quite easily make adjustments to the plot, such as changing the number of bins. Let us change the bins from `50,50` to `100,100: plt.hist2d(x, y, bins=(100, 100), cmap=plt.cm.Reds)` (below, left). Had we decreased the bins, we would get the following (below, right): `plt.hist2d(x, y, bins=(10, 10), cmap=plt.cm.Reds):`

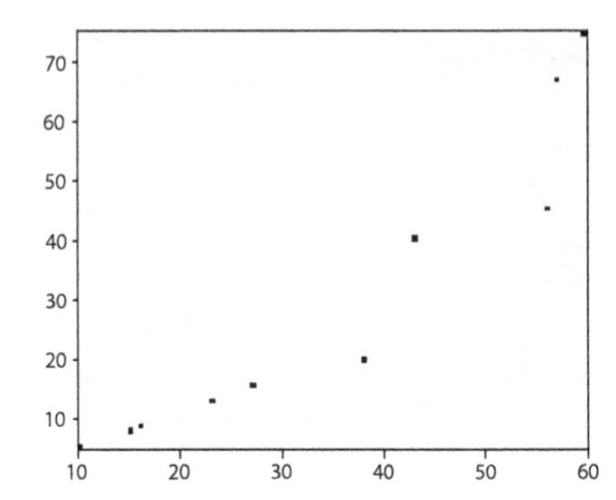

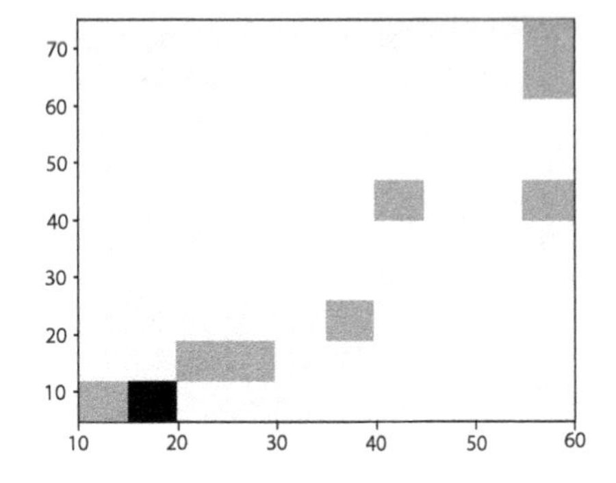

To add labels to the *x* and *y* axes, simply code `plt.xlabel("x")` and `plt.ylabel("y")` as shown earlier, where "`x`" and "`y`" here are just generic labels for convenience. If you were plotting weight by age, then you would rename them `plt.ylabel("weight")` and `plt.xlabel("age")`, and so on. As mentioned at the outset of the book, it is rarely worth your time memorizing code, at least for the purposes of using it for statistical and scientific analysis. What is worth your time, however, is learning how to operate with code, work with it, adjust it, explore it, and see what you can create. Then, as you work more with Python and perform new analyses, you become more experienced, and coding ideas and frameworks become more "available" to you, so at least you have ideas on where to begin. Even professional coders will not have all of this information at their fingertips, which is why their personal libraries are filled with coding books. Daily, they are thumbing through manuals (or the Internet) working with code.

Scatterplots, of course, do not by themselves provide a measure of the **degree of relationship** among variables, they simply reveal the scatter of the data. That is, they give us an idea of the **bivariate variability**, and nothing more. We delay the discussion of measuring bivariate relationships numerically (rather than graphically) to the following chapter where we discuss the Pearson correlation in some detail. The Pearson correlation is worthy of some attention because it is a commonly used measure for assessing the degree of relationship among variables, specifically linear ones. There, we also consider a non-parametric counterpart to Pearson, that of the Spearman rank correlation, to help evaluate relationships that may not be linear in form. We will unpack exactly what makes a correlation "tick" and what can vs. cannot be concluded from a correlation coefficient. As we discuss then, many times too much is made of a scientific result simply because a correlation is observed.

3.5 Correlograms

A **correlogram** is useful for depicting several correlations in the same visual field. As an example, consider the following plot on correlations among features of the **iris** data. For this plot, we use **matplotlib** and produce two versions of the plot. Both functions utilize `sns.pairplot()`. We first import the libraries for generating the plots, specifically `matplotlib.pyplot` and **seaborn**:

```
import matplotlib.pyplot as plt
import seaborn as sns
```

Next, we request Python to load the iris data. The iris data here is contained in **seaborn**, so we are using the `load_dataset` function as the extension to **sns** ("`sns`" because this is what we abbreviated **seaborn** to above). We name the dataframe `df`:

```
df = sns.load_dataset('iris')

df.head()
Out[150]:
  sepal_length  sepal_width  petal_length  petal_width species
0          5.1          3.5           1.4          0.2  setosa
1          4.9          3.0           1.4          0.2  setosa
2          4.7          3.2           1.3          0.2  setosa
3          4.6          3.1           1.5          0.2  setosa
4          5.0          3.6           1.4          0.2  setosa
```

The `df.head()` function requests the first few cases of the dataframe (e.g. `df.head(20)` would give us the first 20 cases, and so on). Let us get on with generating the correlogram. For this, we will use the `pairplot` function in **seaborn**:

```
sns.pairplot(df, kind="scatter", hue="species", markers=["o", "s",
"D"], palette="Set2")
plt.show()
```

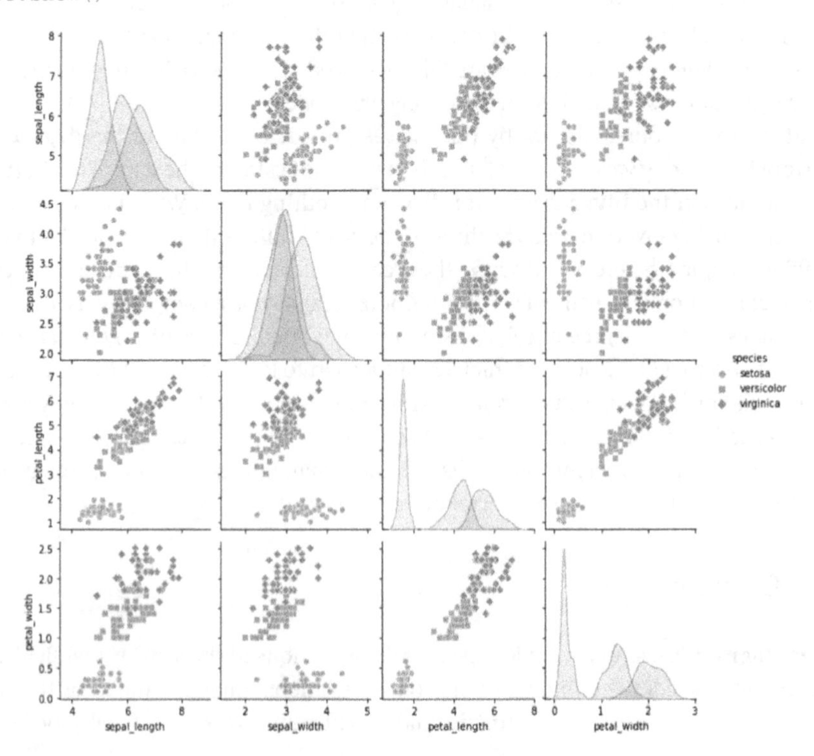

As an exercise (plot not shown), try adjusting the plot further by the following to see what effect these changes have on the plot:

```
sns.pairplot(df, kind="scatter", hue="species", plot_kws=dict(s=80,
edgecolor="white", linewidth=2.5))
plt.show()
```

We can see that included in the correlograms are also univariate distributions along the main diagonal. The options for adjusting graphics such as these are almost endless, far beyond our capacity to detail here. For more details on seaborn pairplot, visit **seaborn.pydata.org**. We will use this plot later in the book when we seek to identify **clusters** in the iris data in our survey of **cluster analysis**.

3.6 Histograms and Bar Graphs

Histograms are useful for plotting univariate data. Typically, a **histogram** implies that what is being measured on the x-axis is continuous, so that the histogram bars are allowed to touch one another. The touching implies the continuity. In contrast, **bar graphs** typically have a small space between bins, to imply that the data are not continuous, but instead **qualitative** or **categorical**. The space indicates that there can be nothing in between the categories. For example, there cannot be anything in between heads or tails on a flip of a coin. You cannot get a result in between (unless it lands on its edge I suppose, but we rule out that possibility). Likewise, if you are Democrat vs. Republican, the bar graph plotting the frequencies of these two categories equally implies a lack of continuity. We know you can be a moderate (i.e. the midpoint between Democrat and Republican), but the point is that a bar graph with these two categories implies that from Democrat to Republican there is no value in between these two categories, unlike the fact that the numbers 0 to 1 on the infinitely dense real line implies a continuum of values. The graph itself can indicate what values are possible along the scale for the given experiment, study, or data description. In practice, the distinction between what is being portrayed in the graphic in terms of continuity vs. discreteness may not always be clear from the given visualization and you may need to familiarize yourself with the data a bit more to learn whether the variable is continuous or not. Histograms and bar graphs are often used interchangeably and the definition we have given may not always be followed to such precision. Formality gives way to practicality in portraying information, so be sure to ask questions regarding what you are looking at when making the continuity vs. discreteness distinction.

In the following plot, we generate a bar graph and specify the heights of the bars as 10 through 100:

```
import numpy as np
import matplotlib.pyplot as plt
height = [10, 20, 30, 75, 100]
bars = ('A', 'B', 'C', 'D', 'E')
y_pos = np.arange(len(bars))
plt.bar(y_pos, height, color=(0.2, 0.4, 0.4, 0.6))

Out[64]: <BarContainer object of 5 artists>
```

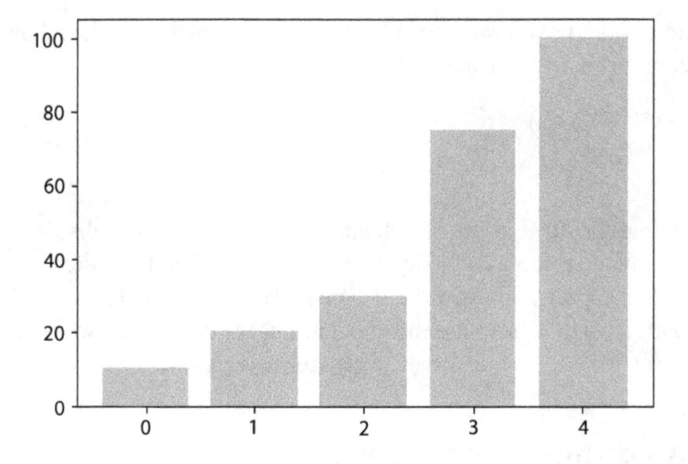

One can easily modify the entries of the plot by adjusting any of the several parameters in it.

3.7 Plotting Side-by-Side Histograms

Sometimes we would like to see histograms next to each other representing distributions of different groups. We can use **seaborn** for this, specifically `distplot`. We use the data **iris** to demonstrate:

```
import seaborn as sns
df = sns.load_dataset('iris')

sns.distplot( df["sepal_length"],  color="skyblue", label="Sepal
Length")
sns.distplot( df["sepal_width"],  color="red", label="Sepal Width")
plt.show()
```

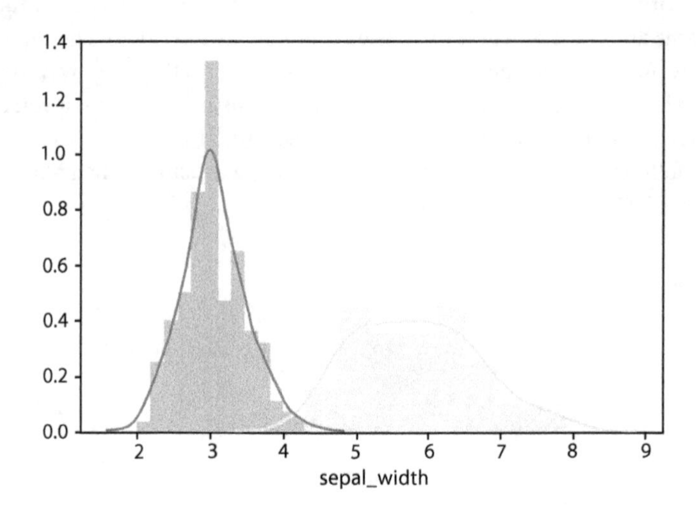

3.8 Bubble Plots

The **bubble plot** allows us to visualize a third variable in the plot while plotting the relationship between two other variables. In the case we consider here, it is read as an ordinary scatterplot, but the size of the bivariate points (i.e. the "bubbles") reveals information about a third variable in terms of **magnitude**. Here we plot a simple plot using random numbers:

```
x = np.random.rand(40)
y = np.random.rand(40)
z = np.random.rand(40)

plt.scatter(x, y, s=z*1000, alpha=0.5)
plt.show()
```

We can easily change the relative sizes of the bubbles by adjusting `s=z*1000`. For example, instead of equal to 1,000, if we size it down to 10, so that we have `s=z*10`, we obtain the plot above on the right. Note that if we make the marker `s` too large, the bubbles become too big and the plot is essentially useless:

```
plt.scatter(x, y, s=z*10000, alpha=0.5)
plt.show()
```

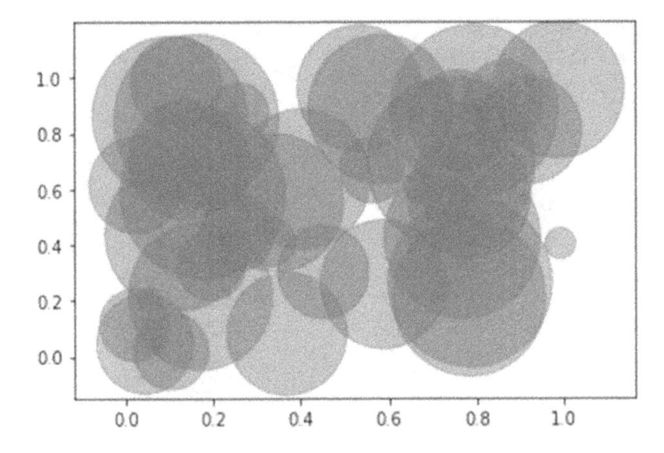

This emphasizes a common theme when generating graphics in software, such that you may have to "tweak" things a few times before you get it "just right" and the graphic communicates what you wish to say in an efficient manner. In the case of the bubble plot, the best size of the bubbles may need some experimenting with before you land on the best option. Novices to data analysis often do not realize the time and energy it can take simply to create even **one** appealing graph when generating results.

Several varieties of bubble plots were especially useful and popular during the COVID-19 pandemic. Figure 3.4 is a plot produced early in the pandemic by Johns Hopkins University showing the number of cases per 10,000 population.

Again, graphing data can help with conjuring up new hypotheses as to the reasons why the concentrations were so high in certain regions but not others. For instance, in the United States, though we know the east coast's population density is greater than in the western states, it is also true that it is quite a bit drier in states such as Montana and Wyoming and the Dakotas. Could this explain the prevalence of COVID-19? Probably not, but the point is that what might appear to be one explanation for the data could quickly change if you consider a different variable or possibility. **One new variable can change everything when it comes to interpreting data or narrowing in on a scientific theory.** Determining true causes or reasons is very difficult, which is why data analysis on COVID-19's spread could not reveal causes. It could only conjure up further hypotheses. As we will see later in the book when we survey ANOVA and regression, even adding a single variable to a statistical model can change the effects of other variables. Hence, when you are seeking an explanation for data or events, you must always try for the most thorough picture you can get. You could have 99 variables at your disposal, yet it is the 100th one that you do not have that is the key to solving the scientific puzzle. Thus, when you model the first 99, whatever model ensues will necessarily be wrong, and if the 100th variable is of vital importance, the model based on the first 99 variables may be entirely misleading without it. We will discuss this issue and more as we progress throughout the book. **Model context matters a great deal.** It is essential that you learn what makes a statistical model "tick" and the extent to which it might (or might not) reflect scientific realities.

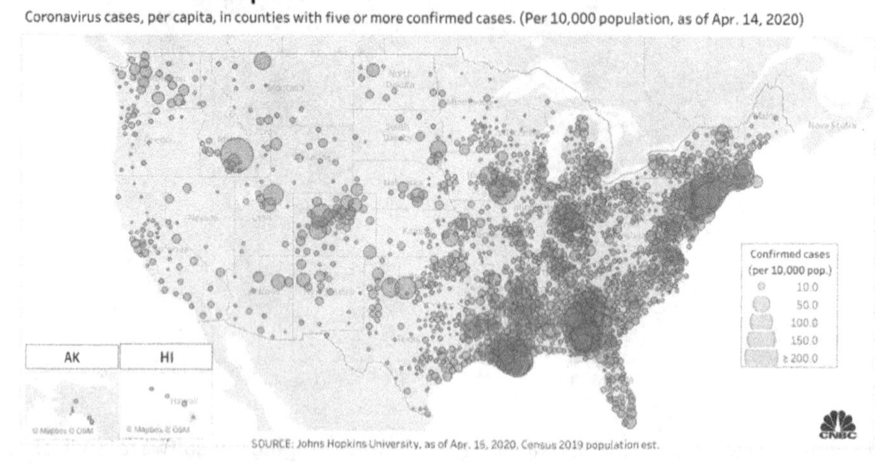

Figure 3.4 Bubble plot of COVID-19 cases across the United States early in the pandemic. *Source: Johns Hopkins CSSE. Used with Permission.*

One weakness of bubble plots is that it may be difficult to decipher the sizes of the bubbles in relation to other bubbles, especially given overlapping bubbles. For instance, in the aforementioned plot, it becomes difficult to know from the plot which areas definitively had the most cases due to the similar-sized bubbles in the plot. Displaying a bar graph would reveal more regarding the **horizon** of the different categories than bubbles do. That is, instead of the bubble plot, one might instead generate a bar graph so that one could more easily see the degradations in density as we move from state to state. That "idea" is virtually impossible to "see" in the bubble plot.

3.9 Pie Plots

Pie plots or pie "charts" are another common and popular graphing device (and a website favorite!), despite the fact that they can be quite problematic when it comes to the clear and effective communication of data. To demonstrate the problems with pie plots, consider the following plot, which we produced using **pandas** with categories `a` through `d`:

```
import pandas as pd
df = pd.DataFrame([8,8,1,2], index=['a', 'b', 'c', 'd'],
columns=['x'])
```

The function we use is `df.plot()`, specifying `pie`:

```
df.plot(kind='pie', subplots=True, figsize=(8, 8))
Out[64]:
array([<matplotlib.axes._subplots.AxesSubplot object at
0x000000001A783550>],
      dtype=object)
```

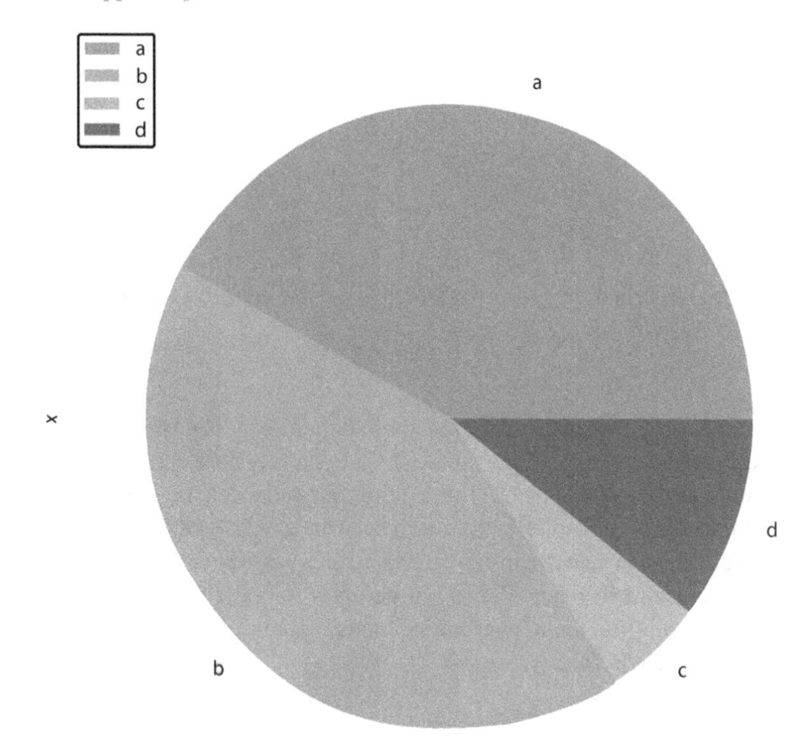

How is the innocent pie graph problematic? In color (you will have to produce it yourself to see the colors), it certainly looks appealing, and we see it in magazines and online data descriptions all the time. However, the plot we just produced clearly reveals the difficulty. Similar to the bubble plot, it becomes difficult to decipher the size of the various "slices" unless they are quite distinct. To convince yourself of this, take a quick two-second glance at the plot and answer the question of which category, a or b, has the most frequency? It is almost impossible to tell this information from a quick glance at the plot, and when the numbers are not included in the slices, the plot is quite useless for making such distinctions. Much more useful would be a bar graph or histogram, where again the **horizon** is more apparent. Then we could immediately see which category has more frequency. Having said that, pie graphs and charts are an all-time favorite and they will likely be a mainstay of visualization techniques (especially outside of professional papers) for years to come. Simply be aware of their drawbacks and limitations when interpreting them.

3.10 Heatmaps

A **heatmap** is a graphic where a measure of magnitude or density is typically indicated in the plot. As an example of a heatmap in Python, let us create one on a random data set. First, we create the data with columns a through e:

```
import seaborn as sns
import pandas as pd
import numpy as np
df = pd.DataFrame(np.random.random((5,5)),
columns=["a","b","c","d","e"])
df
Out[161]:
a b c d e
0 0.304220 0.207444 0.348501 0.381417 0.524326
1 0.163985 0.263075 0.840032 0.592651 0.682820
2 0.265063 0.111662 0.445554 0.370213 0.483865
3 0.337630 0.100143 0.339747 0.579408 0.261271
4 0.056892 0.322099 0.111199 0.915698 0.429183
```

The above is now a matrix of five rows (0 through 4) by five columns (a through e). We now create a heatmap of this data using `sns.heatmap()`:

```
map = sns.heatmap(df)
```

It is easy to match up the heatmap with the above data layout. The smaller the number, the darker the shade. For example, the cell 4-a has a very dark shade. It is associated with a number of only 0.056892. On the other hand, the cell 4-d has a very light shade and is indicated with a much greater number, that of 0.915698. There is no limit to how the shading idea of a heatmap can be created or adjusted to reflect data. Similar plots can be very useful to reveal density of points in a scatterplot as well, where instead of jittering points on top of one another, density is revealed by shade. We demonstrate this using the **iris** data:

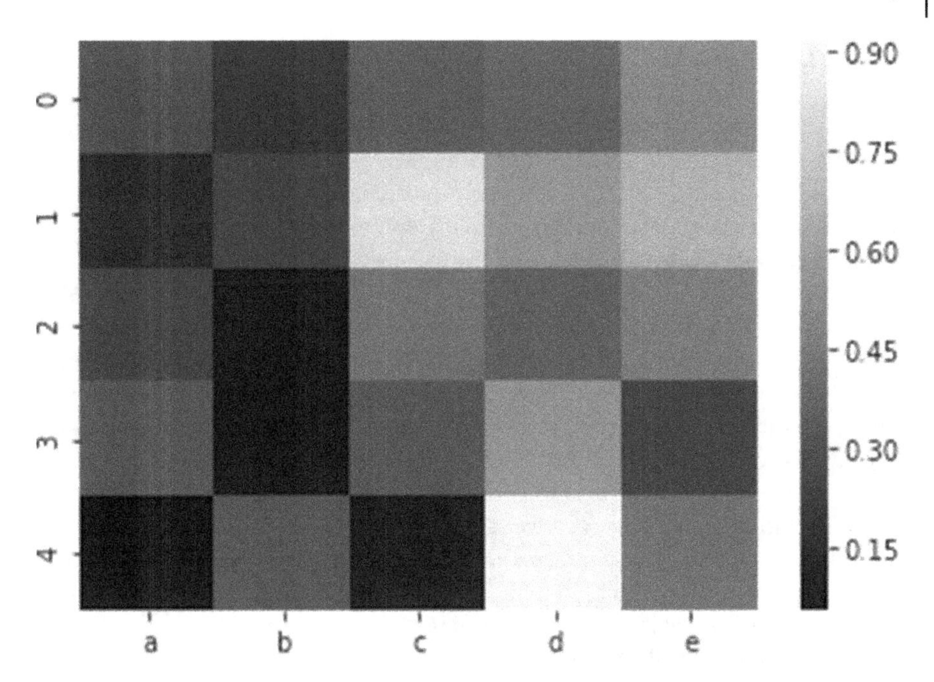


```
import seaborn as sns
df = sns.load_dataset('iris')

sns.jointplot(x=df["sepal_length"], y=df["sepal_width"], kind='hex')
sns.jointplot(x=df["sepal_length"], y=df["sepal_width"], kind='scatter')
sns.jointplot(x=df["sepal_length"], y=df["sepal_width"], kind='kde')
Out[3]: <seaborn.axisgrid.JointGrid at 0x1a284f98
```

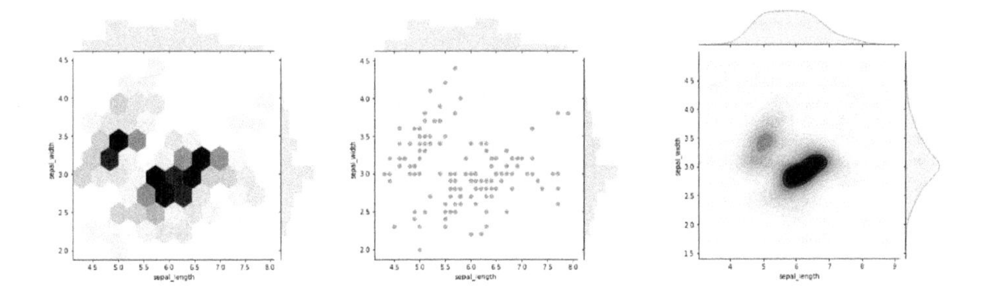

Notice that to generate the different plots, all we adjusted was the `kind = ''` category. For instance, the third plot featured `kde`, which stands for **kernel density estimate**. Which plot of the three is most useful? While they each communicate slightly different information, the leftmost plot (kind = 'hex' for "hexagonal bins") is probably our favorite, as it combines a measure of relationship with density per cell. In all plots marginal distributions are plotted as well, represented by the bar graphs at the top and right side of the first two plots, and corresponding density plots in the third plot.

3.11 Line Charts

Line charts or line graphs are extremely common, and are especially excellent for revealing the course of a variable over **time.** They can be very useful for plotting repeated-measures data to track behavior over time, such as pre-test, post-test, and later timewise evaluations. Consider the line graph of the price of oil since 2019, featuring the historic crash of 2020 where the price of a barrel of oil actually turned negative (Figure 3.5).

What makes line graphs especially appealing is that they have a reasonable **horizon** (unlike those bubble plots and pie charts), meaning that it is (in most cases) easier to see which time points have the highest or least values. Cutting up the x-axis into increasingly more narrow slices is also always a possibility, though it may not always be a good idea because then the categories can get too thinly parsed (for example, suppose the x-axis was by day rather than by month in the oil plot). As always with graphs and plots, it is a trade-off, in that if we try to parse things out too "thinly," we actually lose information, because we are too focused on the case-by-case variation (such as with a table of data with narrow interval sizes). On the other hand, a too "global" portrayal of the data does not allow us to see case-by-case variation nearly as well. **Always aim for a balance of sorts between these extremes to produce the best graph suited to your needs.** No graph will be perfect. It may require a lot of "tweaking" and adjusting before you settle on the graphic that most effectively meets your goals of what you wish to communicate with your data.

We can easily create a line graph in Python using `matplotlib.pyplot`:

```
import matplotlib.pyplot as plt
import numpy as np
```

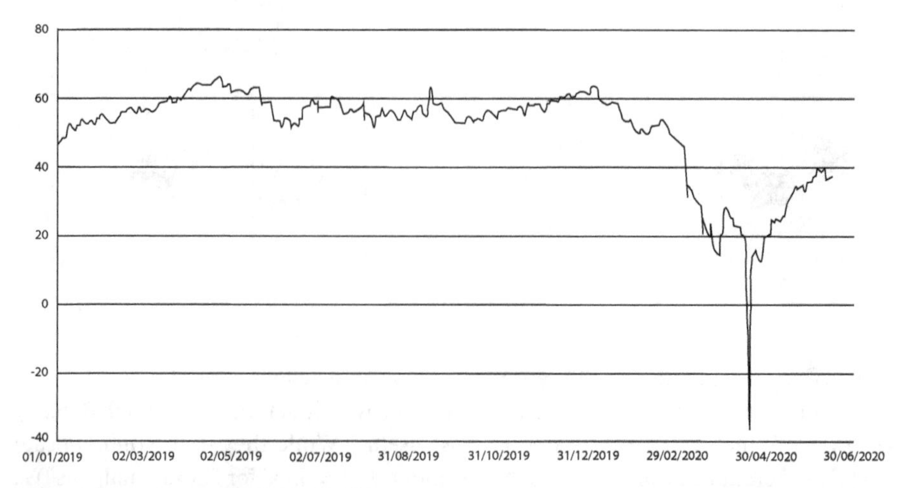

Figure 3.5 The price of oil from 2019 to 2020. The price turned negative during the COVID-19 pandemic. *Used with Permission.*

Let us first create some random numbers from the normal distribution using `np.random.randn()` with a range of 1 to 1,000 on the x-axis:

```
values=np.cumsum(np.random.randn(1000,1))
```

We now plot the resulting data, in which we can clearly see the rise and fall of the data across values on the x-axis:

```
plt.plot(values)
Out[4]: [<matplotlib.lines.Line2D at 0x1a8d89e8>]
```

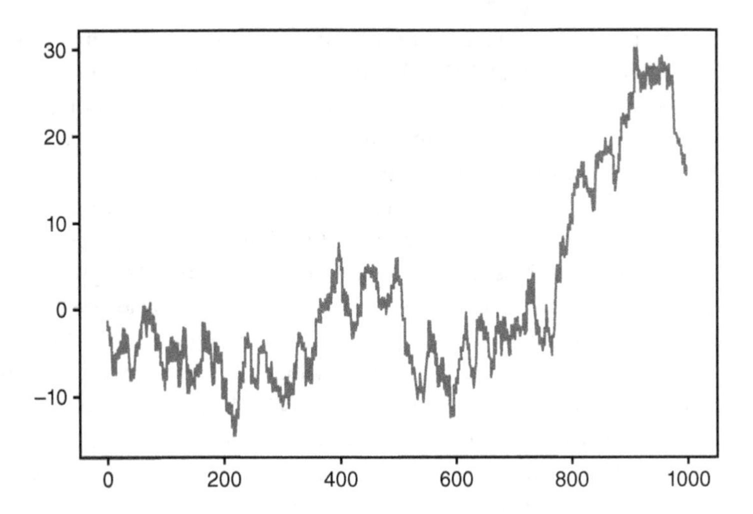

3.12 Closing Thoughts

In this chapter we have merely touched on a few of the possibilities provided by Python when it comes to graphics and visualization. Since the goal of this book is an introduction to applied statistics and data analysis rather than visualization in particular, space has allowed us to simply provide a glimpse into this vast world. Indeed, one can easily spend one's career developing new graphics and ways of visualization (many professionals choose to do exactly this), and combined with the increases in computer software and technology, the sky is the limit. If graphics and visualization interest you, you are strongly encouraged to pursue this exciting field. VanderPlas (2017) provides further detail and specifics on generating powerful visualizations using **Matplotlib** and **Seaborn** in particular, and should be consulted for more information. As always, if you need to produce a fancy graph for your research or publication, it is well worth delving into Python's graphing capabilities to brainstorm the best plot and explore many possibilities, not just one or two. You do not need to be a graphing expert to produce a great graph that is much better than "canned graphs" you can more easily obtain in Excel or SPSS. Open up your Python books, get online, and do some searching; that is what research and computing is all about!

Review Exercises

1. Discuss and explore the impact a **visualization** can have on one's **perception** of a data set. Whether or not the information gleamed from a graphic is correct or not, what are the benefits or dangers of it influencing one's actions and decisions?
2. Why is aiming for **simplicity** in graphics a good strategy? Why is unnecessary complexity not encouraged? (For example, why is a 3-D histogram graphic in MS Excel potentially not a good idea?)
3. Discuss why the idea of a **percentage change** can be very misleading when interpreting data. Refer to the COVID-19 example featured in the chapter and then see if you can come up with an example of two small data sets (of approximately 10 numbers each) to demonstrate the problem with interpretation. Why are raw numbers sometimes preferable over percentage changes?
4. For the population change census data explored in the chapter, compare the **percentage change** to the **absolute change** for Idaho vs. other states. After doing so, comment on why the statement "Idaho is one of the fastest growing states," though technically correct, can nonetheless be somewhat misleading.
5. Elementary as it may seem at first glance, understanding what a **scatterplot** does vs. does not tell you can be challenging. Consider the following two plots relating total confirmed cases of COVID-19 to total COVID-19 tests conducted, presented on both a **linear scale** and then a **log scale**. What are these data telling us, and why does the plot look different across the linear vs. log scales? That is, what is the impact of changing the **scaling** along the y-axis?

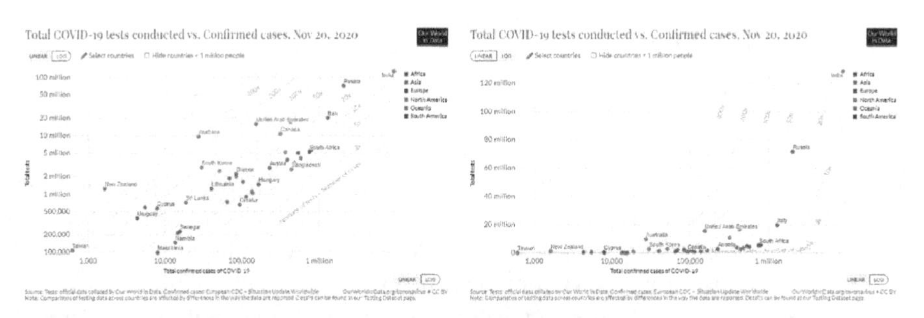

Published online at OurWorldinData.org. Retrieved from: https://ourworldindata.org/grapher/covid-19-total-confirmed-cases-vs-total-tests-conducted.

6. Using the **iris** data, generate a scatterplot in Python with sepal length on the x-axis and petal length on the y-axis. Does the relationship look linear? Estimate the strength of the linear relationship in the plot.
7. Plot a **histogram** of sepal length in the **iris** data and comment on the shape. Is the distribution approximately normal? Skewed? Now, plot the same data but this time with a **boxplot** (use `seaborn.boxplot()`). Compare the histogram to the boxplot. Which is more informative to you, and why?

8. Why are **bubble plots** potentially problematic? For example, for the data featuring the number of COVID-19 cases by country, how might a bar graph be more useful than a bubble plot here? Are there any disadvantages to plotting the bar graph instead of the bubble plot?
9. Derive a very small data set of 5–10 numbers that effectively illustrates the primary problem with **pie graphs.** Why would a "boring" histogram or bar graph be a better choice for your data?
10. When might producing a **heatmap** be useful over simply interpreting numbers in a table? In other words, why not simply look at the numbers in a 2 × 2 table (for instance) instead of plotting a heatmap?
11. Consider the statement, "Graphs, like statistical models, will never reveal the complete picture, and may reveal just enough of it to draw faulty conclusions." Interpret and discuss this statement and invent a situation where **not graphing data** is probably a wise choice.

4

Simple Statistical Techniques for Univariate and Bivariate Analyses

CHAPTER OBJECTIVES

- Understand the Pearson Product-Moment correlation coefficient and evaluate it for statistical significance.
- Understand why the Pearson correlation captures linear relationships only, not nonlinear ones, and hence a correlation of zero does not necessarily imply the absence of a relationship.
- Understand and appreciate what you can vs. cannot conclude from a correlation between variables.
- How to generate a scatterplot in Python, including a scatterplot matrix revealing univariate as well as bivariate distributions.
- How to calculate a Spearman correlation as a nonparametric alternative to Pearson.
- How to compare means using t-tests for independent and paired-samples.
- How to conduct a binomial test as well as a chi-squared goodness-of-fit test on contingency tables when working with categorical data.

In this chapter, we feature a variety of relatively simple tests that can be used immediately without too much background theory to address many of the research questions you may come across over the course of your career. Though in the most complete and formal sense most of the tests in this chapter have an underlying statistical model, they are nonetheless much easier and straightforward than many of the statistical modeling techniques surveyed in the remainder of the book. For instance, techniques such as the **analysis of variance** (ANOVA) or **regression** involve hypothesizing quite intricate models at times, and require a fair bit of expertise when interpreting such models. In this chapter, however, we do not focus on elaborate statistical models, but instead choose to provide some quick and efficient solutions to address basic research questions that often arise. This does not mean that interpreting such tests in the context of research is always easy. Statistics and its application to empirical data can be very **slippery** in that even what might appear to be a very simple test can be very easily misunderstood and misapplied when interpreting data. Never assume that because a test looks easy, its interpretation in the context of science is equally as simple. Simple tests are often the most misused or misunderstood. There is nothing "simple" about a

Applied Univariate, Bivariate, and Multivariate Statistics Using Python: A Beginner's Guide to Advanced Data Analysis, First Edition. Daniel J. Denis.

t-test if you are unaware of what makes the size of t large or small and conclude something from the test that is incorrect.

Tests in this chapter cover much of the kind that would be covered in introductory statistics at the undergraduate level, and include such things as **correlation, non-parametric correlation, *z*-tests for means, *t*-tests**, as well as inferences featuring the use of **categorical** or **count data**, including **binomial tests** as well as **chi-squared goodness-of-fit tests**. The components of these tests form the bedrock of more advanced statistics as well and more complex models.

If you are a true newcomer to statistics, you will have much to learn from these tests. If you are an experienced student or researcher already, all or most of these tests will likely be familiar, yet you are still encouraged to consider each test carefully by embracing the discussion accompanying each one, as there may still be something there for you to learn that may have been missed in previous learning. For instance, tests of correlation are quite simple to conduct. However, what a correlation can actually "tell you" from your data on a scientific level is usually very limited, and a discussion of what correlation does vs. does not provide is essential. Interpretation-wise, **correlation is perhaps one of the most abused statistics computed on data and the most misinterpreted**. Now, that does not mean we have to delve into a deep discussion of the philosophy behind the concept of correlation (that topic could easily fill a chapter in its own right), but it does mean that if one is not at least **familiar** with this discussion, one stands the chance of making quite egregious interpretations of empirical data based on simply computing correlation coefficients! When we conclude something from a statistical test that is methodologically wrong, we are probably better off not computing the statistic in the first place. A false conclusion is often more problematic than no conclusion at all, and such a happening rears its head not only in the social sciences, but also in the natural sciences.

For example, if we found that temperature was correlated with the presence of the COVID-19 virus, and somehow concluded that heat or lack thereof "caused" COVID based only on this correlation, that would be a severe misuse and misunderstanding of what correlation provides or measures, and we would probably do better not computing one at all (if we are to be abusive in the interpretation). In such a case, **our (faulty) interpretation of the statistic is actually misinforming us**. Examples such as this are replete in the research literature where a statistical test is used as backing for a research claim that is simply not supported, and the authors would have done much better not performing or interpreting the test at all. As mentioned earlier in the book, **mediation models** are a prime example of this and of how inaccurate scientific conclusions are drawn based on evidence for statistical (but not necessarily substantive) mediation.

4.1 Pearson Product-Moment Correlation

The **Pearson Product-Moment correlation** is a measure of the linear relationship between variables, and is given by

$$r = \frac{\frac{\sum_{i=1}^{n}(x_i - \bar{x})(y_i - \bar{y})}{n-1}}{s_x \cdot s_y} = \frac{\text{cov}_{xy}}{s_x \cdot s_y}$$

The correlation coefficient here is simply the **standardized covariance**. But what does this mean? To understand any statistical function, you need to look at its components. That is, you need to look at the mathematics of it to see what the formula actually "does." Let us look at the numerator of the quotient:

$$\frac{\sum_{i=1}^{n}(x_i-\bar{x})(y_i-\bar{y})}{n-1}$$

What we have defined immediately above is called the **covariance**. The covariance is the **average cross-product** between variables x_i and y_i. We say "average" cross-product because we are dividing the sum by n–1. Recall that an arithmetic average is typically computed by dividing by n, not n–1. We are dividing by n–1 because we are losing a **degree of freedom**, but nonetheless we notice the quotient is still an "average-like" statistic. The fact that we are losing a degree of freedom does not suddenly make the function "non-average-like." It is still a kind of average. Covariances are used extensively in statistics and statistical modeling, but they come with a potential problem. The problem is that they are **scale-dependent**, which means they will be influenced by the **raw scale** on which x_i and y_i are computed. If x_i and y_i contain a lot of variability, then all else equal, the sum of products may be quite large, since we are multiplying one deviation by another from respective means. Dividing by n–1 does not solve this problem either, since all this is doing is distributing the sum across the n–1 pieces of information that went into the sum. Hence, **it is entirely possible to get a large covariance even if the relationship between x_i and y_i is not terribly strong**. This is the essential point.

The solution, as given in Pearson's r, is to divide by $s_x \cdot s_y$. But why do this? When we divide by $s_x \cdot s_y$, we are **standardizing the covariance**, and hence placing **bounds** on the resulting number between −1.0 and +1.0. Why does dividing by $s_x \cdot s_y$ have this effect? While this can be proved mathematically, for applied purposes the proof is not terribly enlightening. Much more useful is to simply look at what the formula is doing. You might think of $s_x \cdot s_y$ as the total product of deviation **possible** in the data. That is, it is the total "cross-product" that is possible for the two measured variables under consideration. In other words, it gives you the "**play space**" on which to evaluate the observed relationship. It is the baseline covariability available to you for the given data. The numerator of the correlation coefficient tells you how much of this "place space" is due to the actual relationship between x_i and y_i and, hence, the **numerator is a subset of this larger play space**. Its value cannot be greater (in absolute value) than the entire play space since the denominator represents the total cross-variation possible. Hence, the correlation coefficient in this regard is accounting for the total cross-variation observed relative to the total cross-variation possible in the given set of data. In this way, the **Pearson correlation coefficient is, in effect, a ratio**. And because we are generating this ratio, the scale of the variables becomes a non-issue, in that the resulting number reflects a **dimensionless relationship** between x_i and y_i. Since the numerator cannot be greater than the denominator, the bounds on the correlation coefficient are necessarily between −1.0 and +1.0. The sign of the coefficient will be determined by the sign of the covariance, but the magnitude of the (dimensionless) relationship that exists will be determined by Pearson's r.

4.2 A Pearson Correlation Does Not (Necessarily) Imply Zero Relationship

Very important to understand is that **a Pearson correlation of zero in the sample does not equate to there being a zero relationship in the data.** This is a mistake often made by students and even experienced researchers, almost equivalent in nature to that of assuming normal distributions by default when discussing any phenomena. For instance, if I told you the mean IQ of the population is 100, you might automatically mentally assume that half of the population has an IQ less than 100 and half has an IQ greater than 100. However, this is incorrect, unless we knew the distribution of IQs to be normally distributed and/or symmetrical (e.g. for a bimodal distribution this statement would be correct also). When we think of "half below" and "half above," the temptation is usually to automatically incorporate normality. If the distribution is skewed, the statement is false. Of course, for the median the statement is true, but not necessarily for the mean. Another example of how perception is reality, is that if you tell someone shopping for a house that the average home price in the area they are considering is 200k, they might automatically assume that half of the houses are below this price and half above, but this would be an incorrect assumption, given that we do not know the exact form of the distribution. **Always ask about the distribution on which any statistic is computed so you can put that statistic in context of the elements on which it was computed.**

The analogous temptation exists with the correlation coefficient. **A correlation coefficient should never be interpreted without simultaneously visualizing a plot of the two variables.** Now, that is a statement you **should** memorize! Two variables may be related, sometimes strongly so, but not necessarily **linearly**. Pearson's correlation will only reflect the degree (or non-degree) that variables are linearly related. Its magnitude will not account for non-linear trends. If the relationship is curvilinear, for instance, Pearson's correlation, though computable on such data, will not reflect the degree of this curvilinearity. Hence, you could have discovered a very strong relationship, yet it would be missed by a simple computation of Pearson's r. As an example, consider the following plot:

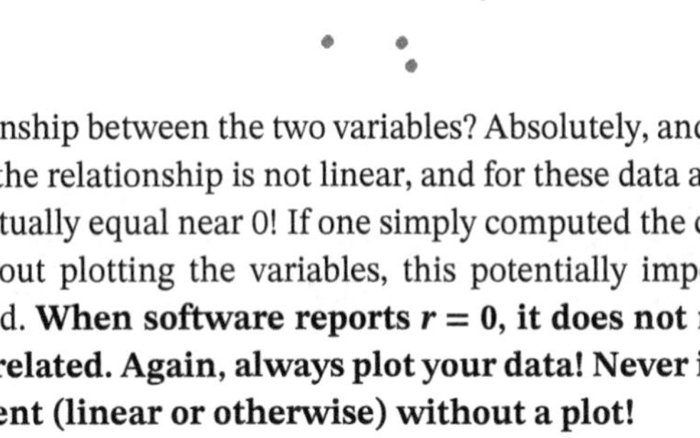

Is there a relationship between the two variables? Absolutely, and quite a strong one at that. However, the relationship is not linear, and for these data at least, the Pearson correlation will actually equal near 0! If one simply computed the correlation via software output without plotting the variables, this potentially important relationship could go unnoticed. **When software reports $r = 0$, it does not necessarily mean variables are unrelated. Again, always plot your data! Never interpret any correlation coefficient (linear or otherwise) without a plot!**

A Pearson Product-Moment correlation assesses the linear relationship among two variables and consists of dividing an average cross-product in the numerator (called the covariance) by the product of standard deviations in the denominator. It is very important to understand that a Pearson r of 0 does not necessarily imply the absence of a relationship between two variables. What it does suggest is the absence of a linear relationship. Always plot your data to potentially detect relationships that are non-linear in form. A coefficient of correlation of any type should never be interpreted without an accompanying plot.

4.3 Spearman's Rho

If the relationship between variables is **non-linear**, one may choose to compute a coefficient that will account for the non-linearity. One coefficient that does this is **Spearman's rho**, named after Charles Spearman who derived the coefficient in 1904. Spearman's rho reflects the relationship between two variables, but is not restricted to that relationship necessarily being linear in form. It may be linear, but it need not be for Spearman's rho to still be quite large. Like Pearson, Spearman's rho ranges from −1.0 to +1.0, but it is distinct from that of Pearson in that it measures **monotonically increasing** or **decreasing** relationships, not simply linear ones. What does this mean, exactly? A plot will help illustrate the concept. Consider the following plot:

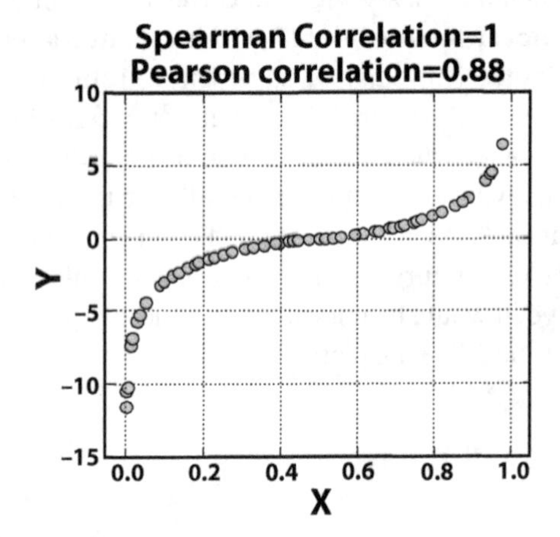

In this plot note that as x_i increases, y_i also increases, even though the relationship is far from linear. The idea of "monotonically increasing" in this case simply means that each increase in x_i pairs itself with an increase in y_i. As we can see from the plot, the relationship is strong, yet Pearson's correlation would not capture this strength. However, according to Spearman, the relationship is perfect and equal to 1.0 because of the monotonicity. It would be a shame to compute a Pearson correlation on such data without a plot and conclude that you have less of a relationship than actually exists. Again, I repeat (I think this is the third time!) – **ALWAYS PLOT YOUR DATA.**

To compare and contrast a Spearman correlation with a Pearson correlation, consider the following data on two generic variables (we are making up the numbers):

```
x = [0, 2, 6, 7, 15]
y = [0, 1, 8, 13, 20]
df = pd.DataFrame(x, y)
df
```

```
Out[26]:
0    0
1    2
8    6
13   7
20   15
```

We first compute Pearson *r* to see what we get. For this, we will use **scipy**, specifically `scipy.stats()`:

```
import scipy.stats
scipy.stats.pearsonr(x,y)

Out[51]: (0.973647485990732, 0.0051149659365227395)
```

We can see that the correlation between *x* and *y* is equal to 0.97 and is statistically significant with a *p*-value of 0.005. That it is statistically significant is not the issue for now. The size of it is, however, and we see that it is quite large and impressive. Now, let us try a Spearman rho on the same data:

```
scipy.stats.spearmanr(x,y)
Out[52]: SpearmanrResult(correlation=0.9999999999999999,
pvalue=1.4042654220543672e-24)
```

We can see that Spearman's rho is equal to 1.0 (rounded up), also statistically significant. But why is Spearman a perfect correlation and Pearson not? Look at the data once more. Notice that for every increase in *x*, there is a corresponding increase in *y*. This is exactly what Spearman's measures, **the monotonic increase in one variable as a function of another**, regardless of whether or not the relationship is linear. With Pearson, it assesses linearity, and hence the correlation will be less. A plot of the data will help shed light on why Spearman is higher than Pearson here:

```
plt.scatter(x, y)
Out[53]: <matplotlib.collections.PathCollection at 0x1c184978>
```

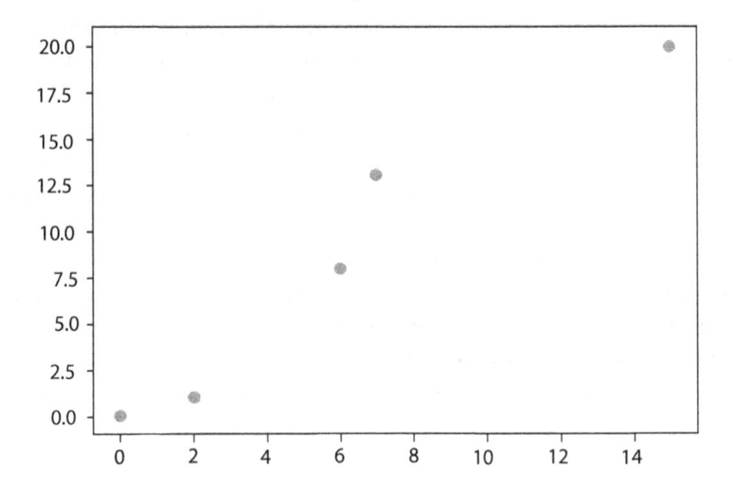

Notice from this plot that the relationship is not perfectly linear. However, an increase in one variable is consistently accompanied by an increase in the other. This is precisely why Spearman and Pearson do not agree. Pearson tries to essentially fit a line through the scatter, which of course does not fit very well since the scatter is a bit messy. Spearman, on the other hand, attempts to assess whether an increase in x goes with an increase in y, which for these data it does so perfectly. Hence, this is why Spearman is larger than Pearson. Which correlation reflects the "true" relationship, Pearson or Spearman? The question makes no sense, since each coefficient is defined differently, and so it assesses different things. It is not a matter of one being a population coefficient and the other not. Both are sample statistics and both can be used to estimate their respective population counterparts. That is not what distinguishes the correlation coefficients here. What does is what we have emphasized: one assesses linearity, the other a monotonically changing relationship. Hence, this highlights an important universal principle when interpreting statistics in general, which is, **what is revealed depends on what is being measured**. Is there a relationship between x and y? The only correct immediate answer to this question is, "Depends on what you mean by 'relationship', in the Spearman sense, yes, quite strong, in the Pearson sense, slightly less so." Interpretations of data are only as meaningful as the functions used to compute them. This is why understanding the formulas and equations is paramount, long before you apply them to data.

As we will see when we study regression later in the book, the same question can be raised as it concerns prediction. "Does x predict y?" It depends on how you define, mathematically, the "idea" of prediction. As we will see, we will usually define it in the sense of least-squares. The "it depends" part is important here. A relationship (or any statistical entity) is only as meaningful as the computation for it. Is there a relationship between COVID-19 and age? It depends on what we are computing to attempt to capture and assess that relationship. The concept of "relationship" is quite hollow without simultaneously including details of the mathematical function used in measuring such a relationship. Formulas and mathematical functions in general tell us first and foremost what we can vs. not conclude about data. Empirical relationships exist only because we have a concept (a mathematical formula) to define them. Otherwise, we cannot measure them. Before correlation was invented by the likes of Francis Galton and company in the 1880s, the idea of a relationship between variables was quite vague. Correlation and regression gave humanity a way of measuring and assessing this "concept." Again, concepts give rise to statistics and mathematics, and these latter are simply precise and very rigorous ways of getting a handle on the underlying concept. Look at the formula for correlation once more and you will see a symbolic representation of a more elusive concept, that of "relationship." However, why stop with simply one way of assessing "relationship?" This is exactly why Spearman's rho was invented, among many other correlation coefficients. **There are different ways of tapping into similar ideas, and in statistics it is no different. Concepts in statistics are philosophical "fuzzy" entities that have been germinating throughout history, until they were eventually defined "rigorously" via symbols (i.e. mathematics).**

4.4 More General Comments on Correlation: Don't Let a Correlation Impress You Too Much!

Virtually everything in nature is correlated to some degree. Understanding this idea is extremely important! This means if you measure almost any variables you could conceive, you will necessarily find a correlation different from zero! Again, pause for a moment to consider how profound this statement is. It means if you go out into nature, take two empirically-derived variables at random, correlate them, **you will find a correction in the sample**. It is a virtual guarantee, which begs the question why evaluating null hypotheses about a zero correlation in the population may seem a bit ridiculous (as many methodologists, not just myself, have long pointed out). What is more, with a large enough sample size, you can reject the null hypothesis for any correlation computed on a sample. Hence, when you see "$p < 0.05$" for a correlation in a research report, it may (and often does) mean very little. You must look further into how large the correlation actually is and whether its statistical significance is mostly a function of sample size. That is, when you obtain a small correlation with your data, you must ask yourself when it is truly that "big of a deal" or if nothing really "is there" at all with your data. Often, researchers like to make "big deals" of very small correlations simply because the correlation comes out to be statistically significant, and these interpretations are often very much misguided if not downright dangerous. Can you imagine correlating a medical treatment for COVID-19 and concluding that the treatment works without having performed a proper experimental design? It is mandatory in this regard to understand the distinctions between a **test of correlation** vs. a **correlational design**. We discuss this in some detail later. Without a good design, you may obtain reasonably-sized correlations, but they may simply not be that meaningful due to profound differences between experimental and correlational research.

For now, the point is that all one can conclude from a statistically significant correlation is that there is evidence to suggest that the true correlation in the population from which the data were drawn is not equal to zero. That's it! A statistically significant correlation does not somehow imply, by itself, that the correlation is large, nor does it even imply the correlation is important. Substantively (read: scientifically), the correlation may be absolutely trivial, and it may even have been **a priori expected**, even if it is statistically significant. **There is nothing at all in nature that should presume that the correlation between variables we study should be zero. Most often, correlations are not zero, but that does not necessarily mean something meaningful or impressive is going on. It simply means most things are correlated.** Hence, when a researcher reports correlations and puts stars next to them to indicate statistical significance, you, as a critical consumer of research, should not automatically associate any measure of importance or even relevance to what has been reported. You simply need to know more about the problem at hand and under investigation to draw more precise conclusions. **The correlation itself is simply a mathematical abstraction**. The computation does not know anything about the "real" empirical objects you are applying the correlation to. This cannot be emphasized enough! What is more, even mediocre correlations still explain very little variance, a concept that will be elaborated on in our discussion of regression models later in the book. For example, consider a correlation of 0.5. When squared, it explains only 25% of the variance. That is, when we square the Pearson correlation coefficient, we obtain a measure of the amount of variance explained:

$$r^2 = \text{proportion of variance explained}$$

So when a researcher simply claims a correlation between two variables, even if the correlation is quite moderate to large (e.g. 0.5) in absolute size, the total variance explained may be only 25%! Confident as they may be that they now understand a "link" between two variables, it is only the humble scientist who acknowledges that 75% of the variance is still a mystery. Indeed, $1 - r^2$ should be called the **coefficient of humility**. As a consumer, you will regularly come across "fascinating" research findings that nonetheless have quite modest effects. This is more to do with research scientists wanting to promote their results than anything necessarily impressive. The exaggeration is a **social phenomenon**, not a scientific one. As Einstein reminded us, even though we know "some" of what there is to know, most of what there is to be known we are still quite clueless about. Einstein was referring to the $1 - r^2$ idea. Marketing and emphasizing small findings should not be a part of science, but sadly it is for the purpose of generating careers (we all have to get paid somehow!). We expect this kind of thing from the popular press, but unfortunately marketing has also made its way into scientific journals for the purpose of career promotion. **Always be critical of what you are reading in even the best of media or journal publications. Is the claim that is being made representative of the actual research, or is at least some of it imbued with clever marketing making extraordinary claims that go far beyond the science?** Good journals (not necessarily always the most prestigious) publish good science, not merely "popular" science.

4.5 Computing Correlation in Python

Python allows us to easily compute a correlation coefficient on random data, and we do so to get us started. We can compute a correlation coefficient on random data via the following:

```
import numpy as np
np.random.seed(1)
x = np.random.randint(0, 50, 1000)
y = x + np.random.normal(0, 10, 1000)
np.corrcoef (x, y)
array([[1.,     0.81543901],
       [0.81543901, 1.        ]])
```

For this data, we see that the correlation is equal to 0.815, a relatively strong correlation in absolute value. We can generate a scatterplot quite easily for these data using **matplotlib**:

```
import matplotlib
import matplotlib.pyplot as plt
%matplotlib inline
matplotlib.style.use('ggplot')
plt.scatter(x, y)
```

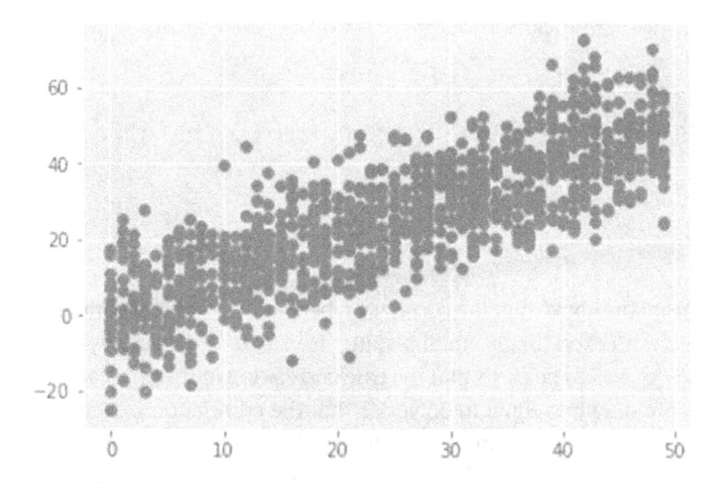

We see that the plot reveals the rather strong relationship between variables. There does not appear to be any visible bivariate outliers in the plot. For these data, a Pearson correlation coefficient is definitely appropriate because we have evidence of linearity in the plot. Even if the linearity were not present, had we theorized a linear relationship between the variables, then computing a Pearson correlation would still be well worth our time, even if to demonstrate the failure of our theory to agree with data.

We now consider covariance and correlation on a more elaborate data set. We consider the Galton data, which we obtained from R (only the first few cases are shown). In this data set are the heights of parents and their grown children. We would like to see if there might be a linear relationship between these two variables:

```
galton = pd.read_csv('Galton.csv')
Out[212]:
     Unnamed: 0  parent  child
0             1    70.5   61.7
1             2    68.5   61.7
2             3    65.5   61.7
3             4    64.5   61.7
4             5    64.0   61.7
5             6    67.5   62.2
6             7    67.5   62.2
7             8    67.5   62.2
8             9    66.5   62.2
9            10    66.5   62.2
10           11    66.5   62.2

galton.head()
Out[226]:
   Unnamed: 0  parent  child
0           1    70.5   61.7
1           2    68.5   61.7
2           3    65.5   61.7
3           4    64.5   61.7
4           5    64.0   61.7
```

```
from scipy import stats
pearson_coef, p_value = stats.pearsonr(galton["child"],
galton["parent"])
print("Pearson Correlation: ", pearson_coef, "and a P-value of:",
p_value)
```

```
Pearson Correlation:  0.45876236829282174 and a P-value of:
1.7325092920165045e-49
```

We can see from the above that the correlation between parent and child height is equal to 0.45, and has an exceedingly small *p*-value. Hence, it is statistically significant, and we can reject the null hypothesis that the true correlation in the population from which these data were drawn is equal to 0. Notice that the correlation here is between what Karl Pearson (famed statistician) would later call **organic variables** of heights. This was indeed one of the first uses of the correlation coefficient back around the year 1888. Galton and others wanted to show that generational characteristics and traits were transmitted genetically (heritability) from one generation to the next. Hence, a correlation here is extremely meaningful. That tall parents are more likely to have tall children (as adults) clues us into something potentially genetic going on across generations. Hence, though all the caveats earlier mentioned about interpreting correlations remains true (in that sometimes they might be quite meaningless), in cases such as with Galton's data, any correlation between generations was truly a meaningful result. The lesson is this: **what determines whether a correlation is meaningful or not is often a function of the variables on which it is computed, and not necessarily the size of the coefficient**. Therefore, the next time you are faced with a correlation coefficient, don't look at the absolute size of the number for a measure of importance. Rather, look at the variables on which it was computed to glean a sense of scientific meaning. If I told you there is a correlation of 0.2 between eating carrots and cancer, that correlation, though small numerically, would likely make you reconsider your vegetable choices. **The correlation is small numerically, but the pragmatic effect is huge**. Why? Because of the physical implications of the established relationship.

Notice that our null hypothesis above is that the correlation in the population is equal to 0. You might ask if that is a suitable null hypothesis, and you would be correct to ask that question. As we alluded earlier, why posit such a null hypothesis if it is virtually impossible in the first place that the correlation in the population could ever be equal to 0? Many excellent methodologists have asked the same question and have been critical about null hypotheses always being about such a particular value. As an example of this, the author once attended a presentation where the correlation between two variables was found to be approximately 0.2, and combined with the sufficiently large sample size, the null was, of course, rejected. The research scientist claimed a discovery! He essentially boasted about a correlation of 0.2. However, what he did not realize is that a correlation of approximately 0.2 for the variables he was working with was, in all likelihood, probably a reasonable **prior expectation** regardless of his scientific intervention. That is, obtaining a correlation of 0.2 was really "no big deal" at all, since even by chance this could have been anticipated. Since a correlation of 0 was automatically assumed under the null, the 0.2 was advertised as being due to his treatment. In all probability, his treatment had virtually zero effect and the correlation of 0.2 was status quo for his research variables to begin with. When you evaluate correlations and other measures blindly, software does not alert you that you are drawing potentially meaningless conclusions. It simply computes things for you. **Always be critical about what null hypothesis is being advanced (and possibly rejected) when evaluating correlation coefficients. Rejecting a null hypothesis that the population coefficient is equal to 0 is usually not very impressive to begin with. You may still get it published, but it may still not be that big of a deal. That the prior expectation is set at 0 on a theoretical level is exactly that, theory, and is usually only for theoretical convenience for the statistical test (i.e. so a theoretical sampling distribution for the statistic can be easily obtained).**

Rejecting a null hypothesis that the population correlation coefficient is equal to 0 may not be a very impressive result, especially if a correlation greater than 0 would have naturally been expected due to chance anyway. Any statistical test can achieve the infamous "$p < 0.05$" given a null hypothesis that is unrealistic. Hence, the statement "I found a statistically significant correlation" may not, from a scientific point of view, mean much at all. Always ask further questions and dig into the problem at hand to understand what is (or not) going on with the data you are studying.

We can obtain some plots of the correlation as well as histograms for both variables univariately through `sns.pairplot()`:

```
import seaborn as sns
sns.pairplot(galton)
Out[261]: <seaborn.axisgrid.PairGrid at 0xfb9dc18>
```

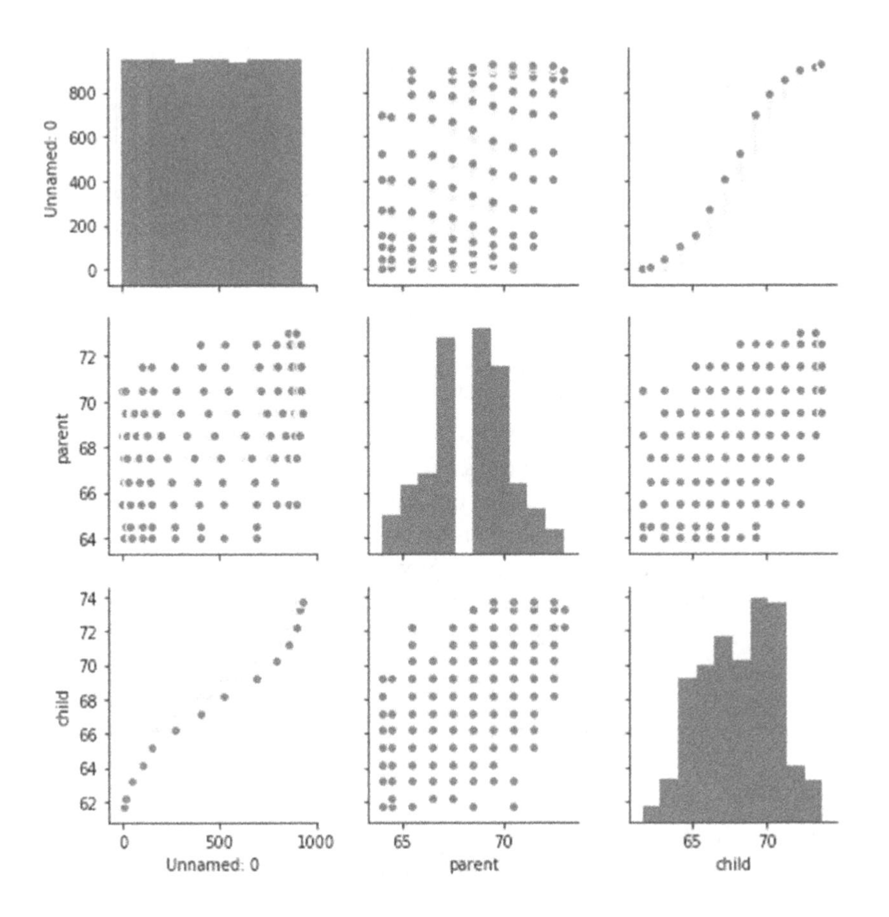

We can see that the correlation for parent and child is represented by the scatterplots in row 2, column 3, as well as row 3, column 2. The remaining scatterplots are of no use, as they simply represent the correlation of each variable with an index variable ("unnamed") that Python produced automatically. Likewise, the plots in row 1, column 3, and row 3, column 1 are of no use. The histograms in row 2, column 2, and row 3, column 3 are useful, as they represent the distributions for parent and child, respectively. We can see that both distributions are relatively normal in shape, even if slightly skewed.

Instead of producing a matrix plot as earlier, we can instead define each variable from the Galton data (printing only a few cases on each variable), and generate a scatterplot through the following:

```
parent = galton['parent']
child = galton['child']
```

This extracts the variables that we will use in the scatterplot. We now build the plot using `plot.scatter()`:

```
columns = ['child', 'parent']
ax1 = galton.plot.scatter(x = 'child', y = 'parent')
```

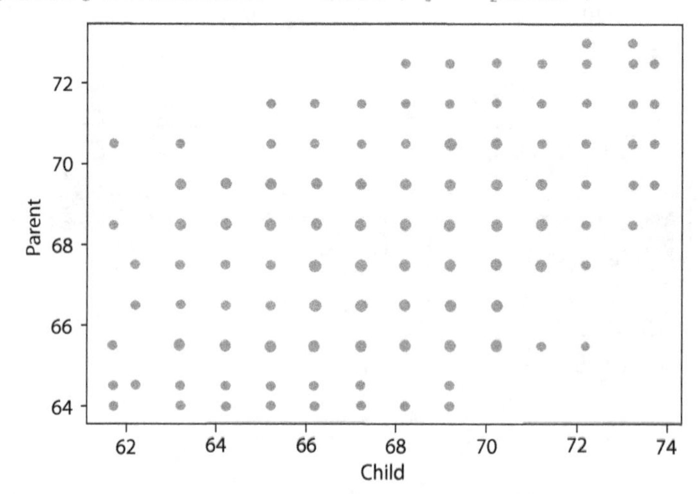

An interesting point about scatterplots is that the data in them represent a subset of the **Cartesian product**. What is the Cartesian product? It is the total of all **possible pairs of points** in the plot. To understand this, imagine in this plot that all points were represented, that is, all possible pairings of child and parent. The correlation would then equal 0, because we would have the full Cartesian product. What the relationship between parent and child represents then is a **narrowing of this product**, such that a subset or subspace of the full Cartesian product set is represented. Indeed, this is precisely what defines a **mathematical relation**, which is a subset of the Cartesian product. For a good discussion of these ideas, see Hays (1994).

4.6 *T*-Tests for Comparing Means

***T*-tests** are used in statistics for several reasons, one of which is as a **test for mean differences**. The reason why we say this is only one purpose for them is because *t*-tests are also used in assessing the statistical significance of regression coefficients and other sample statistics in other contexts. That is, **they are not solely used to evaluate mean differences**. In introductory courses, often students automatically associate *t*-tests with mean differences. However, this is not the case. When we refer to *t*-statistics, we are referring to the *t*-distribution, but the application of this distribution is very widespread, not simply for assessing mean differences.

In a **one-sample *t*-test**, the researcher would like to evaluate the probability that an obtained sample mean could have reasonably been drawn from a particular population with a specified population mean. If the population variance for the problem is known, then performing a ***z*-test** instead of a ***t*-test** would be more appropriate. However, when the variance is unknown and has to be estimated based on sample data, then the *t*-test is the correct test to use. For relatively large sample sizes (e.g. greater than 100 or so), whether one performs the *t*-test or *z*-test will usually make little difference with regard to whether or not the null hypothesis is rejected. The reason for this is because as degrees of freedom get larger and larger, the *t*-distribution converges to that of a *z*-distribution. Still, to be on the safe side, you cannot go wrong with the *t*-test when variances are unknown and have to be estimated from sample data, regardless of how large your sample might be.

A closer look at the one-sample *t*-test reveals how it is constructed:

$$t = \frac{\bar{y} - E(\bar{y})}{\hat{\sigma}_M} = \frac{\bar{y} - E(\bar{y})}{s/\sqrt{n}}$$

where $\bar{y}$ is the sample mean we obtain from our research and $E(\bar{y})$ is the expectation under the null hypothesis, which is the population mean, μ, in this case. Hence, we could rewrite the numerator as $\bar{y} - \mu$. The denominator is the **estimated standard error of the mean**, where s is the sample standard deviation and n is the sample size. We see that the *t*-test is comparing a mean difference in the numerator to variation we would expect under the null in the denominator. We will explore the construction of the *t*-test is more detail in our discussion of statistical power in a later chapter. As we will see then, a large t does not necessarily imply an impressive scientific result, as a large t can be obtained in many cases by simply managing to lower s or increase n or both. The degrees of freedom for the one-sample *t*-test are equal to $n-1$, that is, one less the number of participants in the sample.

We can easily demonstrate the one-sample *t*-test on some fictitious data. Suppose we would like to know whether it is reasonable to assume that the following IQ sample data could have been drawn from a population of IQ scores with mean equal to 100. Our null hypothesis is $H_0: \mu = 100$ against the alternative hypothesis $H_1: \mu \neq 100$. We first enter our data:

```
iq = [105, 98, 110, 105, 95]
df = pd.DataFrame(iq)
```

```
df
Out[38]:
0  105
1   98
2  110
3  105
4   95
```

We can conduct the one-sample t-test using **scipy**, listing first the dataframe (`df`), followed by the population mean under the null hypothesis, which for our data is equal to 100:

```
from scipy import stats
stats.ttest_1samp(df, 100.0)
```

```
Out[41]: Ttest_1sampResult(statistic=array([0.9649505]),
pvalue=array([0.38921348]))
```

The t-statistic for the test is equal to 0.9649505, with an associated p-value of 0.38921348. Since the p-value is rather large (definitely not less than some set level such as 0.05), we do not reject the null hypothesis. That is, we have insufficient evidence to suggest that our sample of IQ values was not drawn from a population with mean equal to 100. Notice that we have not "confirmed" that the population mean is equal to 100, only that we have failed to reject the null hypothesis. There is a difference here. A failure to reject the null never confirms any truth about the null; it merely gives us no reason to suggest it is false. It essentially means "nothing to see here, carry on." If we fail to reject the null hypothesis that older folks get more COVID-19 than younger adults do, it does not mean they get the same amount. It simply means we failed to detect a difference. This distinction is important.

The **two-sample t-test** can be used when collecting data from what we assume, under the null, are two independent populations:

$$t = \frac{E(\bar{y}_1) - E(\bar{y}_2)}{\sqrt{\frac{s_1^2}{n_1} + \frac{s_2^2}{n_2}}} = \frac{\mu_1 - \mu_2}{\sqrt{\frac{s_1^2}{n_1} + \frac{s_2^2}{n_2}}}$$

Notice that in the two-sample case, the mean difference in the numerator is between two population means, μ_1 and μ_2. The standard error is now called the **estimated standard error of the difference in means**, and is composed of both sample variances from each sample as well as a corresponding sample size. We now conduct a two-sample t-test, this time using the previously loaded Galton data. We would like to evaluate the null hypothesis that the mean difference between parent and child is equal to 0:

```
from scipy import stats
stats.ttest_ind(parent, child)
```

```
Out[278]: Ttest_indResult(statistic=2.167665371332533, pvalue=
0.030311049956448916)
```

The obtained t-statistic is equal to 2.167, with p-value of 0.03. Since the p-value is quite small (e.g. less than 0.05), we reject the null and conclude a mean population difference between parent and child. That is, we are rejecting the null hypothesis of $H_0 : \mu_1 = \mu_2$ in favor of the alternative hypothesis $H_1 : \mu_1 \neq \mu_2$. Be sure to note that these hypotheses are about population parameters, not sample statistics.

We can easily evaluate distributions for both variables by obtaining plots:

```
plt.figure(figsize=(10, 7))
sns.distplot(parent)
```

```
plt.figure(figsize=(9, 5))
sns.distplot(child)
```

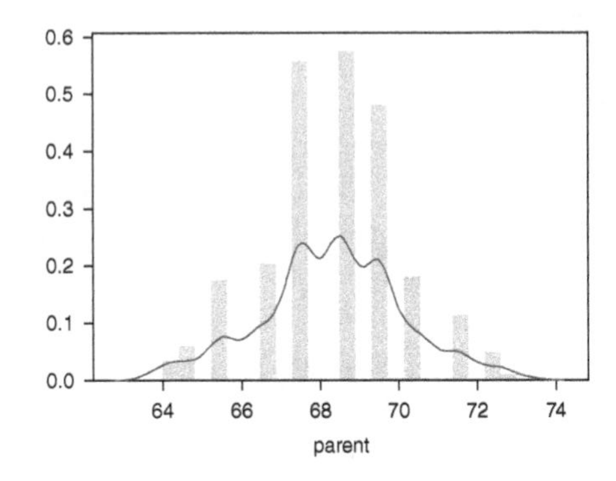

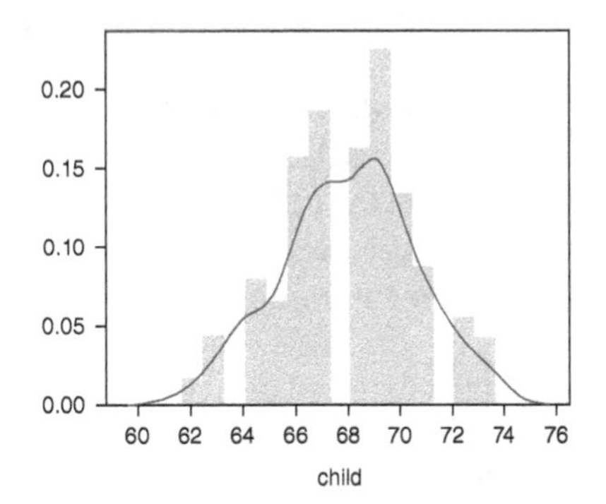

We see from these plots that each population is relatively approximately normally distributed. Hence, we can say that one assumption for the t-test, that of **normality of populations**, is probably at least tentatively satisfied. For the t-test, we also require that the **variances** in each population are equal as well. Informally at least, the variances for each population also appear to be quite similar based on the plots, as the spread along the x-axis appears to be quite similar. However, if you try a **Levene test** on this data using `stats.levene(child, parent, center = 'mean')`, you will reject the null of equal variances. Levene's test is a test for the equality of variances in the population, though its p-value is also quite sensitive to sample size. Use these tests as a guide only.

We can also obtain **boxplots** for both the parent and child:

```
sns.boxplot(parent)
sns.boxplot(child)
```

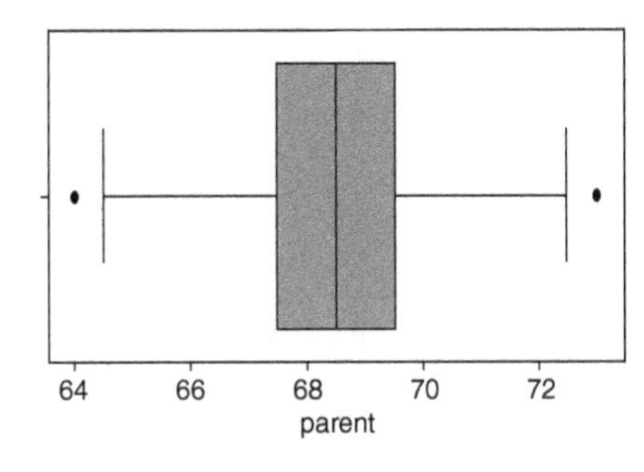

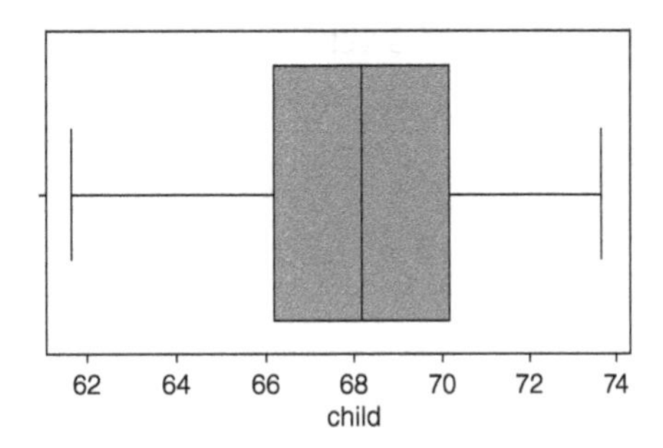

We can see from these boxplots that parent might have a couple of rather extreme scores worth looking into. The center line in each plot represents the median, while the lines enclosing the rectangle are the first quartile (Q1) and the third quartile (Q3). We can see that the distribution for child is pulled ever so slightly to the left, indicating a mild negative skew. If you are not familiar with boxplots and would like a more informative overview, see Montgomery (2005).

4.7 Paired-Samples *t*-Test in Python

Instead of a one-sample or independent-samples situation, sometimes we have data measured on **pairs of observations**. A classic example is asking husbands and wives their happiness ratings for their marriage. The data cannot be presumed to be independent in this case. That is, knowing a husband's rating likely tells us something about a wife's rating (even if the ratings end up potentially being in different directions!). The paired situation is actually a special case of the more general **matched pairs design** where "blocks" of subjects are featured (Table 4.1).

In Table 4.1, each block represents either a natural pairing or one imposed by the researcher. In the case of the husband and wife pairing, "Block 1" would consist of the first married couple. Treatment 1 in this case might be the husband's score, while treatment 2 might be the wife's score. Alternatively, in a different set-up, perhaps treatment 1 is one drug dose and treatment 2 another, and so on. We can see that the matched pairs and block layout is quite general and can accommodate a variety of specific experiments. The key point again is that **within each block, there is a lack of independence between conditions.** In a purely between-subjects design, it is assumed that observations under both conditions are independent. If we generalize this situation to where there are more than two conditions, we obtain results given in Table 4.2, where there are several conditions instead of only two.

We will discuss these designs later in the book in a bit more detail when we feature repeated-measures ANOVA. As we will see, repeated-measures ANOVA is nothing more than a generalization of the randomized block design.

Table 4.1 Matched design.

	Treatment 1	Treatment 2
Block 1	10	8
Block 2	15	12
Block 3	20	14
Block 4	22	15
Block 5	25	24

Table 4.2 Randomized block design.

	Treatment 1	Treatment 2	Treatment 3
Block 1	10	9	8
Block 2	15	13	12
Block 3	20	18	14
Block 4	22	17	15
Block 5	25	25	24

Table 4.3 Learning as a function of trial (hypothetical data).

Rat	Trial 1	Trial 2	Trial 3	Rat Means
1	10.0	8.2	5.3	7.83
2	12.1	11.2	9.1	10.80
3	9.2	8.1	4.6	7.30
4	11.6	10.5	8.1	10.07
5	8.3	7.6	5.5	7.13
6	10.5	9.5	8.1	9.37
Trial means	$M = 10.28$	$M = 9.18$	$M = 6.78$	

Returning to the situation where there are only two conditions, but that they are related by their pairing, a **paired-samples *t*-test** is appropriate. We demonstrate the test on the data given in Table 4.3, in which rats were measured on a learning task across three trials.

Notice that for these data, each rat is measured a total of three times from trials 1 through 3. For our purposes, because we are discussing the paired situation only, we will focus only on the first two trials. Since a rat serving under trial 1 also serves under trial 2, the data are **naturally paired**, and actually, in this situation, are **repeated**. For these data, knowing the behavior of one rat under one trial tells us information about how that same rat might behave under another trial. Hence, we have no reason to believe the trials are independent. We first build the dataframe for trials 1 and 2:

```
trial_1 = [10, 12.1, 9.2, 11.6, 8.3, 10.5]
trial_2 = [8.2, 11.2, 8.1, 10.5, 7.6, 9.5]
paired_data = pd.DataFrame(trial_1, trial_2)
paired_data

Out[66]:
8.2   10.0
11.2  12.1
8.1    9.2
10.5  11.6
7.6    8.3
9.5   10.5
```

We now conduct the *t*-test:

```
stats.ttest_rel(trial_1, trial_2)
Out[69]: Ttest_relResult(statistic=7.201190377787752, pvalue=
0.0008044382024663002)
```

The value of the *t*-statistic is equal to 7.20, with an associated *p*-value of 0.0008, which is statistically significant. Therefore, we have evidence to suggest that the mean difference between the two trials is not equal to 0. That is, we reject the null hypothesis and infer the alternative hypothesis that in the population from which these data were drawn, the true mean difference is unequal to 0. Again, we will return to our discussion of paired data when we discuss blocking designs and repeated measures models more thoroughly later in the book.

4.8 Binomial Test

The binomial distribution is useful for modeling the probability of success on a stochastic experiment that is **binary** and whose events are **mutually exclusive**. The classic example of where the binomial can be used is in coin-flipping, where the experiment or trial can result in either a head or tail. It is a binary event since the only possibilities are one of two categories when the experiment is performed. Other examples include being dead or alive, day or night, young or old, etc. The **binomial distribution** is given by

$$\begin{aligned} p(r) &= \binom{n}{r} p^r (1-p)^{n-r} \\ &= \left(\frac{n!}{r!(n-r)!}\right) p^r (1-p)^{n-r} \end{aligned}$$

where $p(r)$, read as "probability of r," is the probability of observing r successes. The total number of trials is denoted by n. Hence, $p(r)$ is the probability of observing a given number of successes out of a total number of trials (e.g. number of coin flips). Since the events in question are mutually exclusive, the total probability is equal to $p+(1-p)$, where $1-p$ is often denoted by q. Hence, $p+(1-p)=p+q=1$. Returning to our example of how the binomial distribution can be used to answer a research question, let us unpack this a bit further. Suppose we were interested in learning of the probability of a given number of heads on a presumed fair coin. We set $p=0.5$ to denote the **assumed** probability of heads under the null. Suppose we obtain 2 heads out of 5 flips, and we would like to know the probability of obtaining 2 or more heads assuming the coin is fair. This is a binomial setting since:

- The coin variable is **binary** in nature, meaning that the outcome can only be one of two possibilities, "head" or "tail." It is a **discrete variable**. The two events are mutually exclusive, meaning that you either get a head or a tail, and cannot get both on the same flip.
- The probability of "success" on each trial remains the same from trial to trial. That is, the probability is **stationary**. In other words, if the probability of heads on the first trial is 0.5, then this implies the probability of heads on the second and ensuing trials is also 0.5.
- Each trial is **independent** of any other trial. For instance, the probability of a "head" on the first flip does not have an influence on the probability of heads on the second flip, and so on.

These conditions make the binomial distribution suitable for modeling the probability of success (in this case, "heads") on the coin. In Python, we can evaluate the probability of 2 or more heads out of 5 flips as follows, where `stats.binom_test()` is the function for the test, "`2`" is the number of successes out of `n=5` trials, and `p=0.5` is the probability of a success on any given trial:

```
stats.binom_test(2, n=5, p=0.5, alternative='greater')
Out[71]: 0.8125
```

Hence, we see that the probability of obtaining 2 or more heads on 5 flips of a fair coin is equal to 0.8125. But what can we make of this result? Well, if we started out by assuming the coin was fair, where the probability of heads on any given flip were equal to 0.5, then a finding that the probability of 2 heads out of 5 flips were equal to 0.8125 would likely not give us evidence to reject the null hypothesis. That is, what the result is telling us is that on fair coins, **2 heads out of 5 flips happens quite a lot**. It is a common occurrence, assuming we believe in probability theory at all (if we do not, well then, that's a different story). Hence, it gives us no reason to doubt the assumption that the coin is fair. The probability of the data given the null is relatively high.

4.9 The Chi-Squared Distribution and Goodness-of-Fit Test

The binomial test just surveyed for applications can actually be considered a special case of an even more popular test in statistics, that of the **chi-squared goodness-of-fit test**. The chi-squared distribution is not necessarily equivalent to a chi-squared "test" and the two should not be unequivocally equated. In undergraduate applied statistics, chi-square is simply associated with the chi-squared test, but chi-square, like *t*, is simply a statistic with a proper density that can be used in different scenarios to evaluate null hypotheses. In theoretical statistics, the chi-square density pops up virtually everywhere and is associated with the *F* distribution and others.

The **chi-square distribution** is given by

$$f(x)=\frac{1}{2^{v/2}\Gamma(v/2)}x^{[(v/2)-1]}e^{-x/2}$$

where for different inputs of $x>0$, v are degrees of freedom and Γ is the **gamma function**. We can also state the chi-square distribution in terms of the sum of squares of n independent and normally distributed z-scores:

$$\chi_n^2=\sum_{i=1}^{n}z_i^2=\sum_{i=1}^{n}\frac{(x_i-\mu)^2}{\sigma^2}$$

The **goodness-of-fit test** happens to be a statistical method that features the chi-square distribution in evaluating the statistical significance of the test. The goodness-of-fit test features a test on **count data** rather than means such as in the *t*-test or ANOVA. We define the chi-squared goodness-of-fit test as follows:

$$\chi^2=\sum_{i=1}^{r}\sum_{j=1}^{c}(O_i-E_i)^2/E_i$$

where O_i are observed frequencies and E_i are expected frequencies within each cell. The summation notation denotes summing these squared cell differences, $(O_i - E_i)^2$ within c columns and then across r rows. Notice that χ^2 takes squared differences in the numerator for the reason that if we did not square these, the sum would equal 0 and hence the resulting value for χ^2 would always equal 0 regardless of the discrepancy between O_i and E_i. The act of squaring discrepancies or deviations serves to ensure a value for χ^2 different from zero so long as there is at least some discrepancy between observed and expected differences. If the sum of $(O_i - E_i)^2$ is large relative to E_i, then, given the degrees of freedom, χ^2 may become extreme and attain statistical significance. **The null hypothesis then in such a test is that observed frequencies are equivalent to those expected in the population from which the data were drawn**. The alternative hypothesis is that they are not, or equivalently, that there is some **association** between the row variable and the column variable.

Performing a chi-squared goodness-of-fit test in Python is very easy. As an example, let us first consider the case where all observed frequencies are equal in each cell. In the following, we set them all equal to 16:

```
from scipy.stats import chisquare
chisquare([16, 16, 16, 16, 16])
```

```
Out[74]: Power_divergenceResult(statistic=0.0, pvalue=1.0)
```

We see that the p-value comes out to be 1.0, for the reason that when all cell frequencies are equal, there can be no discrepancy between those observed and expected. In other words, the result we are witnessing is the exact result we would expect under the null hypothesis. In this case, the total frequencies available are 16(5) = 80. In this **one-way** ("one-way" because we only have one variable) chi-squared analysis, if there is no discrepancy between cells, then this total frequency should distribute itself evenly across the cells, which is exactly what the above suggests. There is a total of 16 cases per cell, all cells equal, and hence there is absolutely no evidence against the null hypothesis, which is why the p-value is equal to 1.0.

Now we demonstrate the situation where frequencies are slightly **unequal** by adjusting a couple of the cell frequencies downward to 15. Notice elements 2 and 4 in our data in the following are now data points 15 rather than 16:

```
chisquare([16, 15, 16, 15, 16])
Out[75]: Power_divergenceResult(statistic=0.07692307692307693,
pvalue=0.9992790495310748)
```

For these data, we see the resulting p-value is now equal to 0.999, a bit lower than 1.0. This is as a result of the discrepancy in frequencies now from cell to cell. That is, the probability of observing this set of frequencies under the null hypothesis of equal frequencies is still very high, certainly no cause or reason to reject the null hypothesis. The following is an even more disparate distribution of frequencies:

```
chisquare([16, 15, 10, 8, 25])
Out[76]: Power_divergenceResult(statistic=11.81081081081081, pval-
ue=0.01881499899397827)
```

We can see in this case that the observed p-value is equal to 0.018, and hence we deem the result to be statistically significant at the 0.05 level. That is, the probability of observing a discrepancy such as we have observed is very small under the null hypothesis of equal frequencies per cell. Hence, we reject the null hypothesis of equal frequencies per cell.

4.10 Contingency Tables

We just considered the case of the one-way goodness-of-fit test, where we had only a single row. However, the formulation $\chi^2 = \sum_{i=1}^{r} \sum_{j=1}^{c} (O_i - E_i)^2 / E_i$, as mentioned, allows us to generalize to a contingency table with rows and columns. In most practical cases, of course, we have more than a single variable on which we have frequencies. Often research provides us with a contingency table where we have at least one column and one row variable, such as in Table 4.4.

In the data in Table 4.4, the column variable is whether or not a **condition** is present. For this example, the condition can be anything we like. For example, suppose the condition variable is that of whether or not symptoms of post-traumatic stress disorder are present and the row variable of exposure is whether an individual has been exposed or not exposed to combat in Iraq. Or, in the age of COVID-19, the condition could be whether symptoms of the disease are present and the exposure variable that of whether an individual has been exposed via contact tracing (i.e. a measure of whether the given individual has been in contact with those having already been confirmed for the illness). For the sake of example, it does not really matter what we name these variables. The null hypothesis for these data can be framed in at least a couple of ways. One way of stating it is to say that the 50 counts making up the table are essentially randomly distributed across the cells (with consideration of the row and column marginal totals in computing expectations per cell, to be discussed shortly) and that there is no systematic pattern to them. If there were a systematic pattern, it might suggest some cells have disproportionate numbers compared to other cells (again, relative to expectations computed using marginal row and column totals), which would suggest an **association** between the column and row variables.

For example, the expectation for the cell in row 1, column 1 would be computed as the row total of 30 times the column total of 25, divided by the total sample size of 50, yielding an expectation of 15 for the first cell. We see that the observed value of 20 for that cell deviates from 15. The chi-squared analysis will repeat these computations per cell and tally up the squared deviations. If overall observed frequencies deviate from

Table 4.4 Contingency table for a 2 × 2 design.

	Condition Present (1)	Condition Absent (0)	Total
Exposure yes (1)	20	10	30
Exposure no (2)	5	15	20
Total	25	25	50

those expected, this suggests an association between variables. Hence, this is a second way of stating the null, that there is no association between the two variables. Notice that this is analogous to the one-way goodness-of-fit test, only that now we have both a row and a column variable. Otherwise, the concept is exactly the same. To evaluate the null, we will compute a chi-squared statistic and evaluate it for statistical significance.

In Python, we set up our contingency data as follows:

```
import numpy as np
matrix = np.array([[20, 10],
[5, 15]])

matrix
Out[23]:
array([[20, 10],
       [ 5, 15]])
```

To accommodate the test in `scipy.stats()` we organize the data in the following way, and also import `chi2_contingency()` to run the test:

```
from scipy.stats import chi2_contingency
obs = np.array([[20, 10], [5, 15]])

chi2_contingency(obs)
Out[21]:
(6.75, 0.0093747684594349, 1, array([[15., 15.],
        [10., 10.]]))
```

The observed chi-squared statistic is equal to 6.75, with an associated *p*-value of 0.009. Since the *p*-value is quite small, we reject the null hypothesis of no association between the row and column variable and conclude an association between these two variables.

Review Exercises

1. Discuss the nature of a **Pearson Product-Moment correlation coefficient**. What does it seek to measure, exactly?
2. Why should a **plot** of data always accompany the reporting of a correlation coefficient? What is the danger in not including such a plot?
3. Demonstrate with some fictional data why computing the **covariance** is insufficient as a measure of the relationship between two variables. Why is it not dimension-free? Why is computing the correlation the solution to this problem?
4. When should **Spearman's rho** correlation be used in place of that of Pearson? Why might the two computations not agree with one another? Explain.
5. If "**everything is correlated to some degree**," as claimed in the chapter, why bother then with evaluating a null hypothesis that the population correlation is equal to 0?
6. A researcher claims a **scientific discovery** has been made since he computed a correlation of $r = 0.20$ between his variables. What kinds of questions would you ask the researcher in critically evaluating such a claim?

7. Why might rejecting a **null hypothesis** in a test of a correlation coefficient not be that meaningful?
8. Does increasing **sample size** in the computation of Pearson r have any effect on the size of r? Why or why not? Explain.
9. Consider the **one-sample *t*-test** for a mean. How might a large difference in the numerator still not result in a statistically significant value for t? Explain.
10. A scientist claims, "I rejected the null hypothesis for my one-sample *t*-test." What kinds of follow-up questions would you ask next?
11. Demonstrate with made-up data why if you change the null hypothesis for a one-sample *t*-test, the decision on the null will change even if the data remain exactly the same. Use Python to demonstrate your computations.
12. What are the **degrees of freedom** for the two-sample *t*-test? Why are they equal to such? Look at the layout for a two-sample case and see if you can figure out why those degrees of freedom are equal to what they are.
13. How is a **paired-samples *t*-test** different from a **two-sample *t*-test**? Explain.
14. Compute in Python the probability of obtaining 20 heads on 50 flips of a fair coin. What does your result suggest about the coin? Does it "prove" anything about the coin?
15. Explain how the **chi-squared goodness-of-fit test** statistic is similar in some ways to the *t*-test computation. What are the parallels? How are they different?

5

Power, Effect Size, *P*-Values, and Estimating Required Sample Size Using Python

CHAPTER OBJECTIVES

- Understand the nature of *p*-values and their proper use in scientific inference.
- Distinguish between a *p*-value and an effect size and understand why effect sizes are often more important than *p*-values.
- Understand the nature of statistical power and how degree of power relates one-to-one with *p*-values.
- Estimate statistical power and required sample size in *t*-tests to better appreciate how *p*-values, effect size, and sample size relate to one another.

In this chapter, we survey the concepts of **statistical power**, **effect size**, and ***p*-values**. Though some of this information we have already studied earlier in the book in one way or another, the current chapter seeks to unify and explain how these concepts "go together" and influence one another. For example, a researcher interpreting a *p*-value but not understanding statistical power is a researcher who should not be interpreting *p*-values! Yes, understanding how these elements relate to one another is that important! Likewise, the interpretation of effect size, without understanding the role of *p*-values, makes such interpretation incomplete and potentially misguided or misleading. For both the producer and consumer of research, it is essential that these concepts be clearly understood if one is to attempt to produce or interpret any scientific evidence at all. Hence, this chapter is mandatory reading for newcomers to statistics and data analysis, but also to experienced researchers who may not be familiar with how these elements relate to one another or who have never truly understood it. This chapter is extremely important for both such audiences.

5.1 What Determines the Size of a *P*-Value?

In order to understand statistical power and sample size, one must have an excellent understanding of what makes a *p*-value large or small. Otherwise, one will be inclined to potentially draw conclusions from "$p < 0.05$" that are not warranted and attribute

Applied Univariate, Bivariate, and Multivariate Statistics Using Python: A Beginner's Guide to Advanced Data Analysis, First Edition. Daniel J. Denis.

scientific worth to the result of an experiment where there may be none. The good thing about understanding how one inferential statistical test works is that you then have a good grounding in virtually how they all work, and hence, come to know why a *p*-value may be large in one case and small in another.

The easiest and most straightforward demonstration of the mechanics of *p*-values is with the simple univariate *t*-test for means. The principle can be demonstrated just as easily with a *z*-test for means, but since *t*-tests are most often used in research (i.e. we rarely know population variances and have to estimate them), we will use the *t*-test as a prototype for our example. In unpacking the components of the *t*-test, we learn what makes the size of *t*, and hence the size of the *p*-value, large or small. Recall the one-sample *t*-test for a mean:

$$t = \frac{\bar{y} - E(\bar{y})}{\hat{\sigma}_M} = \frac{\bar{y} - E(\bar{y})}{s/\sqrt{n}}$$

Recall that the numerator of the *t*-test, $\bar{y} - E(\bar{y})$, features a difference in means, that of the obtained sample mean minus the expectation of the mean under the null hypothesis. This expectation is given by $E(\bar{y})$, where E denotes the expectation operator. It can be easily shown that the expectation of the sample mean is equal to the population mean, that is, $E(\bar{y}) = \mu$:

$$\bar{y} = \frac{(y_1 + y_2 + \cdots + y_n)}{n}$$

$$E(\bar{y}) = E\left(\frac{(y_1 + y_2 + \cdots + y_n)}{n}\right)$$

It is well known in mathematical statistics that **the expectation of a sum of random variables is equal to the sum of individual expectations**. Given this, along with the above results, we can now write the expectation of the sample mean $\bar{y}$ as

$$E(\bar{y}) = \frac{E(y_1 + y_2 + \cdots + y_n)}{n}$$

$$= \frac{[E(y_1) + E(y_2) + \cdots + E(y_n)]}{n}$$

What is more, since the expectation of each of the individual values y_1 through y_n is $E(y_1) = \mu$, $E(y_2) = \mu, \ldots, E(y_n) = \mu$, we can replace each y_1 through y_n with μ. Hence, we end up with

$$E(\bar{y}) = \frac{[\mu + \mu + \cdots + \mu]}{n}$$

$$E(\bar{y}) = \frac{n\mu}{n}$$

$$E(\bar{y}) = \mu$$

Since the n values cross out, we end up with simply μ. Hence, we have shown that $E(\bar{y}) = \mu$. So, what this means for our *t*-test is that the numerator $\bar{y} - E(\bar{y})$ can just as well be written as $\bar{y} - \mu$, yielding for the *t*-test

$$t = \frac{\bar{y} - E(\bar{y})}{\hat{\sigma}_M} = \frac{\bar{y} - \mu}{s/\sqrt{n}}$$

The numerator thus represents a difference between means, that of the sample from the population. When you think about it, this should be the reason why we are conducting the t-test, because we want to know whether our obtained sample mean differs from the population mean. Is this not the goal of your study or experiment? Now, of course, we already know it will differ to some degree. What are the odds that it doesn't? **Next to nil.** To understand this, consider obtaining the mean IQ of Californians on a sample size of 25. When we compare this to a population mean IQ of 100, we already know there will be at least **some** difference between the sample mean $\bar{y}$ and the population parameter μ. Hence, there being a difference is hardly in question. The true question, as you will recall from our discussion earlier in the book, is whether the difference in the sample is large enough to conclude a true population difference. To evaluate this, we need to compare it to the denominator of the test, which is the estimated standard error of the mean, $s/\sqrt{n}$. Notice what the standard error of the mean is made up of. It has two components. The first component is s, the sample standard deviation, while the second component is a function of sample size, specifically the square root of the sample size, $\sqrt{n}$. Hence, we see that the size of the resulting t-statistic will be determined by three things:

- The actual difference in means observed, that of $\bar{y} - \mu$.
- The size of the standard deviation s.
- The size of the sample taken for the test, given as the square root $\sqrt{n}$.

The reason you did the experiment or study is for the **first of these factors**, not the other two. That is, you surely did not conduct a statistical test to demonstrate that you can obtain a small standard deviation or obtain a large sample size. Though in some cases collecting data to demonstrate a low standard deviation or variance may be a worthwhile research pursuit, it is usually not the reason why scientists conduct studies. You presumably did the test because you were interested in the mean difference in the numerator, but it is clear that **the significance test is not simply a function of this difference**. This fact is extremely important! It is a function of the size of s and $\sqrt{n}$. And, as the size of t gets larger, it becomes an increasingly **unlikely statistic**, such that it lies off into the tail of the corresponding sampling distribution. Hence, it stands that statistical significance, the infamous "$p < 0.05$" can be achieved even for a constant mean difference $\bar{y} - \mu$ by simply increasing the sample size and/or finding a way to ensure s is small.

In most cases, decreasing s can be challenging, but it can be done. For example, if we were to test a treatment on those suffering from COVID-19 vs. those not suffering, we could minimize the variance in our samples by screening beforehand and selecting those from a given age or health background. That way, we are reducing inherent variability in our design and will be able to more easily detect a treatment effect if there is one to be found. In a moment, we will call this ability to detect a treatment difference that of statistical power. **Statistical power** will be the probability of detecting that difference or effect if it truly does exist in the population. Getting a handle on s can be quite difficult, and hence we are usually and most conveniently limited to increasing sample size to increase statistical power. We survey that possibility now.

5.2 How *P*-Values Are a Function of Sample Size

P-values can at times largely be a function of sample size. Be sure this is clear. It means that when you achieve "$p < 0.05$," it may, for all purposes, simply imply that you have collected a large sample. Hence, if you are in the habit of making scientific conclusions and implementing policy or clinical decisions based on the presence of *p*-values alone, you should be very concerned by this, as doing so essentially constitutes **a serious misunderstanding of what the *p*-value communicates**. The influence of sample size has been observed and known since the inception of null hypothesis significance testing, at least since Berkson (1938). The following is what he had to say on the subject. His comments were in relation to the chi-squared test, but they apply equally well to virtually any significance test:

> I believe that an observant statistician who has had any considerable experience with applying the chi-square test repeatedly will agree with my statement that, as a matter of observation, when the numbers in the data are quite large, the P's tend to come out small. Having observed this, and on reflection, I make the following dogmatic statement, referring for illustration to the normal curve: "If the normal curve is fitted to a body of data representing any real observations whatever of quantities in the physical world, then if the number of observations is extremely large – for instance, on the order of 200,000 – the chi-square P will be small beyond any usual limit of significance.
>
> (Berkson, 1938, p. 526)

With Berkson's words in mind, let us look a bit more closely at how this works in the *t*-test. For any difference however small in the numerator of the *t*-test, $\bar{y} - \mu$, increasing sample size will necessarily lead to a larger test statistic. This is not a hypothetical result, it is an arithmetic one (and based on good statistical theory to back it up). That is, the statistic **has** to increase in value for a constant difference $\bar{y} - \mu$ but an increasing sample size. For example, suppose the mean difference in IQ is equal to 1, with a standard deviation of 2. For a sample size of 4, our *t* computation comes out to be

$$t = \frac{\bar{y} - E(\bar{y})}{\hat{\sigma}_M} = \frac{\bar{y} - \mu}{s/\sqrt{n}} = \frac{101 - 100}{2/\sqrt{4}} = \frac{1}{1} = 1$$

Now, suppose we increase sample size to 16, leaving the rest of our *t* computation unchanged:

$$t = \frac{\bar{y} - E(\bar{y})}{\hat{\sigma}_M} = \frac{\bar{y} - \mu}{s/\sqrt{n}} = \frac{101 - 100}{2/\sqrt{16}} = \frac{1}{0.5} = 2$$

Notice that by a simple change in sample size, *t* has increased from 1 to a value of 2. For a sample size of 100, we have

$$t = \frac{\bar{y} - E(\bar{y})}{\hat{\sigma}_M} = \frac{\bar{y} - \mu}{s/\sqrt{n}} = \frac{101 - 100}{2/\sqrt{100}} = \frac{1}{0.2} = 5$$

We can see that, clearly, the value for *t* is increasing as a function of sample size for a constant mean difference. That is, notice that the mean difference in the numerator has not changed at all. It has remained at a value of 1. Meanwhile, the value for t of 5 on a sample size of 100 is easily statistically significant, while for a sample size of 4, the statistic is not nearly as large. If you automatically associate scientific importance with statistical significance, then you can easily and mistakenly conclude a scientific effect for a mean difference of a single IQ point! If based on our discussion so far you are thinking, **"Wow, p-values, from a scientific point of view at least, may be quite meaningless,"** then you are thinking correctly!

As mentioned, noting how the p-value is a function of sample size is not new. In addition to Berkson (1938), Bakan (1966) and other methodologists (e.g. see Cohen, 1990, among a myriad of other methodologists, especially in psychology) have long pointed out the problems with null hypothesis significance testing and how p-values can be made small as a function of sample size. In addition, since the numerator of the t-test in this situation will in practicality never equal 0, that is, $\bar{y} - \mu$ will always be different from zero to at least some decimal place, so long as the denominator $s/\sqrt{n}$ can be made small, **statistical significance is essentially assured.** So, what to do about this? The solution is not to abandon the p-value as some have advocated. The p-value is necessary to secure inferential support for the test statistic. The solution to the problem is to accompany all significance tests with a measure of **effect size. Effect sizes, not *p*-values, are what science is all about.** How does the size of effect influence the degree of statistical power? We consider that issue next.

P-values can quite easily be made almost entirely a function of sample size. Hence, whether you obtain a large or small p-value could depend extensively on whether or not you collected a small or large sample size. That is, statements such as "p < 0.05" do not necessarily reflect a meaningful scientific effect. It may simply indicate that a large-enough sample size was collected that resulted in a small-enough p-value to reject the null hypothesis. To make any sense of any scientific finding that might be present, a measure of effect size should accompany the reporting of p-values. Never, ever, interpret a p-value without an accompanying effect size when needing to draw a firm conclusion from a scientific report.

5.3 What is Effect Size?

Effect size is a general term that takes on different computations depending on the given statistical model or context. It either typically reports a "variance explained" figure or a mean difference of some kind. In most cases, effect sizes in one context can be quite easily translated to effect sizes in another via algebraic manipulation. That is, a Cohen's d, for instance, can be quite easily translated into an R-squared statistic, and so on.

In the context of our t-test, we can easily demonstrate how effect size is determined, and how this determination is not the same as for that of the p-value. Let us start from scratch and recall our t-test:

$$t = \frac{\bar{y} - \mu}{s/\sqrt{n}}$$

Again, as we have emphasized, as a scientist, what interests you most in the formula? You presumably did the study or experiment because you wanted to observe a difference in means. That is, you did the study because you were hoping to see a distance in the numerator, $\bar{y}-\mu$. To use the IQ example once more, if the population mean IQ were equal to $\mu = 100$ and you treated a sample of individuals with a "magic pill" to see if it would make them smarter, yielding a sample mean $\bar{y} = 130$, it is the distance between means 130–100 = 30 that is of most interest, not the fact that you may have obtained a small standard error. After all, we just discussed that the standard error is essentially a measure of how much variance you have in your sample, along with the size of sample used. Hardly of great scientific interest! You did the study because you wanted to see a **difference in means**; hence, any measure of effect size should be primarily about this observed difference, $\bar{y}-\mu$. **It is not the size of the computed statistic that should dominate your scientific interests.** From a statistical point of view, yes, small standard errors are interesting, but in terms of the science, they are not priority. The difference in means in the numerator is what should be guiding your scientific interests.

However, if we simply report the distance of 30 as a measure of effect, it is incomplete. But why? Why is the distance between means not sufficient to quantify the size of effect in this case? It is incomplete because it does not take into account the baseline amount of variation in the population. What this means is that the distance of 30 may or may not be "impressive" depending on how much variability there is in the population, or at least an estimate of it. For example, if IQ scores have lots of inherent variability, then a distance of 30 is not quite as impressive as if the variability is quite low. With very low variability, a distance of 30 becomes much more impressive. Hence, we need a measure that puts the distance in means in some kind of **statistical context**. This is exactly what **Cohen's *d*** accomplishes. Cohen's *d* (Cohen, 1988), a measure of **statistical distance**, is given by

$$d = \left| \frac{\bar{y} - \mu_0}{\sigma} \right|$$

where $\bar{y}$ is our observed sample mean, μ_0 is the population mean under the null, and σ is the population standard deviation. If σ is not known, then s can be used to estimate it. The bars around the fraction mean to take the absolute value of the distance. Now, as a scientist, you naturally have a hypothesis as to the direction of the effect, that is, you are hoping for the IQ case, for instance, that $\bar{y}$ is larger than μ_0 since such would denote the group receiving the magic pill is doing better IQ-wise than what we would expect under the null. However, Cohen's *d* itself does not care much about the direction. And if the distance were reversed such that $\bar{y}$ were equal to 70 instead of 130, you would probably be interested in that result as well, even if it did not align with your theoretical prediction. Hence, this is the reason why we take the **absolute value** and thus ignore the **sign** of the result. That is, Cohen's *d* really does not care about your scientific interests, it simply wants to measure the magnitude difference in means.

What is the influence of σ on the size of effect? We can easily demonstrate this numerically. Using our previous example with a distance of 30, suppose σ were equal to 1. Then

$$d = \left| \frac{\bar{y} - \mu_0}{\sigma} \right| = \left| \frac{130 - 100}{1} \right| = 30$$

Of course, this would be a whopping effect! However, if σ were equal to 30, then

$$d = \left|\frac{\bar{y} - \mu_0}{\sigma}\right| = \left|\frac{130 - 100}{30}\right| = 1$$

Notice carefully that, in both cases, the actual distance in means in the numerator did not change. Only the population standard deviation changed and had a drastic effect on the size of d. This is because under the first scenario, the population standard deviation is much smaller; hence, when we observe a difference in means, it is less likely to occur simply due to natural variation in the population. When variability in the population is rather large, such as in the second scenario, then seeing rather large distances of the type $\bar{y} - \mu_0$ would be expected to be more likely. Thus, this is the reason why Cohen's d will be smaller in such situations.

5.4 Understanding Population Variability in the Context of Experimental Design

The size of σ can be understood as a general concept in scientific experimentation and is not restricted to simply being used in statistical formulae. Rather, it is a central feature of how you should be thinking about research studies, and, even more, how you should plan such research studies. Experiments in physics, for example, usually have quite low values of σ in their investigations. Why is that? It is because **physicists usually have a lot of control over the objects they study**. Molecular movement, for example, is a very precise area of investigation; there is not much variability to begin with regardless of the "treatment" imposed. Medical sciences, in some cases, may also operate in areas that have relatively low population variances. Whether a medication increases or decreases heart rate is likely to be tested on a pool of subjects with a relatively low variance in heart rates, at least practically speaking even due to the natural range of the number of possible beats generated in a given minute. Or, a researcher can purposely screen for participants who have a particular range of heartbeats in the attempt to **factor out** as much of the inherent variability in the data as possible before imposing a treatment. For instance, unless your heartbeat is in the range of 72 to 80, maybe you are disqualified from the study from the start. Only accepting participants within that range will likely afford the experiment more **statistical power** and **sensitivity**, since there is not as much "noise" to filter out when trying to discern the presence or absence of an effect. Then, if the treatment is working or has an effect on heart rate, it can be much more easily detected than if the variance in the population is quite high.

In many other areas of science, however, populations have inherently more variability, and hence our "numerator" in our effect size above will have to be large enough to compensate for this inherent variability to be "noticeable." By analogy, if you have a lot of noise in the water, such that the wind is generating impressive white caps on the lake (i.e. when the waves are so high it shows a white splash), it will take a much larger rock (effect) to generate a splash that is noticeable. When scientists measure people's IQ, for example, the range and variance can be quite high. There is also a lot of

measurement error built into things. **Your job as a scientist is to make the waters as calm as possible before skipping the rock to observe its effect.** If you are okay with stormy waters, then do not expect to see a noticeable splash from your rock, even if the rock is indeed making a splash. In physics, for instance, the waters are usually quite calm. In economics and psychology, on the other hand, often the waters are naturally stormy. Physics may be able to detect effects with very small samples. Psychology and economics? Usually not.

Hence, an experimenter would prefer population variance to be as small as possible, allowing him or her to more easily detect any effect that may be present. All else equal, this will more easily generate effects that are noticeable, and hence Cohen's *d* (or similar computable effect size) will become larger in value. Other than using screening tools and controls, one could also choose a **blocking design** in which nuisance variables are blocked in order to remove variability in the ensuing *t*-test, *F*-test, or what have you. We briefly explored block designs in our previous chapter, and will discuss it further in the following chapter. By blocking, we reduce MS error, allowing MS between to better shine through. **Any method by which within variability is reduced will allow between variability to shine.** This is a principle not only of statistics but also of science in general. If you do experiments (and not simply correlational designs), this concept of between vs. within should always be on your mind! Calm the waters, then skip the rock. If the rock makes a splash, you're bound to see it! If you do not calm the waters first, by way of analogy, you will not see whether the treatment for COVID-19 is working! Calm the waters as much as you can first!

5.5 Where Does Power Fit into All of This?

Having surveyed what makes a *p*-value small, as well as a measure of effect size, we are now well equipped to understand statistical power and how it fits in with all of these components. Recall our definition of power was it being the probability of rejecting the null hypothesis given that it is false. More power means bigger test statistics. Hence, whatever makes the *t* or *F* or whatever test statistic you are using large, you have more power. Given our discussion of *p*-values and effect size, we can now list the determinants of statistical power as it concerns the *t*-test:

- Size of distance in means $\bar{y} - \mu$. All else being equal, the larger the distance, the greater the statistical power.
- Size of σ. All else equal, the smaller the population variance, the greater the statistical power.
- Sample size, n. All else equal, the larger the sample size, the greater the statistical power.

Hence, through the *t*-test, we have seen what makes a test statistic large vs. small, translating to what makes a *p*-value large or small, which translates into the degree of power for the given experiment or study. We can easily see now why increasing sample size is usually, even if typically expensive, the easiest way to boost power. An increase in sample size has the effect of usually decreasing the standard error of the statistic. In the limit (which, if you are not familiar with calculus, crudely means "in the long run" in this case), as $n \to \infty$, the standard error goes to 0, which means, pragmatically, that we now have the entire population. That is, increasing n generates a smaller and smaller $s/\sqrt{n}$, which implies whatever is going on in the numerator of our statistic, $\bar{y} - \mu$, begins to look quite large relative to the diminishing standard error. We also see from our discussion that statistical power is no mystery. It is a function of quantifiable things in our test statistic, and if you understand the determinants of it, *p*-values will forever lose their mystery and you will stop automatically and necessarily associating small *p*-values with anything of scientific import. A small *p*-value may hint toward a meaningful scientific result, but on its own it fails to reveal that much about what is going on scientifically.

Now, we said that increasing sample size usually has the effect of decreasing the standard error and leading to a smaller *p*-value. The qualifier "usually" should not be ignored. When increasing sample size, we in reality do not know what will happen to the variance of the sample, nor do we know what will happen to the distance between means (whether a one-sample test or two) in the numerator. If these figures change, as they may very well due to incorporating new information (sample size) into our design, then it may very well be that the resulting *p*-value actually increases in size with a limited and defined increase in sample size. However, this is extremely rare. In most cases, an increase in sample size overpowers any possible or plausible change in effect size or variance, such that we are almost always guaranteed to observe a smaller and smaller *p*-value. But, if ever you notice a *p*-value increase as you collect more subjects or objects, it may simply be because you are tapping more into the population, and the population is much more variable than your sample data had originally suggested. But in the long run, an increase in sample size will be overpowering to the *p*-value and will eventually drive it to be a very small value. That is, as sample size increases, you will at some point reject the null. As was the case for the *Titanic*, it sinking was a **mathematical certainty**.

5.6 Can You Have Too Much Power? Can a Sample Be Too Large?

Having discussed the determinants of power, it may at first seem to the newcomer that an increase in sample size is a "negative" event, since it would appear to artificially lower *p*-values and make rejecting null hypotheses all too "easy." After all, if Berkson, Bakan and others are correct (hint: *they are*) and *p*-values can be made largely a function of sample size, then maybe we should advise researchers to gather small samples so that the *p*-value is not artificially made small – or at least not collect "too big" of a sample size to accommodate the potential drawbacks of the *p*-value. This idea, while at first glance appears to be plausible, is entirely and utterly **misguided**. We will explain why this is the case in a moment, but first, let us state a fundamental principle of research to get us started:

You cannot have too large of a sample size for your experiment, nor can you have too much power. Period.

Take notice of the above! You cannot have "too big" a sample size! Recall that the goal of a research study is to learn of population parameters. If we had the population parameters, we would not be computing inferential statistics in the first place. That is, we would have no need to compute, for instance, sample means, and could just get on with computing population parameters and be done with it. No inference required! However, we know things usually do not work this way. Populations are usually quite large, if not infinite in size, such as in the case for the behavior of coins in which we can theoretically flip the coin an infinite number of times. Thus, if the goal is to learn of population parameters in the first place, then **this implies that collecting as large of a sample size as possible could never be "wrong," at least not in the sense of statistical inference.** However, what it does mean is that relying on the *p*-value as any indicator of scientific evidence is severely misguided, which is why coupling it with effect size measures is advised. But, to suggest we "appease" the *p*-value by collecting less than large sample sizes is nothing short of ridiculous.

Now, of course, as with all things, there are caveats. While you cannot theoretically go wrong with collecting an increasingly large sample size, **it may still not be worth your while. Enter pragmatics.** That is, thanks to inferential statistics, you may be able to learn just as much, or nearly as much, from a much smaller sample of information than collecting a much larger sample. There is a trade-off of resources you should be aware of, and there is a decreasing **rate of return** for collecting increasingly larger samples.

Let us unpack the above concept a bit with an example. Suppose again we are searching for a treatment for COVID-19 and choose to evaluate our new drug on a sample of 1,000 test subjects. Suppose our power for the experiment is quite high at 0.95. Power ranges from 0 to 1.0, and hence power equal to 0.95 is, in most cases, quite respectable. Now, we have said that you cannot have "too big" of a sample, right? While this may be true theoretically, **you can have too big of a sample size pragmatically**. In other words, while you can double your sample size to 2,000 subjects, such a doubling will probably not tell you that much more than your original 1,000 units. Even if power increased from 0.95 to, say, 0.99, the cost incurred, whether that cost be time, financial expense, or fatigue in collecting and using the larger sample size, may simply not be "worth it." Hence, sticking with the original 1,000 units is probably the most economical choice in giving you the most "bang for your buck" in terms of the scientific experiment. It would be "better" to have the 2,000 subjects, but simply not worth the investment given the rate of return. In other words, you can conclude just about as much from the 1,000 subjects as you can from the 2,000.

The above principle can be easily demonstrated by a survey of **power curves**. Below is an example of a power curve where power is given as a function of sample size. Notice that as sample size increases, the rate of return on power begins to plateau. The shape of the curve is **logarithmic**, which for this case implies that for a sample size of 60 or so, we have reached sufficient power. Collecting more of a sample size will not benefit us much in terms of the inferences we wish to make. To be blunt, collecting a sample size of 500, for example, would be a waste of resources. We can do just as well in terms of our statistical inferences with far less of a sample size. Of course, how we collect the sample is extremely important (e.g.

usually random sampling is advisable), but the point is that the rate of return on collecting larger and larger sample sizes is usually quite minimal past a certain point.

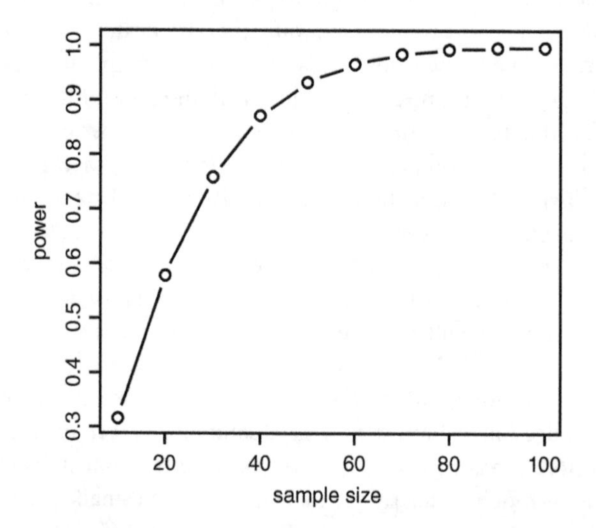

You cannot have "too much" power or sample size. However, at some point, at a practical level, increasing power levels and sample size simply is not worth the investment. That is, the rate of return for seeking higher and higher levels of power is not usually worth the cost of collecting a larger sample, whether those costs be financial or other. Increasing sample size beyond power levels of approximately 0.95 usually affords little benefit in most cases and will incur more expense in the form of time, resources, etc., than is necessary. Aim for a healthy sample size and high power, but there is little need to waste time or resources collecting increasingly larger sample sizes.

5.7 Demonstrating Power Principles in Python: Estimating Power or Sample Size

We can easily demonstrate the above-mentioned principles in Python, showing how sample size and effect size help to determine the degree of statistical power. The question often arises as to whether you should estimate power or sample size. It really does not matter which you estimate, since, as we have seen, one is a function of the other. However, in planning an experiment, it is typical of the researcher to designate his or her degree of preferred power and then learn how much of a sample size will be required to achieve this. Hence, the researcher may set power at, say 0.90, estimate the given effect size, and then solve for sample size. In other cases, the researcher may be restricted by sample size and would like to learn of the degree of statistical power afforded by the fixed sample size. Either way will give you the same estimates. Sometimes researchers have fixed sample sizes and cannot collect more. In such cases, they may simply want to know their chances at rejecting the null and whether the experiment or study is worth doing at all beyond a **pilot study** (which is a study done on a very small number of subjects, presumably to inform the decision to pursue or not pursue a larger-scale study).

As an example, for an independent samples *t*-test, we can use **statsmodels** to provide us with a sample size estimate. For this example, we set **effect size** at 0.8, **alpha** at 0.05, and desire a level of **power** equal to 0.8:

```
from statsmodels.stats.power import TTestIndPower
effect = 0.8
alpha = 0.05
power = 0.8
analysis = TTestIndPower()
result = analysis.solve_power(effect, power=power, nobs1=None,
ratio=1.0, alpha=alpha)
print('Sample Size: %.3f' % result)
```

```
Sample Size: 25.525
```

We see that the estimated sample size is equal to approximately **25 subjects per group**. To be on the safe side, it is always wise to round sample size estimates **upward** to ensure sufficient power. That is, the estimate is for 25.52, but it does not "hurt" to round up in this case. The worst thing that can happen is that we end up having slightly more sample size than we actually need. If we round down, however, we risk having insufficient sample size. Now, in this case, the fraction of 0.525 will make little difference either way, but in principle, **always round up**. The `ratio = 1.0` designates that we have equal numbers of participants in each sample. Though this may change over the course of data collection, it is usually wise to start off assuming equal sample sizes per group.

Now, suppose we adjusted the requested power to 0.9 instead of 0.8. If you are understanding power correctly, this should imply that for such an increase in power we require a greater sample size. Recall that since increasing sample size is one essentially sure way of boosting power, this is what we would expect. Indeed, this is what we find:

```
from statsmodels.stats.power import TTestIndPower
effect = 0.8
alpha = 0.05
power = 0.9
analysis = TTestIndPower()
result = analysis.solve_power(effect, power=power, nobs1=None,
ratio=1.0, alpha=alpha)
print('Sample Size: %.3f' % result)
```

```
Sample Size: 33.826
```

We can see that as our demand for power increases, this is associated with a likewise increase in sample size. In English, what this means is that if you want more of a chance to reject the null, that is, greater sensitivity, you are going to need to collect more subjects.

5.8 Demonstrating the Influence of Effect Size

We can also easily demonstrate the influence of effect size on sample size requirements. For instance, if we hypothesized a much smaller effect (0.2), this should entail requiring a much **larger sample size** to detect it. We keep power set at 0.8, as well as the significance level at 0.05. Let us see what happens to estimated sample size given these parameters:

```
from statsmodels.stats.power import TTestIndPower
effect = 0.2
alpha = 0.05
power = 0.8
analysis = TTestIndPower()
result = analysis.solve_power(effect, power=power, nobs1=None,
ratio=1.0, alpha=alpha)
print('Sample Size: %.3f' % result)
```

```
Sample Size: 393.406
```

We can see that as the effect size drops to 0.2, the required sample size increases to a whopping approximately 393 per group! But why should this make sense? A drop in effect size from 0.8 to 0.2 implies that we are attempting to detect something that is much smaller than before. As an analogy, when the eye doctor asks you to read off the very small tiny letters at the optical exam, you will need a stronger lens to detect it:

A B C

A B C

In the above, the smaller letters require greater sensitivity than the larger letters, since the effect is much smaller in the first case than in the second. That is, **we need a more powerful test to detect the smaller effect than the large**. The tiny letters represent a measure of effect, in which in this case the effect is very small. Hence, you need a stronger "prescription" in order to detect it. If the letters are extremely small, you might even need a "powerful" set of binoculars. Understand the analogy? If the letters were large (i.e. a huge effect), you could probably identify them with no prescription at all. If the Yellowstone bison is directly in front of you (well, run!), you do not need binoculars to see him. You have well sufficient power because the effect (size and proximity of the bison) is quite large! Hence, **the smaller the effect size, all else equal, the greater sensitivity of the test required**. That is, all else being equal, you require more subjects to see something that is hardly there in the first place.

5.9 The Influence of Significance Levels on Statistical Power

We have demonstrated how changing the sample size or altering the anticipated effect size can have a significant impact on power. Conversely, we have seen how adjusting the power level and effect necessitates differing numbers of subjects.

But what about the significance level? A change in significance level will also have a direct influence on statistical power since it necessarily changes the type I error rate. We can easily see why this must be true by studying power curves for distributions, as shown below:

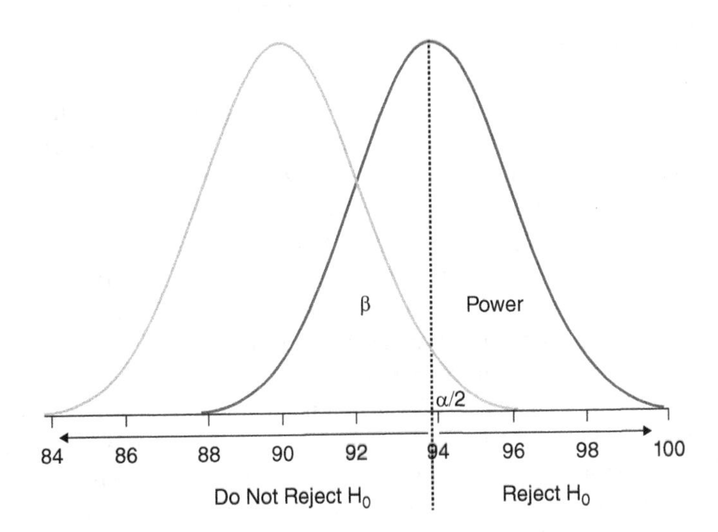

Notice that if we were to move the criterion decision line (the vertical dotted line dividing β, the type II error rate, and power) over toward the left (imagine pushing the line leftward), thereby increasing the type I error rate (i.e. $\alpha / 2$ in the curve), the type II error rate would decrease, and, consequently, since power is the balance of this error rate, power would increase. Hence, if you increase the type I error rate from, say, 0.05 to 0.10, you necessarily increase statistical power. We can easily demonstrate this by using our earlier example in Python. In what follows, we adjust the significance level from 0.05 to 0.20, while maintaining the effect size and power equal to 0.8, respectively:

```
from statsmodels.stats.power import TTestIndPower
effect = 0.8
alpha = 0.20
power = 0.8
analysis = TTestIndPower()
result = analysis.solve_power(effect, power=power, nobs1=None,
ratio=1.0, alpha=alpha)
print('Sample Size: %.3f' % result)
```

```
Sample Size: 14.515
```

We can see that keeping other parameters the same as in our earlier example (where we estimated the number of subjects to equal approximately 25 per group), but adjusting the significance level from 0.05 to 0.20, the required sample size is now only 14 or so, down from 25. But if we just said that this should give us more power, why is power still equal to 0.80? It is equal to 0.80 because we set it **a priori** as such, in other words it was **fixed** at that level. Given that it was fixed there, and the effect at 0.8, there is only one other way to accommodate the new alpha level, and that is via the estimated sample size. This is why the sample size decreased to 14.5.

5.10 What About Power and Hypothesis Testing in the Age of "Big Data"?

With all of the hype around "big data" these days, it is a reasonable question to ask where statistical power fits into this discussion. Big data is a term that connotes a few different things, including the employment of **data engineering** on extremely large data sets. Hence, as the name suggests, when you are working with big data, it implies from a statistical inference perspective that you have sufficient power to make inferences toward populations. However, make no mistake, **inferences are still required**, even if they are not **formalized**. Big data does not somehow make statistical inference and hypothesis testing obsolete in any way, it simply means that we are working with such large data sets that we will usually have sufficient data to make very good estimates and guesses as to the nature of population parameters. Hence, if you have such large data, then calculating statistical power will usually be a waste of time since you will usually have enough power to reject virtually any null hypothesis you put up. The focus must then be on **effect size** and describing or quantifying the degree or extent of variance explained in the response variable or variables you are working with.

For example, when studying data in the COVID-19 pandemic, even if entire populations could not be studied because the data were still coming in day by day, most models had sufficient power or sensitivity since they were based on relatively large sample sizes. Hence, in studies where sample size is easy to come by, such as in big data or areas where samples are extremely large, power is still a concern **implicitly**, but it does not rear its head as much because sample sizes are usually well beyond sufficient. Where power is a much more pressing issue is in **well-calibrated designed experiments**, where the scientist is purposely planning a demonstration to show whether a treatment works or does not. The idea of power as part of null hypothesis significance testing originally and historically arose in the context of performing rather precise experiments and quality control designs, and so when you are building a study from the ground up, you should definitely be paying close attention to statistical power. Most samples in such designs will be potentially difficult and expensive to recruit and hence you definitely do not want to waste resources on collecting more subjects than you may need if indeed collecting subjects is an expensive endeavor. For a big data project in which COVID-19 cases are analyzed on world populations, statistical inference still occurs, but is more **implicit**. Hence, power concerns are likewise more implicit as well. They still exist, however, and inferences are still taking place, again, even if implicitly and we do not bring up *p*-values into the discussion.

What about hypothesis testing? Does hypothesis testing somehow vanish simply because one is using exceedingly large data sets? Absolutely not! Now, from a statistical point of view, yes, rejecting the null hypothesis when you have hundreds of thousands of observations is quite the meaningless result. With such large sample sizes, as we have discussed, null hypotheses will be jettisoned left and right. However, **hypothesis testing is not merely a statistical exercise. It is a scientific one**. Hence, when you reject the null, you are also simultaneously attempting to infer a suitable alternative hypothesis. The principles of hypothesis testing still remain, the only difference being that little if any focus will be on p-values. Most of the focus, as it should be, will be on **effect sizes** or other measures of pragmatic description (such as **prevalence** or **incidence** of disease, etc.). But, rest assured, the principles of hypothesis testing still remain, even in the age of big data.

Statistical power and hypothesis testing are both just as relevant in the age of big data as they were before. The only difference is that since some data sets are so large, usually statistical power is nothing more than an afterthought since it is almost a virtual guarantee that enough participants or objects are being subjected to analysis. And while rejecting a null hypothesis in such situations may no longer be meaningful from a statistical perspective, the idea of detecting a good alternative hypothesis is just as relevant. Hence, big data changes things in terms of software engineering and other facets of research, such as data recruitment, but essential statistical principles are still at play. Even when algorithmic approaches dominate, essential principles of basic statistics and research remain, that of collecting a sample, studying it, then making inferences toward the population.

5.11 Concluding Comments on Power, Effect Size, and Significance Testing

This is as far as we take our discussion of statistical power, sample size, and how these relate to effect size and significance testing. The chapter was only meant as a brief overview and summary of these issues, with a few demonstrations in Python. Short as this chapter may be, it is essential that you clearly understand and appreciate the issues at play if you are at all to interpret p-values, significance tests, and effect sizes. Numerous times, the author has witnessed even experienced and well-published researchers and scientists demonstrate a complete and utter lack of understanding of how these concepts interplay. Such a misunderstanding leads one to make or draw substantive conclusions from research papers that simply do not exist. If you do not understand the contents of this chapter, you should not be interpreting statistical evidence. That is not an exaggeration or an idealistic remark. These fundamental issues are **that** important to grasp. The most common misunderstanding and errors of interpretation are usually the following, and are the ones you should be acutely aware of so that you do not make them:

- Assuming that since the experimental report reads "$p < 0.05$" (or $p < 0.01$, etc.), that somehow this translates into a meaningful scientific result. This is not the case. Without a meaningful effect size measure, "$p < 0.05$" may, on a scientific level, be quite meaningless.
- Not appreciating or understanding that statistical power is intimately related to the size of the p-value, and that insufficient power makes rejecting a null hypothesis virtually impossible, even if the effect size is quite large. If you are interpreting p-values, you need to be acutely aware of how they relate to statistical power. Otherwise, you will likely draw conclusions from your research that are not warranted.
- Failing to appreciate how p-values can largely be a function of sample size, but this is generally not the case for effect sizes. While increasing the sample size to large numbers virtually almost guarantees a rejection of the null hypothesis, the effect size may increase or decrease depending on what the new data have to say.

- Believing that an increase in sample size is always a good idea, in that if you double sample size, for instance, it will result in an equivalent doubling of statistical power or otherwise more sensitivity to detect the alternative hypothesis under consideration. This misunderstanding fails to understand and appreciate the diminishing returns of increasing sample size in relation to power. For a given effect size, at some point, increasing sample size will not afford that much more power, and hence if the sample size is not cheap, then doing so would be a waste of resources.

If any of the principles of this chapter are not clear, you are encouraged to consult good introductory textbooks on statistics targeted toward scientists that further elucidate and explore these topics. Hays (1994) is an excellent reference for all matters discussed in this chapter.

Review Exercises

1. Cite and discuss the determinants of the ***p*-value**. What are the factors that make a *p*-value large or small?
2. Why does the statement "$p < 0.05$" not necessarily imply a scientifically meaningful result? Explain.
3. Why is it the case that for **increasing sample size**, this might not result in a rejection of the null hypothesis? Explain.
4. How does reducing **population variance** contribute to increasing statistical power? Try to explain by highlighting what it means to conduct a quality experiment.
5. Someone says to you that having **too much statistical power** for an experiment is a bad thing. Respond and explain, and correct if necessary.
6. Your boss says to you, "Let's collect an additional 2,000 participants for our study over and above the 2,000 we already have." What kinds of concerns might you have over this plan? Respond to your boss with information that he/she may want to consider further.
7. How does an **increase in sample size** virtually guarantee a rejection of the null hypothesis, yet not necessarily a small or large effect size?
8. Why is the **distance** between the sample mean and the population mean insufficient on its own as a measure of effect size? That is, why is it necessary to divide by the population standard deviation to better contextualize this distance? Explain.
9. Suppose a medical researcher says, "This COVID-19 vaccine has worked successfully on 10,000 individuals we tested it on." Explain why it may not necessarily work on the 10,001st individual using what you know about the principles of statistical inference.
10. Demonstrate using **power curves** for two distributions why decreasing the type II error rate, all else being equal, has the effect of increasing statistical power. Further, can you actively and deliberately decrease type II error in a real research context given a fixed significance level? Why or why not?
11. Using Python, for a one-sample *t*-test, demonstrate the computation for power for an effect equal to 0.0. Does this result surprise you? Why or why not?
12. Estimate sample size in Python for a two-sample *t*-test with both a very small effect size and then a rather large one. Note the difference you observed in required sample sizes. Does this surprise you at all? Why or why not?

6

Analysis of Variance

CHAPTER OBJECTIVES

- Understand the nature and logic of the analysis of variance (ANOVA) from first principles.
- Conceptualize t-tests for means as a special case of the ANOVA model.
- Understand why performing multiple t-tests does not guard against inflating error rates.
- Evaluate assumptions in ANOVA, including normality and homogeneity of variance.
- Conduct one-way and two-way ANOVA in Python and interpret effects.
- Understand the concept of an interaction in factorial ANOVA and why science is mostly about interactions, not main effects.
- How to read an interaction graph in two-way ANOVA.
- Understand the difference between between-subjects ANOVA and repeated measures ANOVA.

The analysis of variance, or "ANOVA" for short, plays a central role in statistical modeling. Not only is it one of the most commonly applied statistical procedures across virtually all of the sciences, but learning ANOVA helps the student of statistics or science better understand the principles behind not only the ANOVA model, but of many other statistical models as well, even relatively advanced ones. In this day and age of high-speed computing, "big data," and all the rest of it, classic ANOVA is sometimes perceived as a "basic" method that is somewhat "antiquated." Nothing could be further from the truth. ANOVA, along with regression of the following chapter, are the backbone statistical methodologies of modern statistics, and one could easily give a full course in ANOVA and still have many topics still left to cover. Our treatment of the method here is quite brief. For the student interested in a much deeper study of the procedure and to learn just how technical ANOVA can get, he/she is encouraged to consult Scheffé (1999) for a thorough, if not challenging, treatment.

Applied Univariate, Bivariate, and Multivariate Statistics Using Python: A Beginner's Guide to Advanced Data Analysis, First Edition. Daniel J. Denis.

6.1 *T*-Tests for Means as a "Special Case" of ANOVA

There are two ways of conceptualizing ANOVA. The first is to perceive ANOVA models as extensions of t-tests for means where instead of only two groups, we now have three or more populations that we wish to compare. The second is to see the ANOVA model as a wider, more complete model and as a special case of the regression model. The second approach provides much more perspective into statistics than the first, but since we have not covered regression yet, our focus will be on understanding ANOVA as an extension of the t-test for means. Once we study regression in the chapter to follow, we will correctly place the ANOVA model in its proper context as a special case of the more **general linear model**.

Recall that in an independent samples t-test, we were interested in evaluating a null hypothesis about population means. In the ideal and prototypical circumstance, we had both a **control group** and a **treatment group**, and our null was a statement that the two samples arose from the same population. Our statistical alternative hypothesis was that they were selected from distinct populations. We expressed this null hypothesis symbolically with the following:

$$H_0 : \mu_1 = \mu_2$$

where μ_1 was the mean for the first population and μ_2 was the mean for the second population. The alternative hypothesis was that the population means were unequal. That inequality could take the form of μ_1 being greater than μ_2 or μ_2 being greater than μ_1. Hence, we could, if we really wanted to, specify the alternative hypothesis as $H_1 : \mu_1 > \mu_2$ or $H_1 : \mu_1 < \mu_2$. Each of these were known as **one-sided alternatives**. However, because we were interested in potential differences in either direction, we instead typically set our alternative hypothesis as

$$H_1 : \mu_1 \neq \mu_2$$

This was known as a **two-sided alternative** hypothesis and allowed for either population mean to be greater than the other. The two-sided alternative merely hypothesized a difference in means, not direction. As we discussed when doing t-tests, it is generally the preferred alternative hypothesis to state as a competitor to the null. In the case of an experiment, you may think and hope the treatment mean will be larger (or smaller, depending on the context of the experiment) than the control mean, but should the opposite effect occur, surely you would be just as interested in rejecting the null. The two-sided alternative allows for a rejection in either case and hence is the preferred statement. Unless you have a very specific reason for performing a one-tailed test, you should virtually always be conducting two-sided ones.

From there, we constructed the t-test:

$$t = \frac{\bar{y}_1 - \bar{y}_2}{\sqrt{\frac{s_1^2}{n_1} + \frac{s_2^2}{n_2}}}$$

The numerator in the independent samples t-test measured the difference between sample means. The extent to which the difference was large, all else equal, was the extent to which t would come out to be large as well. However, the mean difference

$\bar{y}_1 - \bar{y}_2$ was not the only thing going on in the t-test. The denominator also played a critical role and represented the estimated standard error of the difference in means. In the case where sample sizes are unequal, such that $n_1 \neq n_2$, we can pool variances, which yields the following t-test:

$$t = \frac{\bar{y}_1 - \bar{y}_2}{\sqrt{s_p^2\left(\frac{1}{n_1} + \frac{1}{n_2}\right)}}$$

The pooled variance term s_p^2, unpacked, is equal to $s_p^2 = \frac{(n_1 - 1)s_1^2 + (n_2 - 1)s_2^2}{n_1 + n_2 - 2}$ and is essentially a sort of average of the two variances, s_1^2 and s_2^2.

If you understand the logic of the t-test, you will have virtually no problems understanding ANOVA. But what is the logic of the t-test, exactly? We have unpacked the t-test in our earlier discussions including in that of statistical power, where we had to understand the components of the t-test in some detail in order to make sense of what gave a test more power than not. Recall that the t-test is comparing a mean difference in the numerator to an estimated standard error of the difference in the denominator. That is, the t-test is comparing an observed sample difference to variation we would expect, on average, under the null hypothesis. We have already mentioned that if the numerator were large, all else being equal, the value of t would likewise be large. However, what of the denominator? Notice that all else equal, if the denominator is large, then the value of t will be small. This makes sense, because since the denominator reflects the average variability one might expect to find under the null hypothesis, we see the t-test as a comparison of **observed to expected average difference or variation**. Recall in the chi-squared goodness-of-fit test that we compared observed to expected frequencies. Though the t-test deals with different data, the analogy is powerful in terms of understanding what the statistical test actually "does." Hence, even if $\bar{y}_1 - \bar{y}_2$ is large, if the denominator is quite large, then t will not come out very large, since the difference $\bar{y}_1 - \bar{y}_2$ is not that "impressive" when compared to the denominator, $\sqrt{s_p^2\left(\left(1/n_1\right) + \left(1/n_2\right)\right)}$. This is the logic of the t-test, to compare an observed difference to an average difference (in terms of variability) expected under the null.

So what does any of this have to do with ANOVA? Everything! As we will see, ANOVA works in an incredibly similar way. We will see that ANOVA effectively compares a numerator to a denominator, where the numerator also reflects a difference in means, only now, we typically have more than simply two means. The denominator will also reflect more than variabilities within only two groups. Hence, we will see that t-tests and ANOVA are very similar conceptually in what they accomplish. However, why do ANOVA in the first place? Why not simply perform multiple t-tests on our group of means to determine where mean differences lie? There is a very good reason for this, and hence before considering the ANOVA model in detail, we first need to understand its justification and motivation. That is, we need to understand better why performing multiple t-tests on a set of means is typically not a good idea. We will then introduce the ANOVA model as a solution to this problem.

6.2 Why Not Do Several *t*-Tests?

We have said that ANOVA is used when there are more than two means to compare. Well, as we have suggested, the obvious question that arises is why not simply do multiple *t*-tests? For example, a null hypothesis for a four-population problem would look like this:

$$H_0 : \mu_1 = \mu_2 = \mu_3 = \mu_4$$

If we would like to compare means, why not simply evaluate all pairwise mean comparisons via independent samples *t*-tests? That is, why not evaluate null hypotheses such as $H_0 : \mu_1 = \mu_2$, $H_0 : \mu_1 = \mu_3$, and so on, for all pairwise comparisons via independent samples *t*-tests? At first glance, it would seem like an attractive option and save us the trouble of having to learn ANOVA! However, the problem is that, as you will recall, with each mean comparison using a *t*-test comes with it a type I error rate, which is equal to the significance level for the test, often set by convention at 0.05. When we perform several mean comparisons on the same data, this error rate multiplies and hence gets larger with the number of comparisons we are doing. How large it gets is beyond the scope of this book (for details, see Denis, 2021). However, as a very rough approximation, especially for a relatively small number of comparisons, one can think of the error rate as multiplying by the number of comparisons made. This is only an approximation, but helpful in terms of conceptualizing the inflating error rate.

Hence, if we made two comparisons, we would have approximately double the error rate, and so our error rate across both comparisons would no longer be 0.05 as it is for each comparison, but rather near 0.10 across both comparisons. Again, for reasons beyond this book, the error rate is not precisely additive like this, but it does not hurt to think of it in this way, especially when performing relatively few comparisons. The morale of the story is that **when you perform several comparisons on the same data, the theoretical type I error rate gets larger the more comparisons you perform.** Now, one could simply attempt to control the otherwise inflating error rate by using what is known as a **Bonferroni adjustment**, where alpha is split up by the number of comparisons one wishes to perform. For instance, if we are performing three mean comparisons, we could conduct each comparison at 0.05/3 = 0.0167. This would help keep the overall type I error rate down, though at the expense of a loss of statistical power for each comparison. As you can imagine, as the number of means grows in a given problem, performing such corrections on alpha becomes implausible, since the power for any of the given comparisons will be quite low. For instance, for even a six-comparison problem, each comparison would be conducted at 0.0083 (i.e. 0.05/6), which yields hardly any statistical power for each comparison. That is, rejecting the null hypothesis at a significance level of 0.0083 will be quite difficult since alpha is so small. As we will see, performing an ANOVA will be much more efficient.

Hence, we now have the motivation for performing ANOVA, to help control the inflating error rate. ANOVA will allow us to evaluate the null hypothesis while keeping the type I error rate at a nominal level. That is, ANOVA will allow us to learn whether there are mean differences, all the while keeping the type I error more or less at the level of 0.05 for the family of comparisons. In this sense, we can test an **omnibus null hypothesis** without allowing the error rate to inflate as it would when performing several *t*-tests. This is one definitive advantage of performing ANOVA. As we will see, this is not the only reason for ANOVA, and it is a bit more complex than simply the type I error rate. Historically, ANOVA was responsible for providing researchers with a "toolkit" of statistical methods, of which many varieties well beyond the "classic" ANOVA have sprung. ANOVA allows for much more complex modeling that is beyond the scope of this book, and, both historically and currently, had broadened the landscape a great deal as it concerns statistical models. For example, multilevel and mixed modeling, relatively advanced subjects, are direct extensions to "basic" ANOVA models.

6.3 Understanding ANOVA Through an Example

The easiest way to introduce and understand ANOVA is by considering an example with some data. From this example the primary motivation for how ANOVA works on a theoretical level will surface. Consider the data (featured earlier in the book) on achievement scores as a function of which teacher students were randomly assigned. The data appear in Table 6.1.

Let us describe this data in some detail so that we are very familiar with it as we unpack the concepts of ANOVA. With data, especially such as these, which are rather small, you should always inspect and look at the data carefully to become familiar with it. **Always get to know your data regardless of what model you are fitting.** As concerns the current data:

- In each group are achievement scores of students who were randomly assigned to each teacher. For instance, under "Teacher 1" the first student obtained a score of 70, the second student a score of 67, and so on.
- There is a total of four teachers in the data layout.
- The mean of achievement scores for each teacher are given at the bottom of each column. The mean for teacher 1 is 71.00, the mean for teacher 2 is 72.5, and so on.
- Note that the sample means are different from one another. The first two means (71.00 and 72.50) are relatively similar, while the other two means are quite different from both each other and from the first two means. The idea of sample means being different from one another can be thought of as **between-group variability**.

Table 6.1 Achievement as a function of teacher.

Teacher			
1	2	3	4
70	69	85	95
67	68	86	94
65	70	85	89
75	76	76	94
76	77	75	93
73	75	73	91
$M = 71.00$	$M = 72.5$	$M = 80.0$	$M = 92.67$

- We note that in addition to between-group variability, there is also **within-group variability**. For example, in group 1, we note a certain degree of variability, such that 70 is different from 67, which is different from 65, and so on. We note there is within-group variability in groups 2, 3, and 4 as well.

The fundamental question that ANOVA will attempt to address of this data is the following:

Is total variation in the data better explained by between-group variation or within-group variation?

That's it! That is the underlying entire goal of the analysis of variance. The rest is in the details and complexity of each individual model. Applied to our data, ANOVA will attempt to compare **between-group variation** to **within-group variation**. But why do this? Because it is going to help us in evaluating the null hypothesis for ANOVA. The null hypothesis for ANOVA is that all population means are equal. That is, for our data, the null would be

$$H_0 : \mu_1 = \mu_2 = \mu_3 = \mu_4$$

Note that as was true for the independent samples *t*-test, this hypothesis is about **population means**, not samples! Make sure you understand this point. Many students who claim to be familiar with ANOVA and statistical methods in general often still fail to realize that **null hypotheses are about population parameters, not sample statistics.** In our achievement data, we clearly do not have populations. We have samples. Hence, the sample means of 71.00, 72.5, etc., different as they are, would not by themselves justify rejecting the above null hypothesis. Otherwise, the null hypothesis would always be rejected for virtually all data since we always expect to see differences in any samples we can gather. The question for ANOVA is whether the differences we are observing in our sample provides any evidence that the population means are different from one another. That is the question posed by ANOVA. Again, this detail is extremely important! Make sure you understand it. **If you have population data, and hence have knowledge of the population parameters, you are no longer interested in statistical significance!** If you have the population data, a significance test would be meaningless. The only reason for conducting a significance test is because you have sample data and would like to use the sample data to get a picture of what might be going on in the population.

As an example, suppose a department of faculty members is voting on a policy. Suppose there are 20 faculty members who vote. In the actual vote, 11 are in favor, while 9 are opposed. Does it make sense to ask the question of whether the result is statistically significant? Absolutely not! Why not? **Because we have the population data. The population is the faculty.** Unless the faculty comprises a subset of a wider group of faculty members to which we wish to infer, the vote of 11 to 9 is the population result we were seeking to learn about. It just so happens that, in this case, the numbers are quite small, but they are still parameters. Hence, the concept of statistical significance has no meaning here. What may still have meaning is the idea of **measurement error**, but that is different from statistical significance. Measurement error is how precisely we are assessing or evaluating an attribute, so, in this case, it may be worth re-doing the vote to make sure measurement error is minimized and that no faculty members misunderstood the question or the item being voted on, or voted out of fear of not conforming to the group (group conformity in a faculty meeting, who would have guessed that?). Assessing measurement error is well beyond the scope of this book and is a topic more aligned with **psychometrics** than statistical methods in general (though both areas closely overlap). If you have a significant degree of measurement error in your variables, it can severely compromise your statistical analyses.

Let's get our data into Python to have a better look (we print only a few cases):

```
data = {'ac' : [70, 67, 65, 75, 76, 73, 69, 68, 70, 76, 77, 75, 85,
86, 85, 76, 75, 73, 95, 94, 89, 94, 93, 91],
'teach' : [1, 1, 1, 1, 1, 1, 2, 2, 2, 2, 2, 2, 3, 3, 3, 3, 3, 3, 4,
4, 4, 4, 4, 4]}

df = pd.DataFrame(data)

df
Out[211]:
   ac  teach
0  70      1
1  67      1
2  65      1
3  75      1
4  76      1
5  73      1
```

Visualizing our data will help us get a better idea of the distributions of achievement by each teacher:

```
df['ac'].hist(by=df['teach'])
Out[224]:
```

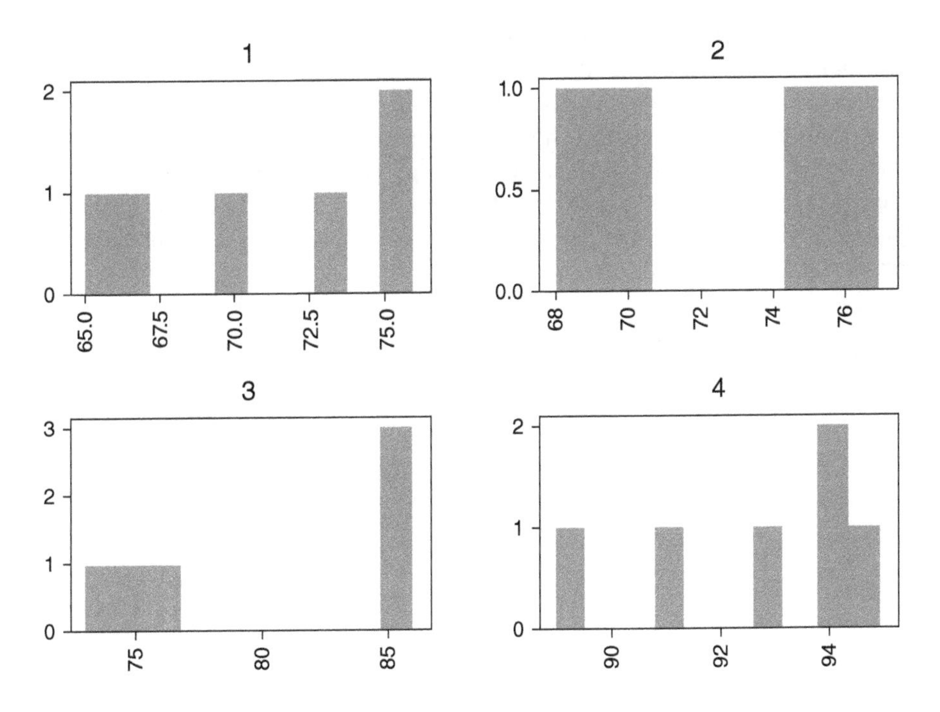

The histograms are a bit "ugly" because there is not a lot of data in each group, but one can still get a rough sense of how much variability there is in each of the teacher groups. Some descriptive statistics will also help. We could get the sum and mean of each variable. First, the sum is computed:

```
df.sum()
Out[228]:
ac         1897
teach        60
dtype: int64
```

Always be cautious about interpreting summary statistics, as depending on the types of variables on which you are computing, the result may be quite meaningless or misleading. For instance, we see that the sum for teach is equal to 60, but this means nothing to us. If you look at our `teach` variable, it is simply summing the following:

$$1 + 1 + 1 + 1 + 1 + 1 + 2 + 2 + 2 + 2 + \cdots + 4$$

However, this corresponds to **levels** of `teach` as a factor and hence the total sum is a useless statistic. On the other hand, the sum of 1897 for `ac` is a bit more meaningful, since it gives us an idea of how scores across teacher groups total up. In the following, we obtain the **arithmetic mean** for both `ac` and `teach`:

```
df.mean()
Out[233]:
ac         79.041667
teach       2.500000
dtype: float64
```

The mean for `ac` is equal to 79.04. This is definitely meaningful, since it gives us the **overall** or **grand mean** of all the data. The mean for `teach` is quite useless, since it gives us the mean factor level. Again, when you ask Python to compute such things as means and standard deviations, while it is easier sometimes to not specify exact variables on which you want these computations (e.g. we could have preselected `ac` by itself had we really wanted to), you simply have to know what to interpret and what to ignore. This applies for virtually all software output. In the following we get the **standard deviations** and **medians**:

```
df.std()
Out[234]:
ac         9.648064
teach      1.142080
dtype: float64
```

```
df.median()
Out[235]:
ac         76.0
teach       2.5
dtype: float64
```

It is easier in many cases to simply request the above statistics and others using `describe()` in Python:

```
df.describe()
Out[236]:
              ac      teach
count  24.000000  24.00000
mean   79.041667   2.50000
std     9.648064   1.14208
min    65.000000   1.00000
25%    72.250000   1.75000
50%    76.000000   2.50000
75%    86.750000   3.25000
max    95.000000   4.00000
```

All of the descriptives for `ac` are useful to us. For example, we see that the minimum value in our data is equal to 65 while the maximum is equal to 95. This can be important information, since if the passing score is, say, 50, we immediately know that nobody failed. Likewise, if we know that the theoretical maximum is equal to 100, then we know nobody got the highest score possible, since the highest score is equal to 95. The 25%, 50%, and 75% values are also useful. These are the **quartiles** of the distribution. That is, 25% of scores in the data set are below 72.25, 50% are less than 76.00, and 75% are smaller than 86.75. Notice that the 50% score of 76.00 is the actual **median** of our data, and agrees with our request for the median earlier using `df.median()`. The `count` field is also of interest, as it tells us there are a total of 24 data points. This can be useful as a quick verification and check to make sure all data points are represented in our summary statistics and that no data omission errors have occurred. For the `teach` variable, again most of the information is irrelevant due to the nature of `teach` being a factor variable. However, the fact that the minimum value is 1.00 and the maximum value is 4.00 confirms for us the range on the factor level from 1 to 4. Such **confirmatory checks** can be useful to make sure Python is reading your data file correctly and that all information is being accounted for. Checking, double-checking, and re-checking those double-checks is what good data management and analysis is all about. Never hesitate to "confirm the obvious," it may surprise you!

6.4 Evaluating Assumptions in ANOVA

Recall that as with virtually any statistical model, it is necessary to ensure that assumptions for the model are satisfied, or at least not so violated that it makes the model inappropriate and inferences misguided. Recall from the outset that assumptions for any statistical model are almost always wrong. **The job is to make sure they are within bounds.**

A first assumption of ANOVA is that of **independence of errors**, which essentially implies that errors within one group are not statistically related to other errors within the group and to errors in other groups (Kirk, 1995). This assumption is essential and can usually (though it does not necessarily guarantee) be considered satisfied by the method of data recruitment and assignment to groups. If you have a reliable means of randomly sampling observations and assigning those observations to groups on a random non-biased basis, it is likely that the assumption will be more or less met. If you put out a flyer for participation for your study and two college students participate and share information about the experiment with one another, independence will be violated to some degree. Whether the **degree of violation** will be enough to thwart inferential tests may be unknown, but it should concern you if you have reason to doubt independence at any point in your experiment or study. Rarely in social studies is independence ever truly satisfied, but the task is to minimize it being violated as much as possible. **Residual plots** can sometimes be used to try to detect a violation of independence, but not without some difficulty. See Denis (2021) for more details.

A second assumption in ANOVA is that of **normality**. Technically speaking, the assumption refers to that of the errors of the model. However, the assumption can also be generally (though not exactly equivalently) phrased as the normality of population distributions in each level of the independent variable (IV). How to verify this assumption? Usually, visual plots of **residuals** are sufficient. For instance, plotting boxplots and histograms will usually suffice in giving you an idea of whether the assumption is more or less tenable. Some researchers (and reviewers of journal articles) like to also supplement this information with an inferential test of normality. The **Shapiro-Wilk normality test** is one such test that evaluates the null hypothesis that the data were drawn from a normal population. The technical details of the test need not concern us here and are not tremendously enlightening. What you should be aware of, however, is that the test has its own issues and can be quite sensitive to assumptions itself as well as sample size. Hence, it is generally recommended that you do not take the test "too seriously," but instead use it as a guide only for assessing normality.

Now, for our data, we have a total of six observations per group. There is realistically no way of evaluating normality per group with only six observations; hence, conducting the Shapiro-Wilk test on each group would be pointless. However, and for demonstration here, conducting the test on the entire data set may nonetheless give us a sense of the **distribution** of the entire data. Again, we are not taking this test too seriously since it is on the entire data set and not by group, but it nonetheless can give us an idea of the global normality of the data (and allow us to demonstrate the Shapiro-Wilk test). To get the test, we import **scipy** and use the `stats.shapiro()` function:

```
import scipy as sp

ac = df['ac']
sp.stats.shapiro(ac)
Out[241]: (0.90565425157547, 0.028417399153113365)
```

The value of the test statistic is equal to 0.90, with a p-value equal to 0.028. Recall that the null hypothesis is one of normality. Hence, since the p-value is quite small, it indicates that we can reject the assumption of normality and infer non-normality. That is, we have evidence to suggest that the sample of data was not drawn from a normal population. Again, with larger data sets the test will be more informative and you should conduct the test **per group or cell** to get an idea of normality, not on the entire data set. And to reiterate, use the test (if you use it at all) as a **guide** only, not a final verdict. We have more to say on assumptions for parametric models later in the chapter, where we discuss this issue at some length and how you should ideally approach it from both a statistical and scientific perspective.

A third assumption that needs to be satisfied when conducting ANOVA is that of **homogeneity of variance**. That is, ANOVA assumes that variances in each population are equal to one another. The null hypothesis is thus

$$H_0 : \sigma_1^2 = \sigma_2^2 = \sigma_3^2 = \sigma_4^2$$

where σ_1^2 is the variance for population 1, σ_2^2 is the variance for population 2, and so on. Since we have a total of four populations, we have a total of four variances to verify. Now, a moment's contemplation will reveal that, just like with virtually any null hypothesis significance test, **the null is dead in the water even before conducting the test!** That is, surely the population variances will not be equal, which we know long before conducting the test! They will be different to at least some degree, to some decimal place, even if they are quite similar to begin with. So why conduct the test at all? Why evaluate an assumption we already know to be false? For the same reason why we do so with means. We know the null is, in theory, a priori false, yet we conduct the test anyhow as a test against an "ideal" model to learn whether the data differ sufficiently to reject the hypothetical model. To evaluate equality of variances, we can conduct several tests, a popular one being that of **Levene's test for the equality of variances:**

```
import scipy as sp
teach = df['teach']
sp.stats.levene(ac, teach)

Out[247]: LeveneResult(statistic=26.34364367460562,
pvalue=5.625609083376561e-06)
```

The value of the test statistic is equal to 26.34. The p-value for the test is very small, suggesting that we can reject the null hypothesis of equal variances. What the p-value is saying here is that the probability of the data we are witnessing in terms of variances is very small under the hypothetical situation of equal population variances. Hence, since the probability of this data is so small, we deem it unlikely that it could have arisen from a model in which the variances are equal. We reject the null hypothesis and conclude that the variances are unequal. However, since ANOVA is quite robust to the equality of the variance assumption and for the sake of demonstration, we choose to push forward with the ANOVA anyhow and interpret it. In practical cases, if you do have wildly different variances, it behooves you to look further at your data in some detail. **Get curious about your data**. There may be a good reason why your variances are unequal and it may be of scientific interest, in addition to statistical, to learn why. For example, if it were found that improvement in COVID-19 symptoms were much more variable for individuals receiving a heavy dose of a drug than a low dose, this in itself may be an interesting result, and may be a reason to follow up on the finding. Inequality of variances may be itself an interesting phenomenon. **Do not simply seek to appease a statistical model with the assumption. A violation of variances could indicate something of scientific interest. Be more curious about your data than simply satisfying assumptions!**

6.5 ANOVA in Python

We now push forward with the ANOVA in Python, exploring a couple ways in which we can obtain output. We also interpret the output in great detail to learn exactly what each piece is telling us. Before conducting the actual ANOVA, it can be useful as a data check to compute `info()` to verify some basic logistics:

```
df.info()
<class 'pandas.core.frame.DataFrame'>
RangeIndex: 24 entries, 0 to 23
Data columns (total 2 columns):
ac        24 non-null int64
teach     24 non-null int64
dtypes: int64(2)
memory usage: 464.0 bytes
```

We now run the ANOVA model using the `ols()` function:

```
import statsmodels.api as sm
from statsmodels.formula.api import ols
model = ols('ac ~ C(teach)', data=df).fit()
```

Notice that we named our object `model` and specified the formula as `ac ~ C(teach)`, where `C(teach)` is telling Python to treat it as a categorical variable. To get the ANOVA summary table, we compute the following, where we also request **Type 2 sums of squares** (you will obtain the same in this case had you chosen `typ=3`):

```
table = sm.stats.anova_lm(model, typ=2)
print(table)
                sum_sq    df          F        PR(>F)
C(teach)   1764.125000   3.0  31.209642  9.677198e-08
Residual    376.833333  20.0        NaN           NaN
```

Let us unpack each component of this summary table:

- The ANOVA has separated components of **sums of squares** into two parts, that due to `teach` and that left over in `residual`. The sums of squares for `teach` are equal to 1764.125, while the residual sums of squares are equal to 376.833. The total sums of squares is therefore equal to $1764.125 + 376.833 = 2140.958$.
- Each sum of squares has associated with it **degrees of freedom**. The degrees of freedom for `teach` are equal to one minus the number of groups. Since there are four groups, the degrees of freedom for `teach` are equal to 3.0. For within, we lose a single degree of freedom per group across all groups. Since there are four groups, this yields a total of 20 degrees of freedom.
- The **mean squares** for ANOVA are not reported in the table, though they are implicit, since they are used to compute the resulting F-ratio. The mean squares, or "MS" for short, are computed as SS/df, which in our case for `teach` are equal to $1764.125/3 = 588.04$. The mean squares for `residual` are equal to $376.833/20 = 18.84$.
- The ***F*-statistic** for the ANOVA recall is computed as a ratio of MS between to MS within, so, for our data, the computation is $588.04/18.84 = 31.212$, which is evaluated as coming from an F distribution on 3 and 20 degrees of freedom.

- The **p-value** resulting from the significance test on F is equal to 9.677198e-08, which when translated out of scientific notation, equals 0.00000009677198. Hence, we can see it is much less than a conventional level set such as 0.05, and hence we deem the F statistic statistically significant. That is, we have evidence to reject the null hypothesis and conclude that somewhere among the means there is a mean population difference.
- Simply because we have rejected the null, however, does not inform us of where exactly the mean differences may lie. The F-test is an omnibus test such that it gives us the overall test, but it does not tell us specifically which means are different from one another. For that, we will require a **post-hoc test**.

6.6 Effect Size for Teacher

Having conducted the ANOVA, we would now like to obtain, in addition to the observed *p*-value, a measure of **effect size**. Remember, *p*-values do not tell the full story and in many cases can be quite useless as an indicator of scientific evidence. Much more meaningful is the size of effect. So how should we compute the effect size in this context? Though a variety of effect sizes can be computed, by far the most common is to simply compute an **eta-squared** statistic, often denoted by the symbol, η^2. Eta-squared is given by

$$\text{SS treatment} / \text{SS total}$$

Since we know that the total variation of all the data is contained in SS total, it makes good sense to take this ratio as the proportion of variation or variance attributable to treatment. In our case, the treatment effect is the teacher effect. Hence, for our data, we compute η^2 to be $\text{SS teach} / (\text{SS teach} + \text{SS residual}) = 1{,}764.1 / (1{,}764.1 + 376.8) = 0.82$. We interpret the number 0.82 to mean that approximately 82% of the variance in achievement scores can be attributed to teacher differences. Is that effect large? In a sense, it depends on the research area. Effect sizes should be interpreted relative to the research area from which they are originating. However, if I told you that depending on which teacher one had in school, it could explain upward of 82% of the variance in their scores, you would probably take a great interest in who teaches your son or daughter! Eta-squared it turns out is a bit biased upward, meaning that the size of the statistic is somewhat of an overestimate of the actual effect size in the population. A less biased estimate of effect size is **omega-squared**, though not discussed here (see Kirk, 2012, for details). In terms of a ballpark estimate, eta-squared usually does fine for describing an effect, and will give you a quick sense of what "happened" (if anything) in the study or experiment.

6.7 Post-Hoc Tests Following the ANOVA *F*-Test

The ANOVA conducted earlier yielded evidence of population mean differences between levels on the IV. However, the F-test does not address specifically which means are different from one another, only that somewhere among the means there is a difference. If a researcher has an idea in advance of conducting the analysis where mean differences are likely to lie, then that researcher has the option of preparing and performing **a priori contrasts**. For example, if the researcher hypothesized that means for teachers 1 and 4 would be different (perhaps they have completely contrasting teaching styles), then planning to do a *t*-test to compare these groups can be set out long before the data are collected and obtained. Comparisons or contrasts such as this are known as "a priori" because they are **planned before looking at the data**. Even if the data are archival, the key point is that you have not yet "looked at them." Thus, when you conduct one or two contrasts, you can be relatively assured that your error rate is not skyrocketing. However, if you look at the means first, then it is most likely that you are performing well more than simply that comparison. Let us unpack this concept a bit further, as it is an important point, as the rationale behind the post-hoc concept can seem at first quite mysterious. Suppose you had no prior theory as to which means might be different from other means for the given study. Having collected the data, you then list the means featured in the data table discussed earlier:

71.00 72.50 80.00 92.67

Now, having looked at the means, and if asked to make one pairwise comparison among them, which one would you make? You would likely choose to compare the first mean to that of the fourth mean. That is, you would compare 71.00 to 92.67. Why? Simply because that is without a doubt your best chance at obtaining a statistically significant pairwise comparison. Since these means are the most disparate, they stand pretty much the best chance at rejecting the null for a pairwise comparison. However, even if you choose to conduct this comparison, how many comparisons did you really make? You did not make just one. You made several more. When you looked at the means, you most likely compared the first mean to the second, the second to the third, the third to the fourth, and so on. That is, **it is likely that you made many more mean comparisons than the one you chose to perform**. Now, you did not actually conduct these comparisons formally, but you did so **mentally**, and it stands that any good test of means should in some sense incorporate these numerous comparisons into its computation. This is exactly what post-hoc tests attempt to do. A post-hoc test assumes from the outset that you are performing more than one or two comparisons. Post-hocs generally assume that you are performing, at least visually, virtually all pairwise comparisons possible on the data.

What is wrong with performing this many comparisons? As discussed earlier with respect to the *t*-test, for each comparison made, there is a type I error rate to be concerned about. And, across the board, the type I error rate will begin to rise sharply. Hence, when you compare means 1 to 4, you are not evaluating this difference at a familywise error rate of 0.05. Instead, you are conducting this comparison at a much higher familywise error rate. Often this error rate is unknown, because nobody knows how many comparisons you actually did do in choosing the comparison of interest. Perhaps you simply did pairwise comparisons. However, you could have possibly also compared means 1 and 2 to mean 3. Hence, a good post-hoc test will attempt to mitigate the overall and inflating type I error rate across the numerous comparisons you may in reality be making. Or, sometimes you may wish to simply "snoop" the data to learn where the differences actually are, and have no prior theoretical inclination as to where these differences may be. In both cases, post-hoc tests are appropriate.

6.8 A Myriad of Post-Hoc Tests

There are a huge number of post-hoc tests a researcher can choose from when shopping for a test. What differentiates them, in general, is how they go about controlling the type I error rate. Some tests attempt to guard against inflation for all pairwise comparisons simultaneously, while other tests do so in an incremental stepwise fashion. Choose one test and you may find statistical evidence for a difference. Choose another and you may possibly not. This is the nature of post-hoc tests.

Now, you may at first ask yourself: "If there is a difference in means, shouldn't any and all post-hoc tests 'find it?' For example, in our achievement data, shouldn't the mean difference between teacher 1 and, say, teacher 3, of 71.0 vs. 80.0, be 'detectable' by any test? After all, it is clearly a difference, so why should it matter what post-hoc test is used to detect it?" If you are asking this question, it suggests you do not quite yet understand how statistical inference works. Recall these differences are on **samples**, not populations. When we observe a difference of 71.0 vs. 80.0, though this is clearly an arithmetic difference in the sample, we do not yet know whether a difference exists in the population from which these data were drawn. This is why we need to do the significance test. We are evaluating whether μ_1 is different from μ_3, which is the question we are asking, not whether there are differences in the sample. Remember, this is how statistical inference operates in general. Unless we have the population parameters at our disposal, we are using the sample information to make a "guess" at what might be going on in the population. Hence, it makes perfect sense that different post-hoc tests will potentially generate different conclusions on the null $H_0: \mu_1 = \mu_3$, for instance. It is how those different tests mitigate or regulate type I error rates and statistical power that is the issue. **A difference in the sample does not necessarily suggest a true difference in the population. This is the precise reason why statistical inference is necessary.**

This is about as far as we explore the logic of post-hoc tests in this book. There are many excellent treatments of post-hoc tests, including Howell (2002), which is highly recommended for a very good overview. For our purposes, we cut to the chase and will stick with recommending one of the most common and reliable tests, that of the **Tukey Honestly Significant Difference** test, or **TukeyHSD**, for short. It does a good job at controlling inflating type I error rates, without sacrificing too much power. If you would like an even more stringent and conservative test (meaning that it will guard even more against type I error, but at the expense of lower power), then the **Scheffé test** is recommended. The Scheffé is often a choice of hardliner scientists who only want to reject the null if there is relatively strong evidence that the null is false. However, as you can imagine, the test often lacks power and hence if you are trying to learn if a treatment for COVID-19 is effective, for instance, the Scheffé test might miss it. We should also mention the Bonferroni adjustment, discussed earlier. Recall that this is when family-wise alpha is divided by the number of pairwise comparisons you are making. This test can sometimes be used as a post-hoc as well. For example, if you wish to make three comparisons at a family-wise error rate of 0.05, then you would be testing each comparison at 0.05/3 = 0.016. As you can imagine, as the number of comparisons increases, power begins to suffer greatly. For example, for 10 comparisons, we would be testing each comparison at 0.05/10 = 0.005, which does not afford us very much power to reject the null for any comparison. Hence, use the Bonferroni adjustment judiciously as a post-hoc test. This actually goes for all post-hoc tests; try out a few of your favorites on your data and see what they say. If they report conflicting results, it may be that the differences you would like to think are present are not quite so strong.

We can easily conduct the Tukey test on our achievement data by the following. We will name the object `post` and request `tukeyhsd()`:

```
from statsmodels.stats.multicomp import (pairwise_
tukeyhsd,MultiComparison)
post = pairwise_tukeyhsd(df['ac'], df['teach'])
print(post)
Multiple Comparison of Means - Tukey HSD, FWER=0.05
====================================================
group1 group2 meandiff p-adj   lower    upper  reject
----------------------------------------------------
     1      2      1.5    0.9 -5.5147  8.5147  False
     1      3      9.0 0.0091  1.9853 16.0147   True
     1      4  21.6667  0.001 14.6519 28.6814   True
     2      3      7.5 0.0335  0.4853 14.5147   True
     2      4  20.1667  0.001 13.1519 27.1814   True
     3      4  12.6667  0.001  5.6519 19.6814   True
----------------------------------------------------
```

Let us unpack what this output is telling us:

- Python tells us it is conducting a multiple comparison of means, where `FWER = 0.05` is the familywise error rate.
- In the actual output are the pairwise comparisons. We see that for group 1 vs. group 2, the mean difference is equal to 1.5, with an associated *p*-value of 0.9. Since the *p*-value is so high, the decision is to not reject the null hypothesis. The output also includes confidence limits on the difference, which for this particular comparison ranges from −5.5147 to 8.5147.
- For group 1 vs. group 3, the mean difference is equal to 9.0, with an associated *p*-value of 0.0091. Since the *p*-value is so small, definitely less than a cut-off value of 0.05, we reject the null hypothesis. This is what is indicated by Python under `reject = True`.
- The remaining group differences are given in the output and are interpreted analogously. The largest difference is between that of group 1 vs. group 4, a mean difference equal to 21.6667 with an associated *p*-value of 0.001. Hence, we can see that for all pairwise differences, the only sample difference not statistically significant according to the Tukey test is that between groups 1 and 2. All the other pairwise differences are statistically significant, suggesting that in the population from which these data were drawn, there are mean differences.

6.9 Factorial ANOVA

Our survey of the analysis of variance thus far has focused on the one-way ANOVA. That is, in our earlier analysis, we only had a single dependent variable (DV) and IV. The goal was to learn of any mean differences on that single IV. In many cases, however, a researcher has more than a single IV and may wish to model them simultaneously. For the example we will use, suppose that instead of being interested in whether there are mean achievement differences across teachers, the researcher was also interested in learning whether there might be mean differences across teachers and textbooks used (Table 6.2). Suppose that in addition to students being randomly assigned to a teacher, they were also randomly assigned to different textbooks, as shown in the following table (notice that "Textbook" is now the new variable in the rows of the layout):

Table 6.2 Achievement as a function of teacher and textbook.

	Teacher			
Textbook	**1**	**2**	**3**	**4**
1	70	69	85	95
1	67	68	86	94
1	65	70	85	89
2	75	76	76	94
2	76	77	75	93
2	73	75	73	91

Thus, the primary question asked by the researcher is now the following:

Is achievement a function of teacher and textbook considered simultaneously?

Now, the wording here matters. The "simultaneously" matters a great deal. Why? Let us demonstrate. Recall in our ANOVA we found evidence for teacher differences. Would our model have detected this same effect had we also included textbook as a factor? Let us find out. In the following, we model both teacher and textbook in the same model:

```
data = {'ac' : [70, 67, 65, 75, 76, 73, 69, 68, 70, 76, 77, 75, 85,
86, 85, 76, 75, 73, 95, 94, 89, 94, 93, 91],
'teach' : [1, 1, 1, 1, 1, 1, 2, 2, 2, 2, 2, 2, 3, 3, 3, 3, 3, 3, 4,
4, 4, 4, 4, 4],
'text' : [1, 1, 1, 2, 2, 2, 1, 1, 1, 2, 2, 2, 1, 1, 1, 2, 2, 2, 1,
1, 1, 2, 2, 2]}
df_2 = pd.DataFrame(data)
df_2
```

```
Out[288]:
    ac  teach  text
0   70      1     1
1   67      1     1
2   65      1     1
3   75      1     2
4   76      1     2
5   73      1     2
. . . . . .

import statsmodels as sm
import statsmodels.api as sm
model_2 = ols('ac ~ C(teach) + C(text)', data = df_2).fit()
table_2 = sm.stats.anova_lm(model_2, typ = 2)
print(table_2)
               sum_sq    df          F        PR(>F)
C(teach)  1764.125000   3.0  30.051216  2.029775e-07
C(text)      5.041667   1.0   0.257649  6.175816e-01
Residual   371.791667  19.0        NaN           NaN
```

Notice carefully that now that we have added the second factor `text`, **the output line for teach is not the same as it was in the one-way model!** The sums of squares for `teach` are the same, the degrees of freedom are the same, but the resulting *F*-statistic is different. In the one-way model, recall our *F* and corresponding *p*-value came out to be 31.21 and 9.677198e-08, respectively. When we include the `text` factor, the *F*-statistic now comes out to be 30.05 with the *p*-value 2.029775e-07. Because the *p*-value is now larger in the two-way than in the one-way model, we have lost some power. But why? More specifically, **why are we seeing one effect in one model and another effect in the other?** Though both are clearly recommending a rejection of the null hypothesis, it is equally clear that the *p*-values are different.

The reason this is occurring is attributable not only to how data analysis works in ANOVA, but how statistical analysis works in virtually all statistical models. The key phrase is "variance partitioning." In any model, we are partitioning sources of variance into components. When we performed the one-way model, we partitioned variance into that attributable to teach and residual. When we do the two-way, we are partitioning variance into that attributable to teach, text, and residual. Though SS total remains the same in each model, **the MS values will change because of the new partitioning**. Hence, the *p*-value for teach is not the same as it was in the original model. Let us summarize the primary outcome of this discussion, and that is this: **As a researcher and scientist, you need to be aware of this principle. When you test and evaluate effects in a statistical model, you are always evaluating them in the presence or context of other effects also included in the model. What might be statistically significant in one model may not be in another.** Of course, the scientific ramifications and implications of this are huge; in one model the treatment has an effect on COVID-19, but in another model, less so! It is vital that you look deeper into how a given model was constructed before interpreting effects. This is also why models in the social sciences can sometimes be explained as simply being **statistical artifacts**, an exercise in variance partitioning, with no clear case to be made that any variable has an effect on any other. As you learn to appreciate how statistical models function, you will no longer look at research findings in the same way, and will look for evidence that the material under discussion (the "stuff" on which the statistics are being computed) has in fact changed due to the treatment, rather than merely be satisfied with a change in statistical coefficients.

6.10 Statistical Interactions

Having above added textbook to the model, the real reason for conducting a factorial ANOVA over two one-way ANOVAs is not simply to add an additional factor. The real reason is usually to seek out **statistical interactions**. What is an interaction? An interaction between factors is said to exist when **the effect of one IV on the response is contingent upon the level of the second factor**. As a COVID-19 example, suppose a treatment for the disease was found to be effective, but this effectiveness was different for males than it was for females. For instance, suppose it worked well for females, but not males. Then we would conclude that the treatment works, but not quite as well for males as for females. This is the nature of an interaction effect. Hence, in our achievement as a function of teacher and textbook layout featured earlier, we are no longer merely interested in the effect of teacher and of textbook independently. We are instead interested in the **cell means** primarily as they give us a sense of whether an interaction may be occurring in our data. For example, in the following (Table 6.3) we specifically emphasize the cell means (each cell enclosed in a rectangle):

Table 6.3 Achievement as a function of teacher and textbook.

	Teacher			
Textbook	1	2	3	4
1	70	69	85	95
1	67	68	86	94
1	65	70	85	89
2	75	76	76	94
2	76	77	75	93
2	73	75	73	91

Let's look at the above cell means very carefully. Notice that, in general, as we move from teacher 1 through 4, there is a general inclination for means to increase. However, this is primarily only true for textbook 1. Look at textbook 2. For textbook 2, though there is a slight increase from teacher 1 to 2, there is then a decrease to teacher 3, then up again to teacher 4. Hence, **the effect of teachers on achievement is not consistent across textbooks.** This suggests the existence of an interaction effect. Interaction plots are by far the easiest way to detect the presence or absence of interaction effects, as we can plainly see in the following plot of the achievement data.

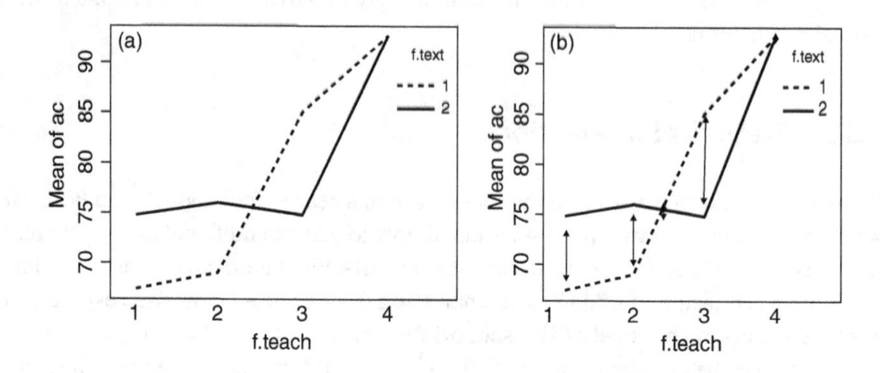

Closer inspection of these plots will reveal that the presence or absence of an interaction is indicated by whether or not the lines are **parallel.** If the lines are perfectly parallel, then there is no interaction in the sample, since the same "story" is being told regardless of which textbook is being considered. As an example of a generic plot in which there is no interaction, consider the following graph, top left cell.

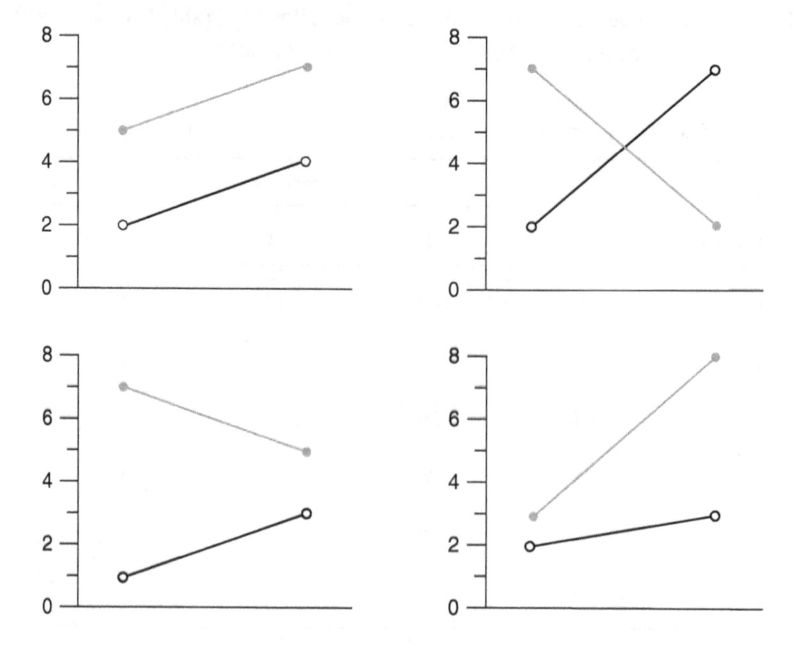

Note that the lines in the top left cell are perfectly parallel. All of the other plots suggest the presence of an **interaction in the sample,** most notably the complete crossing of lines in the top right cell. This brings us to an important question: **Can an interaction in the sample ever not exist?** We consider this possibility next.

6.11 Interactions in the Sample Are a Virtual Guarantee: Interactions in the Population Are Not

There is an important distinction between an interaction in the sample versus one in the population. An interaction in the sample is virtually a guarantee for any data set you may consider. How so? Because just as there is a virtual guarantee of sample mean differences in a *t*-test or ANOVA to some decimal place, **non-parallel lines in an interaction plot will virtually always exist**. Think about why this is so. Do you really think if we plotted an interaction graph in a sample that the lines could ever be perfectly parallel? Hardly ever, unless we were dealing with a very contrived situation or data set, or rounding to whole numbers or something. Otherwise, lines will always be **non-parallel,** at least to some degree in the sample. Of course, our interest, as always, is not in the sample, but rather the population. Whether the sample data suggest the presence of an interaction in the population is what we are truly interested in. For that, we require the significance test on the interaction term. However, in the sample, you will virtually always see an interaction.

The author has observed many times researchers presenting interaction graphs in which there is clearly an interaction in the sample, yet standing next to the graph claiming there is no interaction! Of course, they were right to say they had no evidence of an interaction in the population. However, in most cases, there is clearly an interaction in the sample. Unpacking the distinction here to your audience is of vital importance, but never stand next to an interaction plot in which the lines are clearly non-parallel and say there is no interaction if there clearly is. As with all significance tests, as sample size increases, the *p*-value for the interaction effect will typically get smaller, and hence eventually, if the effect remains more or less constant, you will reject the null hypothesis of no interaction.

6.12 Modeling the Interaction Term

Earlier, we fit the ANOVA model that featured both main effects. However, we did not specifically model the interaction term. We can model the interaction term as follows by including the term `C(teach)*C(text)`:

```
import statsmodels.api as sm
from statsmodels.formula.api import ols
model_3 = ols('ac ~ C(teach) + C(text) + C(teach)*C(text)', data =
df_2).fit()
table_3 = sm.stats.anova_lm(model_3, typ = 2)
print(table_3)
                        sum_sq    df           F        PR(>F)
C(teach)           1764.125000   3.0  180.935897  1.488709e-12
C(text)               5.041667   1.0    1.551282  2.308781e-01
C(teach):C(text)    319.791667   3.0   32.799145  4.574415e-07
Residual             52.000000  16.0         NaN           NaN
```

We unpack each component of the output:

- The two-way factorial has separated the sources of variability into an effect for `teach`, `text`, the interaction of `teach` by `text`, and `residual`.
- We begin by interpreting the interaction term and see that it is statistically significant, yielding a very small *p*-value ($p = 4.574415e\text{-}07$). This suggests that any effect of `teach` on achievement is not consistent across `text`, and that any effect of `text` is not consistent across `teach`. Ideally, **interaction terms should be interpreted first over and above main effect terms.** Why? Because they tell more of the "story" of the data. While it is technically not completely incorrect to interpret main effects in the context of an interaction, in light of an interaction, main effect terms will be misleading as they do not tell you the full story (e.g. if there is a treatment effect and it is only for older people and not young, the main effect term is still correct to say there is a treatment effect overall, but you (and your readers) are being misguided by interpreting the main effect only).
- Both main effects of `teach` and `text` are both statistically significant, yielding *p*-values of 1.488709e-12 and 2.308781e-01, respectively.
- The residual sums of squares for the model are equal to 52 on 16 degrees of freedom. No *F*-statistic or *p*-value is computed for the residual.

It is pedagogical to compare the two-way ANOVA fit earlier, where no interaction was modeled, to the two-way ANOVA we just fit with the interaction term. Notice that the total sum of squares is the same in both cases:

Main Effects Only: $1{,}764.125000 + 5.041667 + 371.791667 = 2{,}140.958334$

Interaction Model: $1{,}764.125000 + 5.041667 + 319.791667 + 52.000000 = 2{,}140.958334$

The only difference in the two models is that SS residual in the first model (i.e. "main effects only") was partitioned further into SS interaction in the second model. That is, notice that the SS residual of 371.791667 in the main effects model is equal to the sum of residual + interaction in the interactive model (i.e. $319.791667 + 52.000000 = 371.791667$). To fit the interaction, the ANOVA "stole" from the residual term, that is, it took from the portion of variance explained, and attributed some of that to interaction. This is how modeling in general works, partitioning sources of variance into components. But then why have the *p*-values changed? The *p*-values are different because the mean squares values have changed as a result of including the interaction term. **The model you test determines the effects that you will or will not see**. This is well worth reminding yourself of repeatedly as you interpret research. It is a core principle of statistical modeling. Hence, when you read a research claim attesting that a given variable has an effect on another, your concern is the model in which the given effect was found, since all models are **context-dependent** in this sense. This principle is true of virtually all statistical models, not only ANOVA. It is likely to be the most important principle you can take away, as a scientist, from this book! We will revisit this issue further when we discuss regression models a bit later in the book.

6.13 Plotting Residuals

Recall that we have mentioned earlier the importance of satisfying model assumptions when fitting any statistical model, and that though we can examine assumptions to some degree by plotting raw data, it is the **residuals** we are most interested in observing for any odd or abnormal behavior. We can easily plot the residuals from our factorial model above using the following **quantile plot**. We fit the plot using **statsmodels'** `qqplot()` function:

```
import statsmodels.api as sm
res = model_3.resid
fig = sm.qqplot(res, line='s')
```

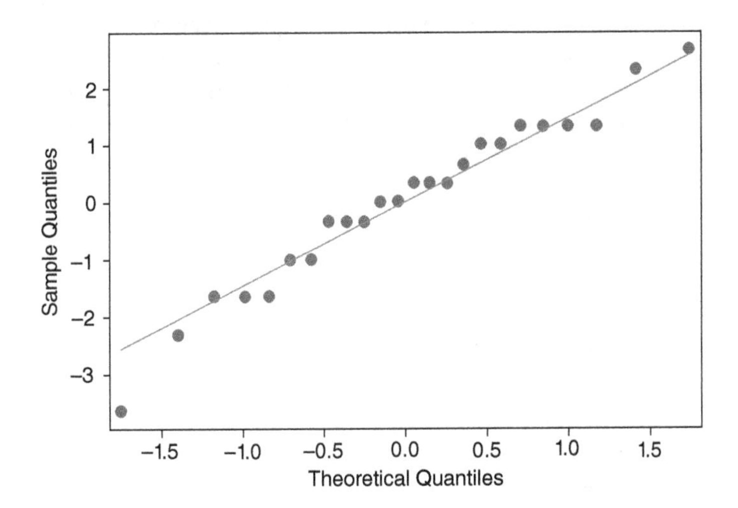

This residual plot plots **sample quantiles** against **theoretical quantiles** that would be expected under a **normal density**. A perfectly normal distribution would be represented by all dots falling along the line. Since dots deviate to some degree, we know the distribution is not perfectly normal. However, the deviation is so slight that this residual analysis is not a serious threat to model assumptions (though the residual in the lower left may be worth investigating a bit further).

6.14 Randomized Block Designs and Repeated Measures

We close our discussion of ANOVA with a brief look at **randomized block designs**, of which **repeated measures ANOVA** is a special case. Our treatment here is very cursory, as entire books are written on repeated measures models and their intricacies in modeling longitudinal data. For a good overview, see Denis (2021), which also features a fairly elaborate discussion of **random effects** and **mixed models**, of which the randomized block and repeated measures models can be considered special cases.

In the ANOVA models we have surveyed up to now, it has been assumed that individuals serving under one condition or level of the IV were different from individuals serving under other conditions. Hence, it was reasonable to expect no correlation among groups on levels of the IV. We randomly assigned individuals to each condition, hoping that any "nuisance" factors would more or less balance out across conditions. If there were a lot of nuisance factors at play, then these would pile up in the error term of our analysis, **MS error**. If you are not randomly assigning to conditions, your design is not actually experimental. Rather, it is more akin to **quasi-experimental** and the conclusions you can draw from it are much more limited. Many statistical models, sophisticated as they are, if used without good experimental design, end up not revealing nearly as much from the data as the researcher might have hoped. Implementing a good design is half the battle to good science. The statistical model is not nearly as important. **Design first. Statistics second.**

There are times where hoping nuisance factors "balance out" in this way is not the most efficient strategy, and we can do better by purposely controlling for one or more nuisance factors by pulling such variability out of our data. To do this, we can **match participants** on one or more nuisance factors. These **matched participants** are known as **blocks**. Recall the example of a matched pairs design layout featured earlier in the book (reproduced in Table 6.4):

Table 6.4 Matched pairs design.

	Treatment 1	Treatment 2
Block 1	10	8
Block 2	15	12
Block 3	20	14
Block 4	22	15
Block 5	25	24

In the table, different participants serve under each treatment, but the **blocking factor** makes the individuals within each block more similar than had they simply been randomly assigned without the blocking factor. That is, data point 10 in block 1 is assumed to be related to data point 8 in block 1. Likewise, data point 15 is assumed to be related to data point 12 in block 2, and so on. What is the advantage of a blocking design? Quite simply, it is to extract variation due to block and remove it from the overall error term for the ANOVA. That is, the blocking factor is removed from the error term and thereby provides a more sensitive (powerful) test of the null hypothesis for the main effect. Blocking designs are especially popular in areas where rigorous experiments are regularly conducted. We do not treat block designs any further in this book. Our purpose in mentioning them at all is that they serve as the stepping stone to a model we do consider next, that of the repeated-measures model. For details on randomized blocks, see Kirk (2012).

Repeated measures is nothing more than a randomized block design in which each block represents a subject, and measurements are taken for each subject under each condition. Hence the reason why understanding block designs is mandatory to understanding the repeated-measures model. Referring back to our randomized block layout, it becomes a repeated-measures design when we replace block 1 with subject 1, block 2 with subject 2, and so on. Recall our rat data featured earlier (see Table 6.5)

Table 6.5 Learning as a function of trial (hypothetical data).

	Trial			
Rat	**1**	**2**	**3**	**Rat Means**
1	10.0	8.2	5.3	7.83
2	12.1	11.2	9.1	10.80
3	9.2	8.1	4.6	7.30
4	11.6	10.5	8.1	10.07
5	8.3	7.6	5.5	7.13
6	10.5	9.5	8.1	9.37
Trial means	$M = 10.28$	$M = 9.18$	$M = 6.78$	

Let us run the above repeated measures in Python. First, we enter the data:

```
rat = pd.DataFrame({'rat': np.repeat([1, 2, 3, 4, 5, 6], 3),
'trial': np.tile([1, 2, 3], 6),
'time': [10.0, 8.2, 5.3,
          12.1, 11.2, 9.1,
          9.2, 8.1, 4.6,
          11.6, 10.5, 8.1,
          8.3, 7.6, 5.5,
          10.5, 9.5, 8.1]})
```

```
rat
Out[410]:
    rat  trial  time
0     1      1  10.0
1     1      2   8.2
2     1      3   5.3
3     2      1  12.1
4     2      2  11.2
5     2      3   9.1
6     3      1   9.2
7     3      2   8.1
8     3      3   4.6
9     4      1  11.6
10    4      2  10.5
11    4      3   8.1
12    5      1   8.3
13    5      2   7.6
14    5      3   5.5
15    6      1  10.5
16    6      2   9.5
17    6      3   8.1
```

There are different options in Python for running a repeated measures. We chose to use `statsmodels.stats.anova` and `AnovaRM`:

```
from statsmodels.stats.anova import AnovaRM
print(AnovaRM(data = rat, depvar = 'time', subject = 'rat', within
= ['trial']).fit())

              Anova
====================================
      F Value Num DF  Den DF Pr > F
------------------------------------
trial 72.6196 2.0000 10.0000 0.0000
====================================
```

The F-stat associated with the ANOVA is 72.6196, evaluated on 2 and 10 degrees of freedom. Clearly, the obtained p-value is very small (e.g. $p < 0.001$, though never **exactly** equal to 0), and hence we can reject the null hypothesis that mean time is equal across trials. We have inferential evidence that the mean on the response variable is not the same across trials.

6.15 Nonparametric Alternatives

In this final section of the chapter, we briefly survey a couple of the more common **nonparametric** (nonpar) tests that may be of use to the research scientist, and demonstrate how such tests can be run in Python. What is a nonpar test? To understand what a nonpar test is, we first need to understand the nature of a parametric test. **A parametric test is one for which before the test is performed, certain assumptions about the population from which the data were collected are assumed to be met and satisfied.** For example, in the ANOVA model just surveyed, we have had to assume such things as **normal populations, equal variances, independence of observations**. These are elements of the test that, ideally, should be true before the test is run, otherwise the results of the test will not be as valid as they could be if the assumptions were met. This is true not only of ANOVA, but also regression and a lot of other simpler statistical tests such as t-tests and so on. In the computation of F-tests, p-values, confidence intervals, etc., it was implicit that these assumptions were more or less satisfied. If they were not, or they were violated to some extent, it could compromise the validity of the test, and hence potentially thwart ensuing substantive conclusions.

Nonparametric tests are tests developed that do not make as many or as **deep assumptions** about populations as do parametric ones. Though some authors define these tests as "distribution-free" tests, other authors categorize them more as "distribution-free-er" tests (e.g. see Kirk, 2007), since they still make some assumptions, but less so than traditional parametric tests. So why is there the necessity for nonpar tests? Simply put, it is because, in most cases, assumptions about populations are either **difficult or impossible to satisfy**, and hence the assumption of an underlying probability model is often unrealistic. It takes only a moment's reflection to realize that satisfying parametric assumptions in a perfect sense exists only in the ideal. That is, in

99.9% of cases, assumptions for virtually all statistical tests will be violated to some degree, and hence at first glance it would seem like nonpar tests should be used in all cases for all data. However, there is no need to be this absolute. Even under conditions of mild to moderate departure of assumptions, parametric tests are usually still advised over their nonpar counterparts, especially if sample size is at least minimally sufficient. However, in cases where violations are more severe, or sample size is quite small, a nonpar test may be more suitable.

When are distributional assumptions virtually guaranteed to be unsatisfied, or at minimum, unknowable? While this can be true for any data, it is especially the case for data on **small samples**. For example, with a sample size of say, six observations, if we cannot assume a priori that the sample was drawn from a normal population, then we have no way of knowing whether normality is satisfied. Plot the sample data you say? Investigating normality assumptions only works in cases where you have sufficient data to make the exploration worthwhile, such that you will actually be able to learn something about the population under investigation in the plot. In the case of six observations (again, to use our example), plotting the data will not tell you much about your distribution other than whether you have at least **some variability** in your small data set. However, if you cannot assume normality before plotting your data, the plot will not really help you either way. Hence, one common use of nonpar tests, historically, has been to use them when sample data are quite small in number and the researcher seeks a relatively quick and easy way to evaluate their data without having to invoke parametric assumptions. Many of the simpler nonpar tests are easy to conduct, simple in structure, and can readily inform the researcher of the presence or absence of a potential scientific effect. Historically, nonpar tests were also quite common before computers, since they allowed a scientist easy computation and to immediately get a glimpse of whether or not their experimental treatment had an effect. It allowed them for a simple way to exclude chance from their findings.

In many cases, one potential drawback (though this depends on what you are seeking to actually learn from your data) to using a nonpar test is that they often work on **ranking data** rather than treating the data as originally continuous. This is not an inconsequential side effect of using a nonpar test, and researchers using them need to be fully aware of how the test is treating their data. It can have fairly significant consequences. For example, recording IQs of 120, 100, and 80 is quite different from ranking them first, second, and third. After all, IQ values of 120, 119, and 118 also hold the same ranking as those of 120, 100 and 80. When a test ranks data and ignores the pseudo-continuous scale on which they were originally conceived, make no mistake, it can sometimes have pretty important consequences for how the results turn out. That is, the response to the question "**Does the treatment have an effect?**" may be different depending on whether one conducts a parametric or a nonpar test. Such differences in substantive and scientific conclusions should definitely be of concern. This is why it is essential to understand the difference between these types of tests so you can make an **informed choice** and interpretation of what the data are actually telling you. Otherwise, you are making conclusions about your data without an appreciation of what the results are based on, akin to making a diagnosis of COVID-19 without understanding how the test works. Your diagnosis is only as good as understanding how the test works and what "ingredients" went into the test to draw such a conclusion.

6.15.1 Revisiting What "Satisfying Assumptions" Means: A Brief Discussion and Suggestion of How to Approach the Decision Regarding Nonparametrics

Before we survey a couple of nonparametric tests in Python, it behoves us to comment a bit further on how we can learn whether assumptions might be satisfied for a particular model. Recall that assumptions of normality, for instance, typically involve those of the **population data** as well as the **sampling distributions** for the test statistic. How can we know whether such assumptions are satisfied? Both normality of distributions and that of sampling distributions can be considered satisfied more or less regardless of what you are seeing in your sample (not "sampling") distributions **if you have reasonable evidence or background knowledge to assume your data were sampled from normal distributions**. For instance, if we sample IQ scores, obtain 25 scores, and are quite confident from prior research or well-known population data that IQ scores in the population are **generally normally distributed**, then even if we observe some abnormality in our limited sample, it does not necessarily mean the assumption of normality needs to be outright rejected. It may simply mean by chance we got a few scores pilling up here or there enough to generate some skew. Now you might ask, "Isn't this what the normality tests surveyed previously in the book are for, to rule out this probability?" and the answer is, they can help, but they are not the end all and be all of things. For one, those tests (Shapiro, etc.) are relatively sensitive to assumptions themselves, and also sample size, so what they report in terms of p-values can change depending on their own deviations from assumptions. Therefore, you should not take those tests as the final verdict, which is one reason why **graphical methods** are often recommended, even by the best of data analysts, instead of relying on such tests. Usually by looking at plots that we generated in previous chapters (e.g. histograms, boxplots, etc.), it usually gives us sufficient enough of an idea regarding assumptions to know whether or not they might be tenable. Inferential tests are usually not even required.

6.15.2 Your Experience in the Area Counts

In deciding whether assumptions are tenable, the question in part also comes down to your own expertise with the material you are working with. When you look at your sample of scores, do you think "Yes, I expected to see this" or do you think "Hmm, that's a bit high on that end of the distribution, no real reason why that should be happening though." If it is the former, then you may be dealing with **naturally** or **inherently skewed distributions** to begin with, which might suggest what you are measuring has likewise an inherently skewed population distribution underlying it. However, if it is the latter, you can probably attribute such deviation to simply "randomness." That is, you may tentatively conclude that you are seeing a bit of a "blip" in the graph in the sample data, but that you have no reason to suspect your data was drawn from a severely naturally occurring **abnormal** (i.e. in the sense of "non-normal") distribution, given what you know about data of this kind in general. That is, why you are seeing "blips" in data becomes important because what you are witnessing in your data, if your data was collected in a sound manner, should be a very small fraction of the actual population distribution, and the assumption for the statistical test is that the population distribution is normal, not the sample data. The sample data is just a potential "sign" that something might be amiss or not amiss. **Sample data should be interpreted with this perspective and context in mind. The particular sample data you achieved should never be considered the entire "be all" and "end all" of things.**

Hence, unless you can think of some specific reason why the sample is abnormal, you may be fine with just attributing it to general variation, and conclude something such as, "It can reasonably be assumed that such data arose from an approximately (emphasis on **approximately**) normal population based on past behavior of similar data" and carry on. Again, you might think of it this way; your sample data is 0.00001%, suppose, of all data that has been collected on similar objects. Most has already been collected (again, I am just ball-parking for demonstration). If most data over the past few years suggest normality or something close to it, even if your particular sample does not quite as much, **and if you have no specific substantive or scientific reasons to think of why normality might not hold in your sample**, then regardless of the aforementioned inferential tests and plots in this case (so long as they are not too extreme), you are probably fine in most cases with the decision to **not** reject normality. That is, if you can assume you are sampling from even approximate normal populations, then the sample distribution you obtained becomes but one "input" to add to that prior knowledge (just like my IQ distribution can supplement my prior already known info on IQs in the population). Again, the sample data is just one indicator of the underlying population distribution; there is no need to take it as the final verdict in totality.

6.15.3 What If Assumptions Are Truly Violated?

To carry our discussion further, let us assume for a moment the assumption is not satisfied and that "blip" (and blips like it in other graphs) truly suggests a skewed distribution in the population. Furthermore, the inferential tests you ran are trying to alert you to this. And, assume, as a scientist, such abnormality does not surprise you either. Let us assume you have heavy skew and you can confidently declare it to be a "real" violation. Now what? If the data are naturally skewed to a significant degree, it could suggest using an entirely different model (e.g. sometimes fitting a Poisson model or other **generalized linear model** works best with highly skewed data, especially if all distributions are coming out with similar skew shapes and it is blatantly clear the phenomena under evaluation is "naturally skewed."). However, if most of your distributions are at least approximately normal or at minimum rectangular-like, that is probably not what is going on here. Some would say you should start performing transformations on the variable to bring it more in line, like taking square roots and such. As an example, consider numbers 4, 9, 16, 25, 49. Notice if we take square roots, we get 2, 3, 4, 5, 7. The distribution is a bit tighter. The 49 (7) is not as "extreme" from the 25 (5) anymore. Logarithms and other transformations can sometimes help as well. However, some (such as myself!) are not a big fan of this approach (i.e. that of generally transforming data). For one, transforming data is meant to **appease the statistical test** so that *F*-ratio inferences, etc., are more valid and that you are not committing more or less type I errors than you think you are when you set your sig. level (e.g. 0.05 or whichever you are choosing). However, we already know (or should know) *p*-values are not the end-all and be-all of things as well, and small or large *p*-values can happen due to sample size, variability, etc., and not just a violation of assumptions. Hence, to all of a sudden be that concerned with the *p*-value being a bit higher or lower than it should be because of a potential assumption violation seems a bit "overkill." What is the solution then? Do as you normally do when interpreting statistical tests, which is **to not take the *p*-value too darn seriously in the first place**! Use it as a **guide**, because you know all the things that it is a function of, not only the assumptions of the test.

Hence, when you get a *p*-value of say, 0.04, and you know your distributions might be a bit "off" more than you would like, incorporate that knowledge more or less mentally into your thought process and use the *p*-value as you should be doing anyway, as a **ballpark indicator** toward your scientific (not merely statistical) conclusion. As you should be doing anyway in any statistical test, you are going to look at **effect size** to see if there is anything of scientific value going on. Furthermore, ANOVA, for example, is quite **robust** to violations of normality and other assumptions, so in most cases you are likely to be fine from a **scientific perspective** because you are using the stats as a **tool** (only) to inquire about a potential scientific effect. That is, you are not trying to "appease" the aesthetics of the statistics and make everything perfectly "formal" and correct. That will never be the case anyway in any model-fitting because assumptions will always typically be "off" to some degree. It is always messy to some extent (as COVID predictions were during the pandemic, recall how most point predictions about deaths and such were way off). Incidentally, this is in part the spirit that has been promoting the "data science" and such movements, to be a bit less concerned about parametric assumptions and instead more concerned with just getting a sense of what the data have to say. In a very real way, the movement away from traditional statistical inference has been to emphasize **effects over formal inference**. That does not in itself obliterate formal statistical inference and *p*-values, it simply sees them for what they are (which good methodologists have been telling us for years anyway).

Check assumptions just as much for the **science** as for the **stats**. If you come across something "weird" in distribution, that may be even more important than just making sure it conforms to assumptions. **Evaluate assumptions because you want to get to know your data, first and foremost, on a scientific level**. Assumption checking might actually suggest new lines of research, you never know, because you have spotted something of scientific value. Follow a similar approach regarding spotting **outliers**. All a statistic can do is tell you if it is extreme statistically. You have to incorporate your knowledge of the material under test to know just how "extreme" it is, since you usually have only a very small fraction of the entire population, and sampling, even if done randomly with the best of intentions, does not guarantee adequate representation of a population. It is just the best we can usually do to allow chance to govern discrepancies we observe.

Also, if you have outliers in your plots (e.g. extremes in boxplots), you may have **substantive reasons** as to why a few of your points are extreme. If you do, then you may want to look at those observations more closely to make sure nothing abnormal was going on with your research units such that they violated expectations for the experiment. You could delete this or that observation, redo your analyses, and observe any difference in findings, but for most data sets, it is probably not worth doing this. Especially with a relatively small sample size on high-variability units, we usually expect more or less messy data anyway. Now if you have 10,000 observations, and two or three are beyond *z*-scores of, say, 3 (or similar), then they might be worth looking at. However, if you have many "outliers" in a much smaller sample, you might start to question whether they are actually outliers, or just "data." Case in point, the number 10 is an outlier in a data set of observations 2, 3, 10, but it is probably not an outlier from a scientific perspective – it is probably just **data.** When the next few data points come in, they might be 12 and 15, making 10 quite ordinary and normal in the updated data set. **Think about your data, do not just look at statistical indicators.** Never be too hasty to consider an innocent data point an outlier and not "belonging" to your data set. **It likely does belong. After all, it was a recorded data point like any other.** If the data recording mechanism was working properly, then you have even less justification for removing it from your data set. Again, the task is not to appease your statistical model; the task is to learn about the objects you are modeling.

So, at long last (sigh!) should you use a nonpar test for your data? It definitely would not hurt to see if you get similar *p*-values, and if the same decision on the null is indicated by both parametric and nonpar tests, you might report the parametric, but also note that nonpar tests did not disagree with the conclusion. In that way, if in reality the nonpar test is the most theoretically appropriate test you should be doing (i.e. in the worst case scenario your distributions are naturally skewed in the population), then at least you are covering your bases. I personally do not see a problem with this approach (some sticklers might insist you should do one test or the other and not both, but in the event of any assumption doubt, trying both I do not see as a problem). If you reject using parametric, but not with nonpar, you may simply report that finding for both (and for your own benefit, study what went into that nonpar structurally, e.g. understand how it utilized ranking and such so that you know for yourself what the test is assessing and how it is different from the parametric). Then, after reporting both, the reader can decide which results they wish to interpret.

But again, to reiterate, the focus should be on the **scientific conclusion**, not the **statistical** here. If the treatment for COVID is working (analogy again), and working well, whether a parametric or nonpar test is used **should not make that much of a difference**, and if it does (i.e. if the effect is typically not that large), reporting both gives your reader information they need to know in evaluating the result for themselves. Also, sometimes the skewness can simply suggest you still should conduct the parametric test, but also supplement your findings with things like **medians** as well as means, so readers can see for themselves how much that extended end of the distribution (for instance) is pulling the mean up or down. That is not changing your model necessarily, it is rather addressing characteristics of your data by reporting less "sensitive" measures that are not as influenced by extremes. Range statistics might be a nice idea (such as the semi-interquartile range) to go along with the median as a more "resistant" or "robust" measure of dispersion, since the standard deviation will be influenced by the degree of skew as well. So in this approach, you are reporting parametric tests, but supplementing them with more resistant summary measures to account for skew, informing your readers of the assumption "imperfection," and switching the issue back to the scientific problem and interpretation of data where it belongs, rather than over-reliance on *p*-values.

Given this discussion, if you still do decide a nonpar test is useful, we now consider a couple of the more popular options. For details on other tests, see Howell (2002) for a succinct summary or Siegel and Castellan (1988) for a more thorough (and excellent) treatment.

6.15.4 Mann-Whitney U Test

When the assumptions of the independent samples t-test are violated or unverifiable, the **Mann-Whitney U test** becomes a viable alternative nonpar option. Recall that the null hypothesis in the independent samples t-test was that both population means were equal. That is, the null hypothesis was

$$H_0 : \mu_1 = \mu_2$$

Given a rejection of the null, the alternative $H_1 : \mu_1 \neq \mu_2$ was inferred. The Mann-Whitney U test evaluates the null that both samples were drawn from the same population, but it does so based on the **ranks of the data** instead of their continuous scores. Hence, instead of focusing on score differences such as that between 8 and 10, the Mann-Whitney test emphasizes ranking differences, where 10 may be "first" and 8, "second." This naturally implies that if the data cannot be ranked in some capacity, then the Mann-Whitney test cannot be used. When is data not "rankable?" Nominal data, recall by definition, cannot be ranked. For example, if your favorite color is blue and mine red, the variable "favorite color" is not being measured on a scale that has a natural ranking. Now, that does not imply the colors could not be ranked by some other measure than favorite color. Indeed, if we examined the physical qualities of color, we may possibly rank them on measures of hue or saturation. The point is, however, is that if the scale you are using is not one with at least a natural ranking quality, then it implies that you are dealing with a nominal scale, and the Mann-Whitney test or other similar ranking tests would make no sense to perform. Hence, data for the test should be **at least at the level of ordinal** such that magnitudes between measured quantities can be inferred and make good sense to infer them.

Be sure to note that your data do not need to be merely at the level of ordinal. They can be interval or ratio, since these scales can be **collapsed** into ordinal scales for the purpose of the test. That is, pseudo-continuity can easily be collapsed into a ranking, but not necessarily vice versa. The point is, however, that your data cannot be "less" than ordinal and has to be **at least ordinal** for the test to make any sense performing.

We demonstrate the test in Python. For this example, we will use the **grade data**, on which grades of passing (1) or failing (0) were recorded along with the amount of time studied for a test (recorded in minutes). We first build our data set:

```
data = {'grade':[0, 0, 0, 0, 0, 1, 1, 1, 1, 1],
'studytime': [30, 25, 59, 42, 31, 140, 90, 95, 170, 120]}
df = pd.DataFrame(data)
df
Out[75]:
   grade  studytime
0      0         30
1      0         25
2      0         59
3      0         42
```

```
4        0        31
5        1       140
6        1        90
7        1        95
8        1       170
9        1       120
```

We next pull out and identify the columns of `grade` and `studytime` from the dataframe `df`:

```
grade = df['grade']
studytime = df['studytime']
```

We import **scipy** to conduct the test, on which we will use `stats.mannwhitneyu()`:

```
import scipy
from scipy.stats import mannwhitneyu

Out[269]: <function scipy.stats.stats.mannwhitneyu(x, y, use_
continuity=True, alternative=None)>
```

We now define our samples and then conduct the test:

```
sample1 = 30, 25, 59, 42, 31
sample2 = 140, 90, 95, 170, 120
stat, p = mannwhitneyu(sample1, sample2)
stat, p

Out[514]: (0.0, 0.006092890177672406)
```

We see that the result is statistically significant with a *p*-value equal to approximately 0.006. Hence, we can reject the null and have evidence to suggest that the two groups were not drawn from the same population. The Mann-Whitney test is quick, easy, and immediately gives you an idea as to whether something might be "going on" with your data without having to satisfy parametric assumptions.

6.15.5 Kruskal-Wallis Test as a Nonparametric Alternative to ANOVA

Recall that when we have several means to compare, conducting numerous independent samples *t*-tests unduly inflates the type I error rate over the family of comparisons. The solution in such a case is the one-way ANOVA. When assumptions such as normality and equal variances are violated, a plausible nonpar alternative is the Kruskal-Wallis (K-W) test to evaluate the null that all samples were drawn from the same population. To demonstrate the test in Python, we recall the achievement data on which we computed an ANOVA earlier in the book. For this analysis, we organize each sample into its own vector. Recall that there were four samples and hence we have four data vectors:

```
x = [70, 67, 65, 75, 76, 73]
y = [69, 68, 70, 76, 77, 75]
z = [85, 86, 85, 76, 75, 73]
```

```
w = [95, 94, 89, 94, 93, 91]
```

We use `stats.kruskal()` to run the test:

```
from scipy import stats
stats.kruskal(x, y, z, w)
```

```
Out[127]: KruskalResult(statistic=16.26653554778557, pval-
ue=0.0009998585717080008)
```

We can see that the p-value for the test is quite small at a value of 0.0009, certainly smaller than some pre-set level such as 0.05. Hence, we reject the null that the samples were drawn from the same population and infer that they were not. However, as with ANOVA, simply because we have rejected the overall null, we do not yet know where the differences may lie. To know this, we need to conduct post-hoc tests to determine which populations are pairwise different from which others. One easy way to do this is to simply run the K-W test on pairwise groups and issue a correction on alpha for each comparison. For example, we will see if there is a difference between groups x and y:

```
stats.kruskal(x, y)
Out[128]: KruskalResult(statistic=0.5247349823321599, pval-
ue=0.46882883249379537)
```

We can see that the test comes out to be non-statistically significant ($p = 0.4688$), suggesting that a difference does not exist between these two groups. We could do the same thing for the other groups as well to determine where differences may lie, testing each comparison at a smaller significance level to keep the overall family-wise error rate low. An even more effective strategy would be to use the **Nemenyi test** as a post-hoc test, available in the package **scikit-posthocs**:

```
pip install scikit-posthocs
import scikit_posthocs as sp
```

```
v = [[70, 67, 65, 75, 76, 73], [69, 68, 70, 76, 77, 75], [85, 86,
85, 76, 75, 73], [95, 94, 89, 94, 93, 91]]
sp.posthoc_nemenyi(v)
```

```
Out[283]:

          1         2         3         4
1 -1.000000  0.974034  0.387602  0.003552
2  0.974034 -1.000000  0.657109  0.016002
3  0.387602  0.657109 -1.000000  0.286205
4  0.003552  0.016002  0.286205 -1.000000
```

We can see that the test reports significant differences between groups 1 and 4 ($p = 0.003552$), as well as 2 and 4 ($p = 0.016002$).

Review Exercises

1. Explain how the **t-test** can be considered a special case of the wider **analysis of variance model.**
2. Why is conducting **multiple t-tests** on the same data set a bad idea to discern pairwise mean differences? What advantage does the ANOVA afford over this? Explain.
3. Consider the **pooled variance** term in the *t*-test:

$$s_p^2 = \frac{(n_1 - 1)s_1^2 + (n_2 - 1)s_2^2}{n_1 + n_2 - 2}$$

   Show that when sample size is equal per group, that is, $n_1 = n_2$, whether one uses a pooled term or the original estimate of variance for the two-sample case, will amount to the same thing.
4. What is the nature of a **Bonferroni** adjustment? If one has only two means to compare, is doing a Bonferroni adjustment useful? What is the drawback of performing a Bonferroni adjustment when one is working with several means?
5. Explain to someone the **logic of ANOVA** from the ground up and from first principles. Start with a data layout and explain exactly what ANOVA seeks to do.
6. R.A. Fisher, innovator of the analysis of variance, once said that ANOVA was nothing more than an exercise in **"arranging the arithmetic"** and nothing beyond that. From a scientific point of view, explain what Fisher might have meant by this statement. What does this mean to you, as a scientist?
7. Discuss what information **eta-squared,** η^2, provides that an *F*-statistic in ANOVA does not. What is the potential advantage of interpreting η^2 in conjunction with *F*? How about interpreting it **instead** of *F*? Would that be a good idea?
8. Discuss the nature of a **post-hoc test.** What is the purpose of a post-hoc test, and how can it be that different post-hoc tests could yield different conclusions? Isn't "a mean difference a mean difference?" How could using a different post-hoc test ever **not** produce a difference? Explain.
9. A researcher decides to perform a **two-way ANOVA** without interaction instead of two one-way ANOVAs. What are some implications for this in terms of interpreting results?
10. What is the purpose of performing a **factorial ANOVA** over two one-way ANOVAs?
11. Your colleague says to you, "In virtually every factorial ANOVA conducted on the planet, there is an **interaction in the sample,** but not necessarily in the **population** from which the data were drawn." Is your colleague correct? Defend or correct him.
12. Run an ANOVA on the **iris** data where sepal length is the DV and species the IV. Evaluate the null hypothesis that mean sepal length is the same across species. If you reject, compute a measure of effect size.
13. Discuss the difference between a **parametric** and a **nonpar** test. What is the rationale for performing a nonpar test and how should this decision be approached? Explain.

7

Simple and Multiple Linear Regression

CHAPTER OBJECTIVES

- Learn and appreciate why regression analysis is so central to virtually all statistical models and how many models are simply extensions of the fundamental regression model.
- Understand the nature of least-squares regression, what it accomplishes, and even more importantly, what it cannot do.
- How to estimate a regression equation when you have one response variable and one or more predictors.
- Decipher and distinguish between R^2 and adjusted R^2, and appreciate the benefits of the latter over the former.
- Understand how multiple regression differs from simple linear regression, and the dangers of not fully appreciating this distinction.
- How to run and interpret simple linear regression and multiple regression analysis in Python.
- Learn model-building strategies such as forward and stepwise regression, and critically evaluate each as a way of performing regression.
- Overview of mediation analysis in statistics, and why conclusions drawn from such analyses are often incorrect and overstep what the data actually can say.

Many statistical techniques used by researchers, in one way or another, even if somewhat indirectly, boil down to an enhanced version of **regression analysis**. Just as techniques such as *t*-tests for means could be conceptualized as special cases of ANOVA, the analysis of variance itself can be considered a special case of the wider regression model framework. Historically, regression analysis arose in the late 1800s, while ANOVA was innovated around the 1920s. And actually, the historical seeds of both techniques were laid much earlier at the turn of the nineteenth century (i.e. early 1800s) by Gauss and Legendre (Stigler, 1986). Though on an applied level they arose for different purposes, on a technical level, as we will see, they are remarkably similar. If you are a newcomer to statistics and research, you may at first glance be surprised to learn that there are literally **books and volumes dedicated solely to regression analysis**. One reason for this is because many, or perhaps even most of the topics and

Applied Univariate, Bivariate, and Multivariate Statistics Using Python: A Beginner's Guide to Advanced Data Analysis, First Edition. Daniel J. Denis.

issues that arise in regression are also prevalent in other statistical models as well. And if you are very familiar with these topics in regression, you will not see them as entirely "new" when encountering other statistical models, even the so-called relatively "advanced" ones. Sometimes researchers, using a "modern" statistical method (e.g. in machine learning, for instance), discuss issues with the method that they mistakenly believe are universal only to that method, without realizing that the issue with the "advanced" statistical method is really a page out of a first-year introductory regression book! If one is unaware of the **foundations** on which what they are doing are based, they are more likely to believe in the "hype" of new statistical methods, without realizing that much of the new methods may consist of ingredients simply "glued together" from other methods. Now, that is not to say new methods are not rightfully "new." They are, and innovators of newer methods are remarkable scholars. It is only to say that the concepts inherent in new methods usually have roots in prior methods. A failure to understand or appreciate this leads one to believe that new concepts (as opposed to the specific technique) are being "invented" with new methods. While this is sometimes true, it is usually not the case.

The dependent variable in regression is often called a "**response**" variable, in the sense that it is "responding" to the value entered by the predictor variable. The **predictor** variable in regression is the corresponding independent variable in ANOVA. It really does not matter which terms you use, but "independent variable" is more times associated with **manipulation**, such as one might have in an experimental context, whereas "predictor variable" is a bit more general in scope. The idea of manipulating an independent variable comes from the mathematical literature, where the function statement $f(x)$ implies that as you purposely change the value for x, the value of the function will be affected (i.e. it will also usually change, unless it is, for instance, an identity function). Hence, when we change the values of the independent variable, the "response" (which in this case are the function values, or "y") will also change. Even in ANOVA settings researchers may speak of predictor variables even when experimental control and variable manipulation is at its highest. **Terminology in statistics or mathematics is not standard, and often you have to read the situation and decipher what is being communicated**. This sometimes makes learning from a new textbook, for instance, mostly an exercise in understanding what notation is being adopted! As Feynman once remarked, "We could, of course, use any notation we want; do not laugh at notations; invent them, they are powerful. In fact, mathematics is, to a large extent, invention of better notations." Feynman meant this more in the spirit that similar concepts can be found and communicated in different branches of mathematics (i.e. mathematics is filled with equivalences across domains), but the point is that to understand a definition or new term in a given field, you have to ask yourself what the concept actually "means" and whether it has been represented by something else in the past or in an allied field.

When more than a single predictor is entered into the regression simultaneously, but still with only a single response variable, the model is that of **multiple linear regression**. Be sure to note we still have only a single response variable in this model even if we have more than a single predictor. If, on the other hand, the model has several response variables and several predictor variables, the model is that of **multivariate multiple linear regression**. Multivariate statistics is the topic of later chapters in the book. These models require some justification for entering several response variables into the same model. As we will discuss when we survey the multivariate analysis of variance, or "MANOVA" for short, the set of response variables entered into the model should make up what is known as a **variate** in statistics, which is, for now, a meaningful, or at least substantively "plausible," **linear combination** of variables. As mentioned, we delay our discussion of multivariate models until later in the book.

7.1 Why Use Regression?

As always, before using a statistical technique, one must first be aware of why one would even want to use it. Often students and researchers use a statistical method, thinking it accomplishes what they have in mind for it, yet not realizing the limitations of what the statistical procedure can actually provide. This is especially true in the case of regression. If the statistical method is used without understanding what actually goes on "behind the scenes," then when interpreting the output of the method a researcher can quite easily make philosophically weak or untrue conclusions of what the actual knowledge is that is provided by the procedure. "Behind the scenes" is what the procedure is actually doing! The **bare bones mathematics and philosophy**. The rest are just words and potential misunderstandings of what a researcher would "like" the method to be doing. These are two different things, and the distinction between them is very important to understand. The bridge between what a statistical method actually tells you vs. what you would **like** it to tell you on a scientific basis is not a philosophically simple leap.

Let us feature an example to demonstrate what we mean. Suppose a researcher specializes in the study of alcoholism and is interested in knowing whether drinking behavior is a predictor of suicide. That is, the researcher would like to know whether, in general, the more one drinks, the higher the risk of suicide. Hence, the researcher frames the question this way:

Is there evidence that alcoholism predicts suicide?

The key word in this is, of course, "predicts." What does the word mean? Now, at first glance, you might think to yourself, "That's a silly question. Everyone knows what predicts means." While that may be true in an ordinary and everyday language sense of how the word is used, in statistics and science, ordinary "English" will not suffice. **In statistics, the word "predict" only means as much as how it is defined mathematically.** Otherwise, it has no meaning. Hence, if the researcher runs the regression by having alcoholism as the predictor variable and suicidality as the response, and then draws a conclusion of "alcoholism predicts suicide," the statement is hollow for that researcher unless there is an understanding of how the statistical procedure that had been run operationally defines "predict." Otherwise, the researcher is simply running a regression analysis without even understanding what the procedure even does beyond associating words to it such as "predict!" What is more, if the audience is not aware of what "predict" means on a deeper level than some arbitrary semantic meaning, very little in the way of true communication is happening. "Alcoholism predicts suicide" and nobody in the room knows what it means! Insanity! The software does not know either, as it is just doing its job at computing. To understand the words you use in a statistical context, you need to have at least some appreciation for what the statistical methodology is doing "behind the scenes," otherwise the communication is hollow. You may still get the publication, but you will not appreciate what you yourself are even publishing. This is why understanding statistical methods for scientists is extremely important. As discussed earlier in the book, this is one danger of associating statistics and research solely with software analyses. You have to understand what the software is doing, otherwise, run the regression as you may, you will be at a loss to understand its output at any level other than associating imprecise **verbal labels** to what you are reading. The good news is that for topics like regression, and many other statistical techniques, understanding what the procedure does and how it works does not often require advanced training. What it does require, however, is careful study of the method's principles along with a discussion of what can vs. what cannot be concluded from the method. As highlighted throughout the book, far too many statistical methods are used to draw research conclusions that are simply not warranted. Understanding the statistical method allows one to better appreciate the limitations of what one has computed, and be in a better position for interpreting the research result accurately.

The following are a few very powerful and recent samples and examples of why understanding the statistical tool is so important and vital to good interpretation of the ensuing scientific claim:

- In the COVID-19 pandemic, it was often said by scholars in the area as well as media types that **models are only as good as the assumptions that go into them**. That is, one can make predictions of COVID-19 deaths as much as one would like, but unless the assumptions and other a priori "inputs" to the model are reasonable, then the predictions will necessarily be faulty. We heard this repeatedly on television. If someone uses regression, or any other statistical model, yet does not understand how influential assumptions can be to the model, then they will necessarily misinterpret the findings of the model and mislead those they publicize the model to.
- In the death of **George Floyd** in 2020, the question was raised as to whether being black, in general, is predictive of mistreatment by police, now and historically. Papers examining and seeking to demonstrate a predictive relationship will likely be using (and have used in the past) a statistical model (often a version of regression) to do it. Hence, when a conclusion is made, "Black citizens are more likely to be victim of police brutality than are white citizens," it is imperative that the consumer of such research be able to disentangle what is meant by "predictive" in this case. What does it mean, precisely, in the sense it was used in the regression? If the researcher or consumer of such research is unaware of the underlying principles that govern the statistical model that presumably led to the conclusion (and how those elements could be adjusted on a technical level), then that researcher will be unable to critically evaluate whether the conclusion is correct or incorrect. That is, you will be unable to evaluate the claim that is being made. **If you want to be able to evaluate the claim, you need to understand what the statistical method is doing.** There is no way around this work. What does an educated and informed conclusion in this regard sound like? Something like this: "There is evidence that color predicts police mistreatment ... in the least-squares sense where squared errors in prediction are minimized." This is an example of a conclusion interpreted by someone who understands how a statistical model can impact the conclusion. As a scientist, you should be aiming for making, understanding, and paraphrasing such conclusions if you are to be an intelligent consumer and producer of research.

- Is a poll conducted three months prior to a presidential election predictive of the final election results? As we saw in 2016, most polls were not predictive (well, they were predictive, only of a different result). Understanding polling and the statistics on which the poll is constructed is much more complicated than the succinct conclusions drawn from television personalities. You have to understand what "predictive" means in this sense, and that knowledge comes from knowing how the statistical model works and what conclusions can be drawn from it. **If you want to know what the microscope is detecting, you need to understand a bit about the microscope, not only that which you are seeking to observe.**

Hence, it may sound like an obvious statement, but it is imperative to understand how regression predicts responses if one is to use the word "predict" after performing a regression on some data! Otherwise, it is all "shirt and tie, and no substance." Again, the good news is that in depth mathematics are not absolutely required to understand the most important features of what makes regression "tick" and all the assumptions that go along with it. Now, that does not mean that more mathematics or philosophical justification is not always a good idea. For instance, knowing more of the "underground" of **matrix algebra** will only lead to more enlightenment down the road, just as understanding the logical precursors that must be in place before prediction is entertained for a given research problem will foster a richer understanding. Mathematically, knowing how to pump out derivatives and integrals, for instance, is not nearly as important as understanding how regression works on an intuitive level, while also not ignoring the technicalities.

7.2 The Least-Squares Principle

Though regression analysis may use one of several methods of estimating parameters, by far the one that dominates is **ordinary least-squares**. In fact, much of the history of regression analysis dates back to the origins of the least-squares principle in one form or another. Many early astronomers, for instance, worked on ways of minimizing error in predicting astronomic phenomena (Stigler, 1986). Though the mathematics behind least-squares can get rather complex, the principle is very easy to understand. When we say "least-squares," what we mean is minimizing (i.e. "least") a quantity. The quantity we are wanting to minimize is a squared quantity (i.e. the "squares" part of the principle). But, what quantity do we want to minimize when making predictions? Remarkably, it is analogous to making predictions without least-squares, and we have already seen it before. If you follow our development here, you will have a solid and intuitive understanding of least-squares.

Recall the sample variance discussed earlier in the book:

$$s^2 = \frac{\sum_{i=1}^{n}(y_i - \bar{y})^2}{n-1}$$

In computing the variance, recall that, in the numerator, we are summing squared deviations from the mean, but it is entirely reasonable to ask **why we are summing deviations from the mean, and not any other number**? Why we are squaring deviations is well understood, so that the sum does not always equal 0 (i.e. if we did not square deviations, then $\sum_{i=1}^{n}(y_i - \bar{y})$ would always equal 0 regardless of how much variability we have in our data; try it yourself with some sample data), but why deviations from the **mean** in particular? Believe it or not, the answer to this question helps form the basis of least-squares regression, and the answer is this: because **taking squared deviations from the mean ensures for us a smaller average squared deviation than if we took deviations from any other value, such as the median or the mode** (Hays, 1994). That is, taking a sum of squared deviations minimizes the squared error. The theory behind this result is beyond the scope of this book (see Hays, 1994, for details), but suffice it to say it can be quite easily justified and is on a solid footing. The important point is this, that the variance generates a quantity that minimizes the squared error in prediction. Hence, if you guess the mean from a set of data, you are guaranteed to experience the lowest average squared error possible when you make predictions using this principle. The variance, in effect, provides a least-squares solution, but under the condition of either no predictor, or, if there is a predictor in the model, that the correlation between the response and the predictor is equal to 0. This important conceptualization of the variance paves the way to understanding least-squares as it is applied to ordinary regression analysis, which we now discuss and develop.

7.3 Regression as a "New" Least-Squares Line

To understand regression, it is first helpful to start with the case where there is zero linear correlation between the predictor and the response. If such were the case, then in order to minimize error in prediction overall, the best value to predict would be the **arithmetic mean**. Hence, if the correlation is equal to 0, or there is no predictor at all, the best prediction for the response in the sense of keeping errors to a minimum would be the mean of the response variable. Even under this circumstance, we would want to know how much error we could expect if we made predictions repeatedly using the average. In other words, we need a measure by which to index the degree of error we would, on average, expect to experience if we were to use the mean as a predictor. The answer to this question is the **standard deviation**, which recall is the square root of the variance. That is, under the case of zero correlation, our best prediction would be the mean, and the degree to which we could expect to be "off" in our prediction can be indexed by the standard deviation of the response variable. However, if we do have a correlation between the predictor and the response, then it follows that if we were to use the standard deviation as our error in prediction estimate, it would not adequately account for the average error in prediction. We could still use it as our estimate, but it would not be nearly as useful as if we measured error around a "new" regression line, which is the least-squares line we will fit to our data.

Consider the following two plots. The first plot (left) is one where there is zero correlation between the predictor and the response. In such a case, the arithmetic mean is our best predicted value. In the second plot (right) is where there is a correlation. Consequently, the mean is no longer our best predicted value and instead of using a horizontal line in our prediction, we will "tilt" the line upward. This new line is our regression line, and the degree of error around the line will be indexed by the standard deviation of observations around the regression line. This standard deviation of scores around the regression line has a special name, which is called the **standard error of the estimate**, and is an estimate of the population standard deviation of scores around the population regression line. Hence, we can see that the least-squares regression line constitutes a "tilting" of the original line where the arithmetic mean was our best prediction under the condition of zero correlation.

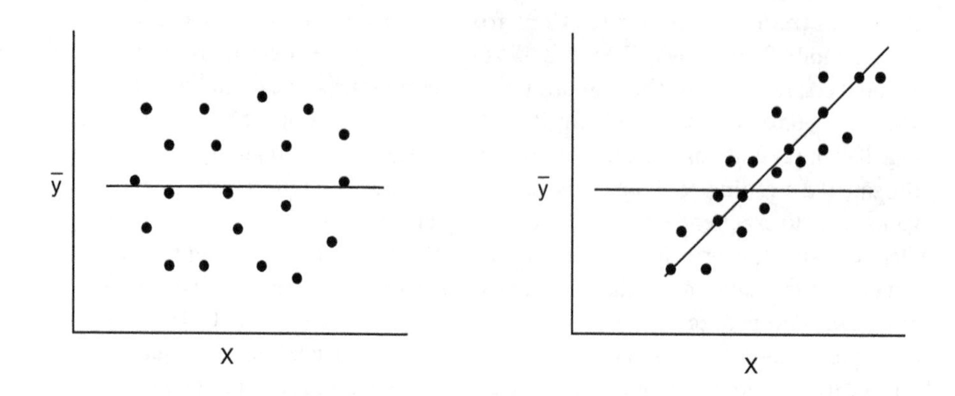

7.4 The Population Least-Squares Regression Line

To obtain the tilted line, we need to know first and foremost how to compute a line. As you may recall from high school mathematics, a line is made up of an **intercept** and a **slope**:

$$y_i = \alpha + \beta x_i$$

where y_i is the observed value along the line, α the intercept of the line, β the slope, and x_i is the value for the predictor variable that we are inputting into the equation to compute the function and obtain y_i. Now, if we were doing mathematics, this would be the end of the story. That is, if we were not fitting the line to a messy swarm of data, the above equation is all we would need. **This is where statistics differs from mathematics.** In statistics, we are fitting the line to a messy swarm of data, and hence, in addition to the linear function above, we need to append a quantity that indicates the reality that in most of our predictions using the line, we will be "off" by a certain degree. This chance of "error" is designated as ε_i and represents the difference between the observed value in our data and the predicted value using the linear function. When we append the error term to our equation, we have

$$y_i = \alpha + \beta x_i + \varepsilon_i$$

This is the **population regression line.** It assumes we have the actual population data, which we typically will never have, and hence we will have to estimate the parameters α and β using sample data. The parameter ε_i, as mentioned, is the difference between observed and predicted values, equal to $y_i - y_i'$, where y_i' is the predicted value for a given input of x_i into the equation. It is estimated as well, though we are much more interested in estimating α and β. Note that α and β are not indexed by i. This is because, for any given population, they are considered **constant**, in that they are the only intercept and slope we are fitting for the given population model. The response variable, y_i, along with the error term, ε_i, are indexed by i to indicate that there will be a different value for each value of x_i entered into the equation. Hence, we can say that the first part of the equation, $\alpha + \beta x_i$, is the **deterministic** component of the model, while the second part, ε_i, is the more **random element** of the model. This random portion of the model is sometimes referred to as the **stochastic** portion of the model. Hence, we see that the regression line, at least in structure, is no different than a regular line from elementary mathematics, only that it now includes an error term to account for being fit to messy data. That is, we can make predictions for y_i, but we can expect our predictions to be off to some degree, which is indexed by ε_i. The next question is how to compute sample estimates for α and β so we can use the equation for prediction. We consider that problem next.

7.5 How to Estimate Parameters in Regression

What we have defined thus far is the population regression line, but we have not discussed how to estimate parameters that make up the equation. Why estimate parameters? Well, if we actually had the true population regression line, we would have no need to estimate anything. For example, if I deemed my population to be equal to one hundred data points, computed the regression line on such data, and were not interested in making inferences toward a larger set of observations (obviously, since I have already deemed my regression equation to be that of the population), then we would have no need to estimate parameters. However, as usual in science and statistics, we are rarely if ever working with population data. Instead, we are working with sample data. Hence, values of α and β will be considered as **unknowns** and will have to be estimated from sample data. What we need now are equations that will tell us exactly how to compute the **sample estimators**.

The estimators for α and β will be denoted by $a_{y \cdot x}$ and $b_{y \cdot x}$, respectively, where the subscript notation $y \cdot x$ indicates that we are predicting y based on x. So, how do we compute these sample quantities? Though the derivation on why they are computed the way they are is well beyond the scope of this book, the most important thing to recognize for our purposes is that these quantities must be computed in a very particular and exact way. Hence, when we show you the formulas for computing them, from an applied perspective, it is enough to realize that these formulas were developed via **differential calculus** so that they ensured a particular condition was satisfied. What condition do they guarantee? They guarantee that if we compute them in this way, we can be assured that the **sum of squared errors around the regression line will be smaller than anywhere else we may have chosen to fit the line**. This is the least-squares principle we discussed earlier. These are called the **least-squares solutions** for this reason. If you compute them in this way, you get the **least-squares regression line**. If you compute them in any other way, you do not. Since we wish to minimize the sum of squared errors around the line, it is essential then that we use these particular formulas. In other words, this is why these formulas "are the way they are," because they ensure we are minimizing a function, in this case, the least-squares function. In general, many formulas are the way they are in statistics because they **minimize** or **maximize** a function. They derive from topics in a field of mathematics called **optimization** where functions are maximized or minimized, given that particular and well-defined constraints are in place during the process. Keep that in mind when you see formulas and ask yourself why the given formula is the way it is. It is usually the way it is for a very good reason, which is often one of optimization criteria.

The **intercept estimator** $a_{y \cdot x}$ is computed by

$$a_{y \cdot x} = \bar{y} - b_{y \cdot x}\bar{x}$$

while the **slope estimator** is computed by

$$b_{y \cdot x} = \frac{\sum_{i=1}^{n}(x_i - \bar{x})(y_i - \bar{y})}{\sum_{i=1}^{n}(x_i - \bar{x})^2}$$

Again, these are referred to as the least-squares solutions for the regression line. If you compute these equations on sample data, you have obtained intercept and slope values that will ensure that when you fit them to data, you have minimized the sum of squared errors around the regression line. That is, you will be guaranteed to have minimized

$$\sum_{i=1}^{n} e_i^2 = \sum_{i=1}^{n} [y_i - (a + bx_i)]^2$$

which is the unpacked version of $\sum_{i=1}^{n} e_i^2 = \sum_{i=1}^{n} [y_i - y_i']^2$ since $y_i' = a + bx_i$, and the **residual error** is equal to "observed minus predicted," or $y_i - y_i'$. Without delving too much into the details of each equation, we can nonetheless get a good feel for what each term "does" in terms of its computation. For instance, look at the equation for the intercept: $a_{y \cdot x} = \bar{y} - b_{y \cdot x}\bar{x}$. Notice we are starting out with the sample mean, $\bar{y}$, and subtracting out $b_{y \cdot x}\bar{x}$. Why start out with the sample mean in the computation? We start out with $\bar{y}$ because under the condition of zero correlation between x_i and y_i, it follows that $b_{y \cdot x}$ will also equal zero. This makes sense, since if you picture a flat slope, this also indicates, as already discussed, that the linear correlation between the two variables is also zero. Therefore, when $b_{y \cdot x} = 0$, then it follows that

$$\begin{aligned} a_{y \cdot x} &= \bar{y} - b_{y \cdot x}\bar{x} \\ &= \bar{y} - (0)\bar{x} \\ &= \bar{y} \end{aligned}$$

On the other hand, when $b_{y \cdot x} \neq 0$, then $a_{y \cdot x} \neq \bar{y}$, which reflects the fact that the line has been "titled" to better fit the observed data. Since $b_{y \cdot x}$ may be greater or less than 0, this will also be reflected in the computation for $a_{y \cdot x}$. We can appreciate then why $a_{y \cdot x}$ is the way it is, even if we have not proved it mathematically. Only a mathematical proof provides ultimate justification. What we have provided is but a breakdown of why the formula might make sense to us on an intuitive level.

How about $b_{y \cdot x}$? If you look at this formula a bit more closely, how it is computed will also make good sense. Again, we are not justifying or proving the formula. Instead, we are looking to see what the formula is "communicating" to us. What is it trying to tell us? We will now look at it in more detail:

$$b_{y \cdot x} = \frac{\sum_{i=1}^{n}(x_i - \bar{x})(y_i - \bar{y})}{\sum_{i=1}^{n}(x_i - \bar{x})^2}$$

We can see that in the numerator is a sum of product deviations, $\sum_{i=1}^{n}(x_i - \bar{x})(y_i - \bar{y})$, while in the denominator is a sum of squares for the predictor. The numerator incorporates the response variable, while the denominator does not. The extent to which the numerator is larger than the denominator is the extent to which $b_{y \cdot x}$ will be different from 1.0. Notice that $b_{y \cdot x}$ is actually communicating a ratio of **cross-products**, since we can unpack $\sum_{i=1}^{n}(x_i - \bar{x})^2$ into $\sum_{i=1}^{n}(x_i - \bar{x})(x_i - \bar{x})$, and hence we rewrite $b_{y \cdot x}$ as

$$b_{y \cdot x} = \frac{\sum_{i=1}^{n}(x_i - \bar{x})(y_i - \bar{y})}{\sum_{i=1}^{n}(x_i - \bar{x})^2} = \frac{\sum_{i=1}^{n}(x_i - \bar{x})(y_i - \bar{y})}{\sum_{i=1}^{n}(x_i - \bar{x})(x_i - \bar{x})}$$

If both sums are the same, then the slope is 1.0. If the amount of cross-product sum in the numerator is equal to 0, then this implies

$$b_{y \cdot x} = \frac{\sum_{i=1}^{n}(x_i - \bar{x})(y_i - \bar{y})}{\sum_{i=1}^{n}(x_i - \bar{x})^2} = \frac{0}{\sum_{i=1}^{n}(x_i - \bar{x})(x_i - \bar{x})} = 0$$

regardless of how much variation is in the denominator. Of course, if there is no variation in the predictor variable, then $\sum_{i=1}^{n}(x_i - \bar{x})^2 = 0$ and since division by 0 is **undefined** (not "zero," but rather undefined), the slope in this case would also be undefined. The formula for $b_{y \cdot x}$ is literally answering the question of how big one cross-product is in the numerator relative to another cross-product in the denominator. The cross-product in the numerator incorporates both y_i and x_i; the cross-product in the denominator only features x_i.

In summary, then, least-squares regression theory ensures that if we compute the intercept and slope by the above equations, we will, on average, minimize errors in prediction when using x_i to predict y_i. If you understand that, you now have a pretty good grasp of what "prediction" means, in the least-squares "sense" of the word. Estimation of parameters can also be performed using a technique called **gradient descent**, which can be especially useful for large-scale "big data" problems. For most common data analyses, least-squares, or a derivative of it, will suffice. For details on gradient descent, consult Bishop (2006).

7.6 How to Assess Goodness of Fit?

Once we have fit our regression line using the least-squares principle, the question that needs to be answered next is how well does it fit? While a least-squares line is guaranteed to minimize the sum of squared errors, simply minimizing a function does not, by itself, guarantee a good fit to the data. By analogy, a football team might work hard to minimize making errors on the field, but this does not in any way ensure the football team will be a good team. How "good" the fit of the regression line is, is unrelated to the minimization criteria imposed on it to compute the least-squares estimates. As an example, consider the following two plots of bivariate points:

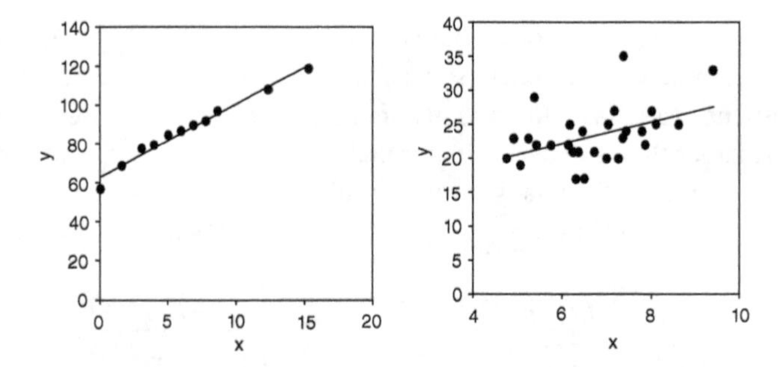

In both cases, the least-squares line is accomplishing the same thing, that of minimizing error. However, the fit of the regression lines to the data in each case could not be more different! Notice in the first plot (left), the fit is quite strong around the regression line. In the second plot (right), the fit is quite poor. What, then, will determine how well a least-squares line fits the data? This is not a statistical issue, it is a scientific one. **The quality of data and the strength of the relationship is what will determine the fit, nothing else.** In both plots, the regression lines are accomplishing the same task, that is, they are both **minimizing the sum of squared errors**, and are each doing it equally well! What differentiates the plots is the degree to which the data conform to the regression line. Hence, what we need is a measure of **model fit** or equivalently, a measure of the **average error** around the regression line. There are many measures we can use to assess model fit. We will survey a couple of them now.

7.7 R^2 – Coefficient of Determination

Undoubtedly, the most popular measure of model fit in a regression is that of R^2, often called the **coefficient of determination**, for the reason that it measures the variance in the response variable accounted for or "explained" by knowledge of the predictor variable, or, in the case of multiple linear regression, predictor variables. R^2 is defined as

$$R^2 = \frac{\sum_{i=1}^{n}(y_i' - \bar{y})^2}{\sum_{i=1}^{n}(y_i - \bar{y})^2}$$

A look at R^2 helps us appreciate what it is accomplishing. In the numerator, notice we are taking squared deviations of predicted values from the mean value of the response, $\sum_{i=1}^{n}(y_i' - \bar{y})^2$. This is contrasted in the denominator to simply a sum of squared deviations of observed values from the mean of the response. Hence, what differentiates the numerator from the denominator is whether we use y_i' or y_i when subtracting the mean, $\bar{y}$. The extent to which the two values agree will be the extent to which R^2 is equal to 1.0. On the other hand, the extent to which the numerator accounts for lesser variation is the extent to which R^2 (in the limit) will approach 0. Notice that, typically, R^2 cannot be negative in value, since both the numerator and denominator must be positive quantities.

Having defined R^2, what does it tell us on a scientific level? That is, we know statistically how it is defined, but how does that translate into drawing scientific conclusions?

The truth is that when we say R^2 accounts for "variance explained," this is all it accomplishes. It does not claim to do anything more than that. Like the correlation coefficient, it cannot pretend to know the nature of your variables or whether there were any experimental controls implemented in your research. R^2, like any other statistic, is an **abstract measure** applied to data and hence the fact remains that if you want to draw strong conclusions from your research, the design of your research is still paramount. **Statistics do not save you from poor design!** And so, even if we obtained an R^2 near 1.0, any kind of directional force or "impact" of one variable onto another cannot be inferred by this figure alone. One needs a good research design in order to draw such claims (and even then, it is usually a methodological challenge).

Conveniently, R^2 can also be defined as the **squared correlation between observed and predicted values from the model**. If you think about what this means for a moment, it should make good sense. Recall that the purpose of a statistical model is to explain or account for variability. If the model does this well, then it should be "reproducing" the data, or, at a minimum, **regenerating a linear transformation of the data**. If predicted values align perfectly with observed values, the model, from a statistical point of view at least, is a definite "winner." Hence, the extent to which observed and predicted values line up is the extent to which the model fits the data. And so, this interpretation of R^2 is quite convenient and useful as well. The goal of precise mathematical modeling is generally to see how well a function rule can be obtained that, in a strong sense, duplicates the observed data. The function rule can be thought of as a "narrative" for "explaining" the observed data. Function rules that do this well fit the data very well. The coefficient of determination is a good way to first assess how well that narrative fits. It does not imply it is the "only" narrative that could have explained the data. It only means that the current narrative does a nice job. In theory, one can have an infinite number of narratives explain the same data, something sometimes researchers fail to acknowledge in reporting their model as if it is "the" only model that could explain the observed data. Be on the lookout for researchers who buy into their theory a bit too much. Many other models or narratives may fit the data just as well. Especially with social research, **one narrative that fits is likely one of many that could fit.**

7.8 Adjusted R^2

Though R^2 is a useful descriptive statistic of how much variance is being accounted for in the sample, it has a pitfall in that it is **biased upward** as an estimator of the population effect size. That is, R^2 will more times than not lead to an overly "optimistic" estimate of the true population effect size. This will be true especially for small samples. For larger samples, the bias is much less, and eventually disappears as sample size grows without bound. What is more, R^2 **can typically only increase as one adds more predictors to a model, even if the contribution of those newly added predictors is minimal.** This is something extremely important to understand. Hence, if I have a regression model with three predictors with a given explained variance and then add a fourth predictor, R^2 will either stay the same or increase. It cannot decrease.

So why is this a problem? The problem is not so much statistical as it is, once again, scientific. In short, it means that **one can continually add predictive power to one's model by simply including more predictors**. But surely a model with 1,000 predictors, as an extreme example, is not very useful on a scientific level. If we can explain nearly as much of the variance with, say, 10 predictors, then the more **parsimonious** model must be preferred. Adjusted R^2 tackles this issue of parsimony head-on, rewarding models that are more parsimonious than not. Consider how it does this through a look at its formulation:

$$R^2_{\text{Adj}} = 1 - (1 - R^2)\left(\frac{n-1}{n-p}\right)$$

where n is the number of data points, while p is the number of parameters fit by the regression. This includes the intercept term. Hence, in a model with one predictor, p will equal 2. The "active" or "working" component in R^2_{Adj} is thus the ratio $(n - 1)/(n - p)$. Notice that for constant n, as p increases, the denominator $n - p$ gets smaller, and hence the ratio $(n - 1)/(n - p)$ gets larger. How does this affect the size of R^2? In the formula, we will have to take the product $(1 - R^2)((n - 1)/(n - p))$ first before subtracting, due to the order of operations (i.e. multiplication comes before subtraction). The value of $(1 - R^2)$ is the amount of variance **unexplained**, since it is the balance of R^2. Therefore, the larger $(1 - R^2)((n - 1)/(n - p))$ is, the more we are subtracting from the "best fit" value for R^2 that is represented by the starting point of the formula, 1.0. Hence, what will keep R^2_{Adj} larger than not is if we are subtracting a smaller value than not for $(1 - R^2)((n - 1)/(n - p))$. But, as p gets larger and larger, $(n - 1)/(n - p)$ also gets larger, which means that $(1 - R^2)((n - 1)/(n - p))$ gets larger, and hence, all else equal, larger p given a more or less constant original R^2 gives us a smaller R^2_{Adj} when we subtract from 1.

Now, at first glance, based on our argument above, you may come to the tentative conclusion that increasing the number of parameters is a "bad thing" overall, since it seems to have a negative impact on the size of R^2_{Adj}. However, this is not entirely true, and therein lies the genius of R^2_{Adj}. If the new parameters increase R^2 by a relatively substantial amount, the increase in explained variance will be "worth" adding the new parameters. This is how R^2_{Adj} can be considered a **penalized** version of R^2, because if you add new predictors that are not worth it, **you get punished**. Hence, models are rewarded that are more **parsimonious** than not, which accords well with how science should work. This is how R^2_{Adj} is of paramount importance from a **scientific** perspective, as well as statistical. **Parsimony is what science is all about**. Simple explanations are always considered better than more complex ones if they provide equivalent explanatory power. If you start adding every predictor under the sun with the hopes of simply increasing variance explained, R^2_{Adj} says, "Nope, not on my watch," and will have the effect of decreasing by an increasing amount the original R^2. R^2_{Adj} is not the only penalized statistic in regression. **Akaike's information criterion** (AIC) is another such measure that attempts to reward more parsimonious models over less parsimonious ones. We do not discuss AIC here. Interested readers are encouraged to consult James et al. (2013) for a good introduction.

7.9 Regression in Python

As an example of regression in Python, and at the same time to demonstrate a data management task, we will import a data file originally saved in SPSS. The original data file is named `iq_data.sav`. We can use `pyreadstat` to read it into Python. We will name the new data file by the name of `df`. We print out only a few cases of the file:

```
pip install pyreadstat
import pyreadstat
df, meta = pyreadstat.read_sav("iq_data.sav")
df
Out[321]:
    verbal  quant  analytic  group
0     56.0   56.0      59.0    0.0
1     59.0   42.0      54.0    0.0
2     62.0   43.0      52.0    0.0
3     74.0   35.0      46.0    0.0
4     63.0   39.0      49.0    0.0
5     68.0   50.0      36.0    0.0
6     54.0   54.0      29.0    0.0
7     56.0   52.0      57.0    0.0
8     51.0   46.0      65.0    0.0
9     49.0   39.0      61.0    0.0
10    66.0   55.0      69.0    1.0
```

To begin, we first obtain a plot of `verbal` against `quant`. To do this, we will extract `verbal` and name the new variable `"y"` and also extract `quant` and name that variable `x`:

```
y = df["verbal"]
x = df["quant"]
import numpy as np
import matplotlib.pyplot as plt
plt.scatter(x, y)
Out[327]: <matplotlib.collections.PathCollection at 0x1f4057b8>
```

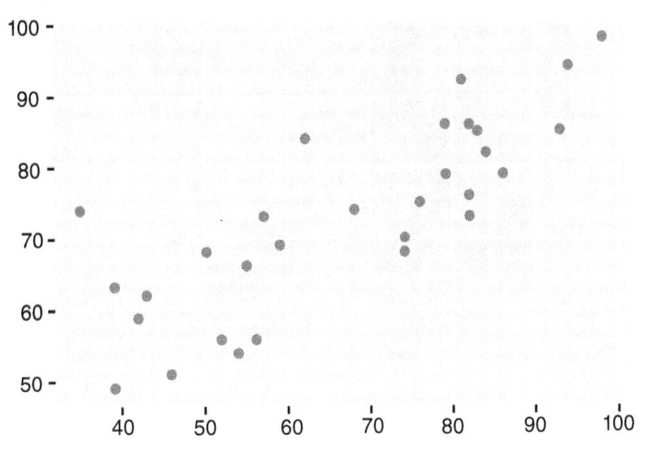

We notice a somewhat moderate but linear relationship between the two variables. That is, an increase in verbal, on average, seems to be paired with an increase in quant. Of course, the plot is very bare, and so it would be nice to add some labels to the respective axes. This can be accomplished using the following (updated plot not shown):

```
plt.title("Scatterplot of Verbal on Quant")
plt.xlabel("Quant")
plt.ylabel("Verbal")
```

In Python, we will now run the regression of verbal on quant. When we say "verbal ON quant," we mean verbal is the response variable and quant is the predictor. First, we add a constant to our predictor variable using statsmodels.api:

```
import statsmodels.api as sm
x = sm.add_constant(x)
```

The constant in the regression equation is the same for everyone, which is a value of "1" and may be readily verified (we print only the first three cases):

```
x
Out[30]:
    const  quant
0     1.0   56.0
1     1.0   42.0
2     1.0   43.0
```

We are now ready to run the regression using sm.OLS, which stands for "ordinary least-squares":

```
model = sm.OLS(y, x).fit()
print_model = model.summary()
print(print_model)
                            OLS Regression Results
==============================================================================
Dep. Variable:                 verbal   R-squared:                       0.653
Model:                            OLS   Adj. R-squared:                  0.641
Method:                 Least Squares   F-statistic:                     52.68
Date:                Tue, 09 Jun 2020   Prob (F-statistic):           6.69e-08
Time:                        14:02:26   Log-Likelihood:                -103.07
No. Observations:                  30   AIC:                             210.1
Df Residuals:                      28   BIC:                             213.0
Df Model:                           1
Covariance Type:            nonrobust
==============================================================================
              coef    std err          t      p>|t|      [0.025      0.975]
------------------------------------------------------------------------------
const      35.1177      5.391      6.514      0.000      24.074      46.162
quant       0.5651      0.078      7.258      0.000       0.406       0.725
==============================================================================
Omnibus:                        1.230   Durbin-Watson:                   1.244
Prob(Omnibus):                  0.541   Jarque-Bera (JB):                1.020
Skew:                           0.431   Prob(JB):                        0.600
Kurtosis:                       2.732   Cond. No.                         263.
==============================================================================
Warnings:
[1] Standard Errors assume that the covariance matrix of the errors
is correctly specified.
```

There is a lot going on in the above output, with enough statistical theory underlying its output to easily fill an entire book on regression. We summarize only the most salient and important features of the output:

- We confirm that the dependent variable (`Dep. Variable`) is that of `verbal` and that the model was estimated by ordinary least squares. The number of observations in the data is equal to 30. The degrees of freedom for `residual` are equal to 28, computed as the total sample size minus the number of predictors (which in our case is 1), minus 1. That is, $n - 1 - 1 = 30 - 1 - 1 = 28$. The degrees of freedom for the model (`Df Model`) are equal to 1, since there is only a single predictor. Had there been two predictors, this number would equal 2.
- The `R-squared` for the model is equal to 0.653, and is the proportion of variance in the response accounted for by the predictor. That is, knowledge of quant accounts for approximately 65% of the variance in verbal. Adjusted R-squared is equal to 0.641 and, as expected, is slightly less than R-squared. However, since the model is not overly complex (we only have a single predictor, after all), Adj. R-squared is essentially almost equal to R-squared.
- The *F*-statistic for the model is equal to 52.68, with an associated probability of 6.69e-08, which translates to a *p*-value of 0.0000000669. Hence, the model is easily statistically significant, which indicates that the predictor is affording prediction of the response over and above chance. As we will see, since there is only a single predictor in the model, this *p*-value will agree with the *p*-value obtained for the *t*-statistic on `quant` that we will survey shortly.
- **AIC** for the model is equal to 210.1. We see that the **BIC** value is only slightly larger. AIC and BIC, like Adjusted R-squared, are penalized fit statistics. Though we do not discuss them in detail in this book, for applied purposes, you should know that **smaller values than not are preferred**, as both of these measures penalize for "too complex" of a model. Typically, BIC penalizes a bit more harshly so we expect it to be a bit larger compared to AIC.
- Next in the output we come to the coefficients for the model. We see that the intercept (`const`) is equal to 35.1177 with standard error equal to 5.391. When we divide 35.1177 by 5.391, we obtain the *t*-statistic of 6.514. We see that it is statistically significant ($p = 0.000$), with confidence limits of 24.074 and 46.162. Recall what the constant represents; it is the predicted value for the response when the value of the predictor variable is equal to 0, since $y_i = a + bx_i = a + b(0) = a$. It usually is not of immediate interest, in that we are much more interested in the value for the coefficient for `quant`, the predictor in the model.
- The coefficient for `quant` is equal to 0.5651, and we see it is also statistically significant ($p = 0.000$). This is the slope for the regression equation and is interpreted as: **for a one-unit increase in quant, we expect, on average, a 0.5651 unit increase in verbal, given the model under test**. Now, the model "under test" is a simple linear regression, but this statement is still required. When we study multiple regression shortly, we will see that when we include additional predictors into the model, the coefficients for other predictors typically change. Hence, "given the model under test" is a statement you should include regardless of the complexity or simplicity of the model. **Any effect in a model is always evaluated relative to the model, never in a vacuum.**

- The remaining pieces of the output are not of immediate concern and hence we do not interpret them here.

This completes our discussion of simple linear regression, and we now move on to consider the more popular multiple regression model. As we will soon see, interpreting multiple regressions, while having some similarities to simple regression, includes additional complexities. Understanding these complexities is very important. Interpreting effects in a multiple regression is not the same as interpreting them in a simple regression, and we will discuss why this is so in some detail.

7.10 Multiple Linear Regression

In the simple linear regression just conducted, we only had a single predictor. Often, however, a researcher wishes to model the relationship between several predictors and a single response simultaneously (recall we included simultaneous independent variables in factorial ANOVA as well). When we include more than a single predictor into a regression model, the model is that of **multiple linear regression**. Technically, what we are doing is modeling the response as a function of a **linear combination** of predictor variables.

At first glance, you may think multiple regression is a simple extension of simple linear regression, and whether a researcher chooses to do one or the other is of little consequence. This impression, however, is misguided. When one performs a multiple regression, there are numerous issues that arise that simply did not exist in simple linear regression models, or, if they did exist, were not as apparent and "in your face" as for multiple regression models. In multiple regression, we cannot avoid those issues and must instead confront them head on.

7.11 Defining the Multiple Regression Model

Where the simple linear regression model was defined as

$$y_i = \alpha + \beta x_i + \varepsilon_i$$

the multiple linear regression model is given by

$$y_i = \alpha + \beta_1 x_{1i} + \beta_2 x_{2i} + \cdots + \beta_p x_{pi} + \varepsilon_i$$

A few things to note when comparing the two models:

- Both models feature only a **single response variable**, y_i. That is, whether one performs a simple linear regression or a multiple linear regression, there is still only a single response. If there were more than a single response, the model would be **multivariate** in nature.
- There is only a single intercept, α, in both models.
- Whereas the simple linear regression model features one predictor with an associated regression weight, βx_i, the multiple regression model features several predictors, each

with associated regression weight, $\beta_1 x_{1i} + \beta_2 x_{2i} + \cdots + \beta_p x_{pi}$. While the weight in simple linear regression is simply a coefficient for x_i, the weights in multiple regression, because there are several of them, are best referred to as **partial regression weights** for each predictor x_{1i} through to x_{pi}. Notice that each predictor variable is still indexed by an "i" following p, to indicate the ith value of the given predictor.

- Both models, whether simple or multiple regression, have an **error term** associated with them, ε_i. The error term is still subscripted by a single "i" in the multiple regression model as it was in the simple regression model, since there is only a single error possible regardless of how many predictors one has in the model. This is important to understand. Simply because we have included more than a single predictor does not mean we are no longer obtaining a single prediction of the response. We are still obtaining a single prediction, only that the response we are predicting is based on a more extensive linear combination of predictor variables. Hence, there is still only a single error term, which implies there will be only a single **residual** of the form $y_i - y_i'$ as well.

It cannot be emphasized enough how multiple regression is different from simple regression, and in this introduction to multiple regression, it allows us to introduce some of the primary features of statistical modeling in general. We have already mentioned it in passing, but now it is time to name it as a fundamental principle of statistical modeling. The following is what we like to call the **universal principle of statistical modeling**, no matter how simple or complex the model:

All models are context-dependent.

It would not be an exaggeration to say that every research paper that features statistical modeling should begin their report of their statistical results by including the above sentence. **Context-dependency** is the idea that whatever model you evaluate, it is always the **model** you are testing, not individual predictors within that model. For example, if I set up a model to predict depression, and include anxiety and social isolation, the regression weight obtained on both anxiety and social isolation cannot be interpreted separately from each other. That is, if a regression weight of 0.70 is obtained for anxiety, that coefficient should only be interpreted in the **context** of having social isolation also in the model, and not distinct from it. This is equally true for the effect of social isolation, in that it can only be interpreted in the context of anxiety. So, if anxiety is statistically significant, that statistical significance "exists" in the context of also having included social isolation, and not **necessarily** distinct from it. Likewise, the obtained estimate of the regression weight is also context-dependent. That is, it exists in the context of the other variable(s). The regression estimates are not independent of the model tested and do not exist in a vacuum from it. For instance, if we only included anxiety in the model, the coefficient for anxiety might change from what it was when social isolation was also included. If a researcher or consumer of research is unaware of this context-dependency of the model he or she is generating or consuming, then there lies a serious risk of misinterpretation, thinking that anxiety is a predictor of depression, **period**. This may not be the case. Anxiety is a predictor given the context in which its coefficient is estimated. Never forget that!

How can we solve the context-dependency issue? Given our discussion above, you may at first attempt to argue that if you want to avoid the issue, simply conduct simple linear regressions and not multiple. However, this idea is misguided because even in a simple regression, **you are still testing the model**! The fact that you are excluding predictors from your model is still an act of making your model context-dependent, only that in this case, you are not including variables that may be relevant to the model. In brief, **you cannot escape the issue!** It will haunt you regardless of the model you fit, even if very simple. Statistical models do not necessarily reflect reality, even if they are often advertised to reflect just that. They partition variability and estimate parameters, that is all. The rest of the substantive conclusions must come from **good research design** and/or quality variables that you are subjecting to the regression in the first place.

7.12 Model Specification Error

The idea of context-dependency is allied to the idea of specifying the correct model. All statistical models, including regression, assume that what you are fitting is a model that includes the most relevant predictors to the response. If the model you are fitting is grossly incomplete, such that important and influential covariates are lacking from the model, then not only will the parameter estimates featured in the model be incorrect, any conclusions about the predictive power of the model will likewise be wrong. Of course, very few, if any models can be perfectly specified, which is in part why George Box infamously once said, **"All models are wrong, some are useful."** The problem of **omitted variables** is sure to plague any regression model you fit, even if only on a theoretical level in some cases. For instance, perhaps a given treatment is a strong predictor of COVID-19 recovery, but there will always be more variables that could be added to the model to help improve on its **specification**. That does not imply that we should not take the COVID-19 regression model seriously that was fit, only that we appreciate the fact that it is still an **incomplete model**. The model you fit will be wrong, even if it fits the data perfectly. Don't feel bad, as Newton's models were wrong too, even if they explained (and still do) a great deal of variability. **Science (if done ethically) is a very tough game.** Be prepared to be wrong all the time! However, if you can explain variability and perhaps theorize on causal mechanisms, you are doing pretty well! Don't despair!

The moral of the story is this – when you fit a regression model, you should always try to ensure it is as well-specified as possible. Otherwise, parameter estimates in the model and conclusions drawn about the model effects will be biased and wrong. When interpreting any statistical model in the literature, the first critical question you should ask yourself when reading the article is whether the model is **correctly specified**. If it is not, or the author does not address the specification of the model in some detail, then the resulting model may not be worth interpreting at all, as it can potentially misguide more than it guides. A good research paper should, in some sense, address the model specification issue. In the physical and biological sciences, the goodness of the specification is usually implicit and requires less backing. In the social sciences, however, such as with economic or psychological variables, model specification should be rigorously defended before the statistical analyses are carried out and interpreted.

7.13 Multiple Regression in Python

We now demonstrate a multiple regression, this time using both `quant` and `analytic` simultaneously as predictors of `verbal`. In this example, we refrain from interpreting *p*-values and simply focus on obtaining the relevant coefficients. Recall our IQ data:

```
import pyreadstat
df, meta = pyreadstat.read_sav("iq_data.sav")

df
Out[169]:
    verbal  quant  analytic  group
0     56.0   56.0      59.0    0.0
1     59.0   42.0      54.0    0.0
2     62.0   43.0      52.0    0.0
```

First we import the required packages:

```
import pandas as pd
from sklearn import linear_model
import statsmodels.api as sm
```

We then bin together the predictor variables along with the response. Note that `X` is now composed of both `quant` and `analytic`:

```
X = df[['quant', 'analytic']]
Y = df['verbal']
```

Next, we define the regression object:

```
regr = linear_model.LinearRegression()
regr.fit(X, Y)
Out[180]: LinearRegression(copy_X=True, fit_intercept=True, n_
jobs=None, normalize=False)
```

At this point, the regression has been performed and we are ready to extract parameters from it. We now extract the intercept fit by the model:

```
print('Intercept: \n', regr.intercept_)
Intercept:
 32.17094152928748
```

We see that the intercept for the regression equation is equal to 32.17. Next, we obtain the partial regression coefficients, the first for `quant`, the second for `analytic`:

```
print('Coefficients: \n', regr.coef_)
Coefficients:
 [0.42814931 0.16964027]
```

The coefficient for `quant` is equal to 0.428, while the coefficient for `analytic` is equal to 0.169. How each of these is interpreted is very important. At first glance, it may be tempting to interpret `quant` in this way:

For a one-unit increase in quant, we can expect, on average, verbal to increase by 0.428 units.

This interpretation is wrong! But how can this interpretation be incorrect? It sounds so consistent with how we interpreted the regression slope in simple linear regression. And therein lies the problem! **The nature of the coefficient is different in multiple regression than it was in the simple linear regression**. That is, we cannot interpret it in the same way. Since the coefficient appears in the context of another coefficient, it is imperative in our interpretation to provide the "context" statement that we initially provided in the simple linear regression. There are two ways of doing this. Either one of the following two interpretations would work:

For a one-unit increase in quant, we can expect, on average, verbal to increase by 0.428 units, given that the model under test has analytic also in the model.
For a one-unit increase in quant, we can expect, on average, verbal to increase by 0.428 units when holding analytic constant.

Either one of these statements will adequately describe the model results. Notice that in both statements, the context is given for the model. Without either of these, the interpretation is literally wrong.

We can obtain predicted values for our model by designating specific values for `quant` and `analytic`. For instance, suppose our new input for `quant` is equal to 20, while our new input for `analytic` is equal to 25:

```
new_quant = 20
new_analytic = 25
print('predicted verbal: \n', regr.predict([[new_quant, new_
analytic]]))
```

```
predicted verbal:
 [44.97493447]
```

We see that given the above inputs, our predicted value for `verbal` is equal to 44.97.

7.14 Model-Building Strategies: Forward, Backward, Stepwise

In this section, we survey approaches to model-building in regression. What does it mean to "build a model?" There are essentially two steps to model-building: (1) determining the set of predictors you will consider to be candidate predictors for the model and (2) adopting a method by which those predictors will be entered into the model. Despite all of the statistical advances and sophistication in model-building over the years, and many algorithms and such developed to aid in the process, the most important step to model-building remains choosing candidate predictors for consideration into the model. A **candidate predictor** is a predictor you have measured or obtained and will consider it for possible and potential inclusion into a model you are building. Without the careful selection of these candidate predictors, any model you build will necessarily be problematic and possibly very misleading. Hence, this necessarily presents a problem:

There is no statistical method that will ensure you have selected the correct or most complete set of candidate predictors.

That is, sophisticated as model-building has become over the last century and despite all of the advances in software, the first step to model-building is not a statistical one at all! Rather, it is a **scientific** one. No statistical algorithm can ensure you have considered the most useful and relevant of predictors. You, as a scientist, must be "on top" of this from the start. For example, for the case of predicting COVID-19 illness, suppose that a researcher designed a model in which pre-existing conditions as a predictor in the model were not considered. The failure to even record this information or include it into the model would constitute a major **specification error**, and the model would be wholly incorrect regardless of which model-building algorithm was then used. If the set of candidate predictors is wrong, or incomplete, using any technology afterward to build the "best" model will simply not work. Hence, as a scientist, you must ensure that before you build a model, you are including the most complete and accurate set of predictors possible before you set the algorithm into motion. That is, **the selection of candidate predictors is the most important step in all of statistical modeling building**. It is also the least technical. Once you have selected your candidate predictors, the second step is consideration on how to enter them into your model or how to designate an algorithm to search for the most "important" (in some sense) predictors. Of these, there are several options: simultaneous, hierarchical, forward, backward, and stepwise regression (among others). We briefly consider each of these.

In **simultaneous regression**, otherwise known as **full entry** regression, the researcher, after determining a suitable set of candidate predictors, enters all of these predictors into the model at once. This type of regression should be considered your "default" choice in most of your data analyses. That is, unless you have a good reason for not entering all of your predictors at once, then most regressions you perform should follow this approach. Sometimes researchers like to preselect the predictors they enter into their model, and enter each individually or as a group at each stage of the model-building process. This is known as the **hierarchical approach** to model-building. For instance, suppose we already know that pre-existing heart conditions can predispose one to COVID-19. However, we would like to know over and above this whether gender is an important predictor. We could build the model by first entering a variable representing pre-existing heart conditions (to keep it simple, though definitely likely medically incorrect, suppose we are entering beats-per-minute), then enter gender as the next predictor. When we examine the effect of gender, we are studying the predictive impact of gender over and above that of pre-existing heart conditions. Hence, the hierarchical model gives us a convenient way to observe the variance explained by adding predictors to our model. It needs to be recognized, however, that regardless of whether you enter predictors in a hierarchical fashion or full entry, the final model results will be the same. That is, to use our example, whether you first enter pre-existing conditions or include them directly with gender, the final model will give you the same results. Hierarchical is preferred, however, in some cases where the researcher would like to observe and take account of the **incremental variance explained** by adding one or more predictors.

The idea of adding predictors into a model after others have already been included is sometimes referred to as **controlling for other predictors**. However, it needs to be emphasized that this is **statistical control** only, not **experimental**. There is no true control unless you have an experimental design that features such control. All we are doing when we observe the effect of one variable in the presence of others is partialling out variability, nothing more. Statistical control does not somehow serve as a panacea to experimental control, nor is it a justifiable way to get out of performing rigorous experiments. Beware of misinterpretations of statistical control. Statistical control is not that big of a thing, scientifically speaking.

Hierarchical regressions form the basis for a procedure some researchers sometimes perform, that of **statistical mediation**. In a mediation model, as shown in the following diagram, an independent variable is hypothesized to predict a response variable, but such a relationship is said to be "mediated" by another variable (the so-called "**mediator**"). As the story goes in mediation, if the original relationship between the IV and DV disappears, or nearly so, then varying degrees of support for mediation are said to exist.

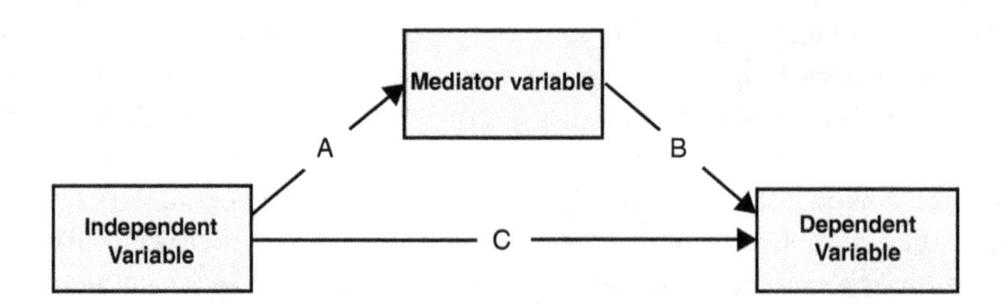

Of course, the establishment of statistical mediation in no way justifies, on its own, a conclusion of a **physical process** between the IV, mediator, and DV, since as we can clearly see, statistical mediation is, at most, simply an example of hierarchical regression. Now, if a substantive argument can be advanced that "true mediation" on a physical substantive level is occurring, then statistical mediation might serve as support for that argument. Too often, however, researchers falsely believe that the establishment of statistical mediation implies or somehow justifies a physical or scientific "process" between variables. This is grossly incorrect. **No statistical model, on its own, can justify any process that does not exist before the statistical model was applied.** Still, and even if regularly misunderstood, mediation analysis is quite common, especially in psychology and related areas. If the **a priori assumptions** governing the variables are tenable, mediation analysis can indeed be quite useful on a scientific level. But if such assumptions are not tenable, then all that is being performed for certain is a hierarchical regression (i.e. variance partitioning) and possibly nothing more. Unless of course there is an underlying process we are simply not aware of yet, but it would still be a gigantic leap to say the independent variable "goes through" the mediator to the response in any sense. All we can really say is that the mediator "accounts for" the relationship between the predictor and response variable, but given that many variables can "account for" bivariate linear relationships, the substantive conclusion may simply still not be that impressive if not strongly governed by theory or substantive context (e.g. working with biological or physical variables). In a real sense, all correlations are "spurious," which implies they are "due" to one or more other variables, but if you are performing correlational analyses, you can only conclude this in a correlational sense. An experiment would be required to detect any true "due to a third variable" idea and attribute any **causal** or **impact** mechanism. Hence, the researcher that concludes, "We found evidence for a mediating relationship," may simply have found evidence that variance can be successfully partitioned among variables, which is not all that ground-breaking. For extensive instruction on mediational analyses, see MacKinnon (2008).

7.15 Computer-Intensive "Algorithmic" Approaches

Thanks to advances in computing power in the last 40–50 years, algorithmic approaches to model selection have become quite popular. Now, let us be sure to understand what this means. A computer does nothing more than obey the commands it is instructed with. Hence, saying that a computer is "performing" model selection for you is a hollow and meaningless statement. What is choosing the model is the code and instructions the user has programmed into the software. This code and instructions are based on, you guessed it, "anticlimactic" (i.e. it won't make cable news) mathematics. Hence, if you want to truly understand algorithmic approaches to model selection, **you need to study the underlying machinery**. You need to understand the sticks and stones. You need to understand what the computer is actually "doing" when it is selecting the model. Of course, we could just test every conceivable model to see which fits best, so-called **best subset regression**, but for p predictors, that would entail testing 2^p models. For instance, for the case of $p = 30$ predictors, the number of models possible is equal to $2^p = 2^{30} = 1,073,741,824$. Choosing to adopt an algorithmic approach to model building may help, and "shrink" the task down to something more manageable, though in managing complexity, there is certain to be an associated cost. Whenever you attempt to trim a problem down to a manageable size, you always lose information along the way.

In **forward regression**, the algorithm searches among the candidate predictors (remember, it cannot identify predictors you do not introduce for consideration to begin with) and selects that which has the largest **bivariate correlation** with the response at some pre-designated alpha level, such as $\alpha = 0.05$. After selecting this first predictor, the model searches among remaining predictors for the one having the highest **squared semipartial correlation** with the response, which essentially translates to the predictor having the greatest statistically significant increase in variance explained in terms of an F-statistic. The algorithm continues in this fashion until no other predictors meet entrance criteria. Key to the forward algorithm is that **it will not remove predictors**. That is, **once a predictor is in the model, it remains in the model**. As we will see, this is contrary to stepwise regression in which predictors are allowed to be removed at each step. In **backward elimination**, the process is reversed, such that the model starts out with all candidate predictors and removes at each step predictors that contribute the least to the model.

In **stepwise regression**, the model first searches for predictors for inclusion in the spirit of forward regression, but then at each step also considers predictors for removal. For example, once the second predictor is entered, the first predictor is re-assessed for statistical significance given the inclusion of the second predictor. If it is no longer statistically significant, it is removed from the model. Stepwise continues in this fashion until no predictors are entered or removed. While the entrance criteria is typically set at an alpha level of 0.05 or so, the removal criteria must be at a level greater than this (e.g. 0.10); otherwise, the algorithm could engage in cyclic behavior of adding and removing predictors without any progress to model fit.

7.16 Which Approach Should You Adopt?

The decision of which model-selection procedure to adopt is thus not an easy one, and as a researcher, you need to be well aware that a given approach may yield different results than adopting a competing approach. Does this problem sound familiar? It should. It shows once again that the model you fit and interpret is contingent upon decisions made along the way. Not only then is model fit determined by what predictors you do or do not include, but it also depends on the methodology used in model selection. A stepwise approach may yield an entirely different model than a forward or simultaneous regression, and the researcher generating such research as well as the consumer consuming such research needs to be intimately familiar with these issues. Otherwise, they are likely to mistake a given model as "the" model that represents the data. It is never "the" model. Rather, it is a model determined by all the inputs we are discussing. **Subjectivity is part of the model-building process**. The moment you decide to build a model, you will, to some extent, be making subjective decisions. So long as you are clear and upfront about these decisions, that is fine, but you need to communicate them to your audience to ensure you are being **authentic** and **ethical**.

One should also be aware that though stepwise approaches may appear attractive at first, they come with their own set of difficulties. For one thing, parameter estimates are usually biased when using an incremental F-test as used in such approaches (Izenman, 2008). Hence, computed standard errors may not be entirely realistic, leading the model to include or not include given predictors. Among students, stepwise regression is often a "first choice" when deciding on a model-building algorithm due to its perceived complexity and "coolness" factor (everyone wants to do complex things), but it is usually not the best choice. The best choice is usually the anticlimactic simultaneous regression where decisions about predictors are made based on a researcher's knowledge of the literature, along with hypothesized outcomes. As with many things in statistics, **it is often the simpler tool that is the most effective**, and hence unless your research is nothing more than a fishing expedition, simultaneous regression will usually meet your needs.

Utility should also be at the forefront of what predictors to include. If a predictor is too expensive to recruit from subjects (for instance, maybe some blood chemistry tests are exceedingly expensive to obtain), then you may be better off going with a simpler predictor that accounts for as much or nearly as much variance in the response variable. A model explaining slightly less variance may nonetheless be a "better" model if its candidate predictors are more easily obtained. If it were found that heart disease could be predicted by rocketing people into space and observing circulatory behavior in microgravity, strong as a predictor this may be (I'm just making this up), it would have virtually zero utility as a predictor (here on planet Earth) since obtaining that information on a sample of individuals would be virtually impossible, if not extremely expensive. **Good predictors explain variance, but they are also manageable, obtainable, and "do-able" for the project. That is, good predictors have good utility.** Once you bring modeling down from the ivory tower of abstractness, these "pragmatic" things matter a great deal. Try convincing a top business CEO of the most "theoretically correct" model and you may be surprised at the response you get! Pragmatics matter!

7.17 Concluding Remarks and Further Directions: Polynomial Regression

In this chapter we have merely scratched the surface on the topic of regression analysis. Regression can easily take up a full year course, and even then we would still be only scratching the surface! For an excellent and thorough treatment of all things about regression, as well as extensions to the regression model into the generalized linear model and all its possibilities, see Fox (2016). Draper and Smith (1998) is also a classic reference and should be consulted. Denis (2021) also features a more thorough chapter on both simple and multiple regression, as well as interactions within a regression framework.

In this chapter, we have assumed that relationships among response variables and predictors could be modeled sufficiently as a **linear** trend. In many cases, however, **non-linear** trends may provide a better fit, both statistically and in line with one's substantive theory. In those cases, **polynomial regression** may be a suitable option, where one can flexibly change the degree of the polynomial away from degree 1 to learn what best fits the data. This chapter has been about first-degree polynomials (otherwise known as "lines"); that is, $y_i = a + bx_i + e$ featured an input variable x_i understood to be raised to the first power. An example of a more complex polynomial is that of $y_i = a + b_1x_i + b_2x_i^2 + e$, where now the predictor is raised to a higher exponent, in this case "2," making the model a **quadratic** one. Such models are still considered **linear in the parameters** since b is still to the first power. However, the curve of these models is different and may necessitate fitting a different polynomial other than a linear one.

From a statistical and computational point of view, fitting polynomial models is not too difficult, and is quite similar to fitting ordinary linear regression models. However, from a scientific point of view, fitting polynomial models can quite easily be **abused** as the analyst attempts to account for as many data points in the plot as possible, thereby potentially "**overfitting**" the model to the data. As an example, consider the following plot:

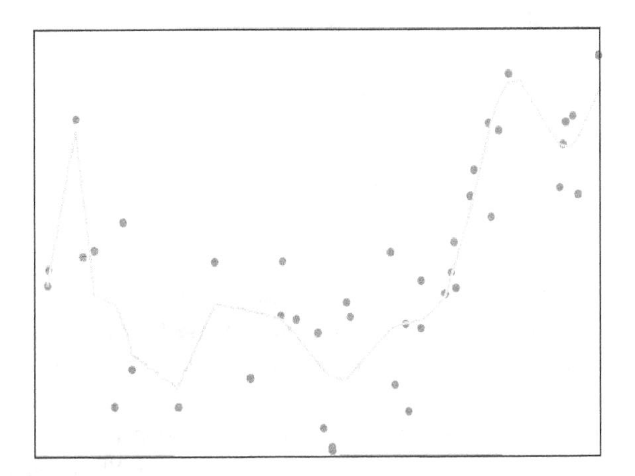

This is an example of overfitting a model to data, since the curve is attempting to essentially account for as many data points as possible. What could possibly be wrong with this? Isn't the goal of modeling to best account for the data? While from a statistical point of view the curve might fit quite nicely, from a scientific point of view it is unlikely that a replication of the study or validation of the model on new data (i.e. "cross-validation") would generate such a tight model fit. Without knowing the material under investigation in the plot (i.e. without knowing the "stuff" on which the science is being performed, that is, the raw objects or material), one cannot adequately assess the extent of the overfit, at least not on an intuitive level. Statistically, one could separate the data into a **train** and **test** set, and attempt to validate the model to see how well the curve fits on the test data. Slightly "**under-fitting**" a model may be preferred to "over-fitting," especially from a scientific perspective, so that whatever model is fit, is fit with a bit more **generalizability** to new data. By way of analogy, if you are a clinical psychologist, over-fitting your diagnosis to a particular client makes it a **case study**, not a generalizable phenomenon. Again, these issues cannot truly be understood without recourse to the scientific material under investigation. In sum, anyone can "connect the dots" on a plot, but that is not what mathematical or statistical modeling is about, just like explaining why one individual contracted COVID-19 is hardly an explanation for what causes the disease. **Generalizability**, as well as **simplicity** are always the goals, not to fit a ridiculous model to demonstrate a failure to relate the statistics you are computing to the scientific data you are modeling. For further details on polynomial regression, James et al. (2013) is an excellent introduction.

Review Exercises

1. Discuss the goal of **regression analysis**. Is it to make predictions? If not, what is it for?
2. Given an example (not featured in the chapter) in the age of COVID-19 where a **simple linear regression** may prove useful.
3. What is the **least-squares** principle? What does least-squares regression guarantee? What can it not guarantee?
4. What does it mean to **estimate parameters** in regression? Under what circumstance would parameter estimation not be required? Is such a circumstance realistic? Why or why not?
5. Consider the formula for the **slope** coefficient for the regression of y_i on x_i:

$$b_{y \cdot x} = \frac{\sum_{i=1}^{n}(x_i - \overline{x})(y_i - \overline{y})}{\sum_{i=1}^{n}(x_i - \overline{x})^2}$$

   Relying on your knowledge and intuition behind what the formula is accomplishing, rewrite the formula for the slope where x_i **is regressed on** y_i instead. What changes would you make to the equation and why? Why do such changes make good sense in terms of what regression analysis accomplishes?
6. Distinguish between an **error** and a **residual** in regression analysis. Why are they not one and the same thing? Do we ever have access to errors? Why or why not?
7. Why is it the case that the **coefficient of determination**, or R^2, cannot typically exceed a value of 1.0 in either simple or multiple linear regression? Justify that it cannot.
8. Your colleague, after conducting her research study, tells you, "The coefficient obtained in my regression is equal to 0.90." Would you consider this an impressive result? Why or why not? What other questions might you ask her?
9. For a constant value of R^2, what happens to the size of R^2_{Adj} as p increases? What about as n increases? Justify your argument using the formula in each case.
10. Consider the **Forest Fires Data Set** from the **Machine Learning Repository** (https://archive.ics.uci.edu/ml/index.php). Regress area burned on the temperature

variable in Python. Is temperature a good predictor of how much land will be burned in a fire? How much variance is explained?

11. What is the effect of lowering the estimated **standard error** for a predictor in regression on the ***p*-value** associated with the resulting *t*-statistic? All else being equal, are small standard errors a good thing or a bad thing for your research project? Why?
12. Consider the **Challenger USA Space Shuttle O-Ring** data set housed at the **Machine Learning Repository**. You can learn more about the data by visiting the site. Regress the launch temperature (degrees F) on leak-check pressure (psi) and comment on the predictive relationship, if any.
13. What is meant by the statement "**All models are context-dependent**" and why this is possibly the most important take-away from learning statistical models from a scientific point of view? Unpack and discuss the implications of this statement and why it is so important.
14. Discuss the difference between **hierarchical** and **stepwise regression**. How are they not the same? Is stepwise regression ever appropriate? If so, when?
15. Why are **mediational analyses**, especially in questionable scientific contexts, a potential philosophical quicksand when it comes to interpretation? Discuss.

8

Logistic Regression and the Generalized Linear Model

CHAPTER OBJECTIVES

- Learn why dependent variables that are not continuous pose challenges for traditional ordinary least-squares regression analysis and why the generalized linear model is better equipped to deal with such variables.
- Why logistic regression is well-suited to deal with binary response variables, either naturally-occurring or artificially created.
- Understand the difference between the odds and the logarithm of the odds, known as the logit.
- How to run a logistic regression in Python and interpret the most relevant output.
- How to run a multiple logistic regression in Python and interpret odds for a variable in the context of other predictors.

Thus far in this book, we have considered a number of different models and statistical methods such as *t*-tests, ANOVA, regression, among others. While at first glance it may not be apparent, the fact of the matter is that most of the models we have considered thus far have assumed the response variable to be more or less **continuous** in nature. For example, when we modeled verbal scores as a function of quantitative scores, we assumed that the verbal variable was drawn from a population distribution that is more or less continuously distributed. That did not imply that verbal had to be normally distributed. It did not. However, it had to have **sufficient distribution in its scores** such that we could at least think of it as measurable along a continuous scale.

Recall the definition of continuity as defined by mathematicians discussed earlier in the book. Though calculus and real analysis define continuity quite formally and very rigorously, the essence of continuity is that any score for a variable is possible along the infinitely dense range of the variable. For example, for an IQ variable to be continuous in the range of, say, 0 to 200, it implies that any score within these limits is theoretically possible. That is, one could obtain a score of 95.02748 as a possible IQ, just as one could obtain a score of 110.08375, and so on for an infinite number of decimal places. And as discussed earlier in the book, true continuity in research variables

Applied Univariate, Bivariate, and Multivariate Statistics Using Python: A Beginner's Guide to Advanced Data Analysis, First Edition. Daniel J. Denis.

usually has to be **assumed** rather than the variable actually possessing a true continuous distribution. Why is this the case? Simply because **true continuity in its deepest theoretical sense implies that any particular value in such a distribution has a probability of zero of occurring.** This is the case since if you reach into a distribution deemed continuous, and if the variable is actually truly continuous, then it should be impossible to actually "catch" or "grab" any particular value. Akin to snorkeling in the weeds under water (imagine you are wading through the weeds as a deep-diver would), it becomes impossible to grab any of the weeds when you reach for a single one. The flow of the water has the effect of moving the particular weed aside just as you reach for it. In other words, it is impossible to catch or secure a particular value. Hence, when we dive into a distribution of a continuous mathematical variable, we likewise cannot secure a particular value. More formally, we say that **in the limit, the probability of any particular value is equal to 0**. As before, the idea of limits is beyond the scope of this book, but you can think of a limit in this case that as the "slices" toward the particular value get smaller and smaller, that is, as we "narrow in" on the distinct value, the probability of that value being obtainable **shrinks** toward zero (i.e. it has a limiting value of zero). In an analogous sense, as you walk toward a flat wall, the distance between you and the wall gets smaller and smaller, until in the limit, any distance greater than zero has a probability of zero. This is a powerful way to conceive of the limit concept.

An even more intuitive definition of continuity and one that everyone is familiar with is that of not lifting one's pencil off the page as one draws a straight line from one endpoint to another. The line is deemed continuous if at no point along the line is the pencil lifted, as shown:

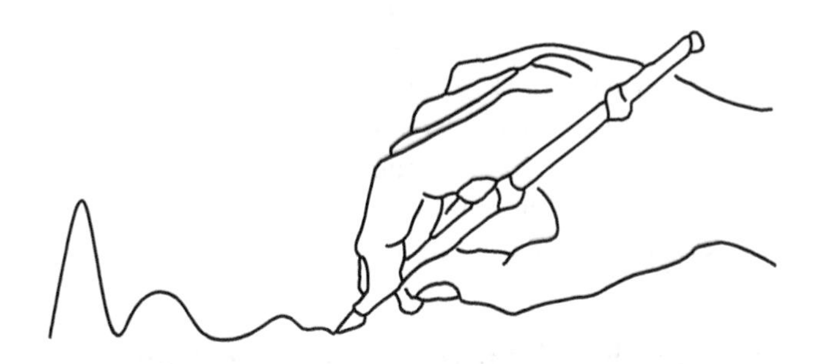

As discussed earlier in the book, whether one is working with continuous or non-continuous variables is a **major decision node** to deciding which analysis one should perform. For instance, if you ever work with a statistical or methodological consultant, you may spend a great deal of time in the planning phase of your experiment or study on the **measurement problem**, which is all about deciding how best to assess what you are seeking to measure. Measurement should not be taken for granted, and especially in sciences that are less "hard," such as psychology, education, sociology, economics, and others, the measurement problem should require a great deal of your attention. Far too often students and researchers skip the measurement problem and proceed directly to performing statistical analyses on empirically measured variables that, when critically evaluated, likely do a poor job at assessing the underlying construct(s). This is not the complete fault of researchers, since many constructs in such sciences are inordinately difficult (or even impossible) to measure and often require self-report to measure them, which is laden with difficulty since people are not always (seldom?) the best judge of their own behavior.

To carry on with our example, if you are to measure something like intelligence, how best should you measure it? Among the several issues you will encounter will be whether to treat the response (dependent variable) as either a continuous variable (at least on a theoretical level) or a non-continuous one. For instance, will you seek to measure individuals' precise scores on IQ or will you attempt to simply **classify** them into "low," "medium," and "high" IQ? How you **operationalize** the measurement of your variable will likely have a huge impact and influence on not only what statistical model you choose for your analysis, but also regarding what conclusions can be drawn or not drawn from the research. In fact, **one reason why some studies find effects and some do not may have less to do with whether a phenomenon truly exists in the population and more to do with how the variable was operationalized!** This itself should make understanding statistical modeling paramount to your work. Though a complete discussion of this sort is well beyond the scope of this book, you should never take the results of a study at "face value." Instead, always "dig" into how variables were **measured**, **defined**, **operationalized**, etc., as this could play a huge factor in determining whether the study yielded fruitful findings. How one study measures a variable may not even be remotely similar to how another study measures it. Hence, when you read the headline, "Those with higher IQ are more likely to earn more money," or some such, that statement or conclusion is only as good as the measurement that went into the study. **Be critical**. Even in medical and biological fields debates over treatment effectiveness from study to study can be quite difficult to discern as a result of the myriad of methodological factors influencing the given study. As we alluded to earlier in the book with regards to the pandemic of 2020, what does the number of deaths "due to COVID" actually mean? How are those dying **from** COVID identified versus those dying **with** COVID but ultimately due to a different underlying condition? How is the **causal mechanism** identified, if at all, regarding the cause of death? These questions are important! Complex statistical analyses on poorly conceived or measured variables is at best meaningless and at worst unethical and misleading. We explore this question a bit further in what follows.

8.1 How Are Variables Best Measured? Are There Ideal Scales on Which a Construct Should Be Targeted?

As mentioned, a full discussion of measurement is well beyond the scope of this book. For our purposes here, we wish only to comment a bit further and deeper on the measurement issue given that the model we will consider shortly will assume a binary-distributed response. However, before assuming our variable binary, the question naturally arises on how best to measure variables in general. As mentioned, this is a central problem in fields such as psychology and other social sciences where the measurement question is, or least should be, the most difficult but urgent methodological issue to resolve.

Should one attempt to measure variables on a continuous scale or should you "bin" them into categories? While there is no definitive and easy answer to this question, as a rough guideline, in most cases **measurement on a continuous scale is probably**

the best ideal to shoot for. Now, there is a good reason why "rough guideline" is mentioned here. In some cases, measurement on a continuous scale can do more harm than good and is definitely not appropriate, even if at first glance it may appear as an attractive option. In most cases, however, it is usually best to side with continuity or at least "pseudo-continuity" than implement a dichotomous or polytomous choice. Yet for some scales, such as **Likert scales** in which respondents are asked to rate their opinion or preference by selecting one of a range of whole numbers, one must be cautious about extracting too much meaning out of difference choices if the phenomena you are measuring has a great deal of measurement error or "random shocks" (i.e. chance) to it.

As an example, one that can be generalized to many Likert scales, suppose you would like to measure how much a person enjoys pizza. While you can informally ask people how much they enjoy it, you know that seeking scientific measurement of the phenomenon is a wiser approach. Hence, you decide to construct a **Likert scale** with values ranging from 0 (dislike pizza) to 10 (enjoy pizza immensely, eat it all day). Now, what do these numbers mean? In the complete abstract sense, there is no debate. That is, the real line with integers on it clearly designates a scatter of different numbers. However, you are not doing mathematics. You are doing science, and things get a whole lot more complicated (in this sense only) fast. For example, if a participant rates "0" while another rates "10," you can probably be quite certain, assuming your participants are not lying, that one person enjoys pizza more than the other. It seems reasonable that the person circling 10 is a pizza lover, while the one circling 0 is likely not. We could probably be quite certain of this distinction.

However, if one person rates 6 while another 7, can you say the same thing? Can you even draw a similar conclusion but to a lesser degree? While numerically they have circled different choices, **whether one actually enjoys pizza more than the other should not be assumed simply because one rated higher than the other**. Doing so would constitute a major mistake and is one reason (among many) why social science is sometimes looked up with suspicion. It may be that these two individuals actually enjoy pizza to the same exact degree in reality, but may have differing **response styles**. By "response styles" here, we mean some people like to circle higher numbers than others to express their pizza-liking, even if they like pizza to the same degree (or even less) as the person circling a lesser score. Given this, how then could we ever know for certain who enjoys pizza more? Probably only through **behavioral observation**. That is, frequency of pizza eating over a given period of time or other behavioral measure would probably provide much more realistic and precise insight into true pizza preference than actually doing something quite trivial and asking subjects to respond to a questionnaire and assuming an abstract numerical system represents true differences on a substantive level. **Self-report, in general, is an excellent way to measure how an individual responds to a questionnaire. It is not necessarily a measure of the phenomenon you are seeking to learn about, unless that phenomenon is primarily about how people respond to questionnaires**! Case in point, the 2016 presidential election!

If you remember anything about social measurement, the above should be it. Never, ever assume that the behavioral action of someone responding on a questionnaire necessarily correlates to what you are ultimately seeking to measure. Even lying aside (good scales will have lie scales built in to attempt to detect dishonesty, to varying degrees of success), human beings are not necessarily **aware of their own behavior**, which is why you are presumably studying them in the first place. People might also **project** in the opposite direction. For instance, for individuals trying to overcome their attachment to money, they may minimize how much they value money on a questionnaire. The truth is that the person likes money a great deal, but in an effort to distance themselves from the addiction, responds with more of their "**intention**" than with what is actually true in reality. Such confounds exist aplenty in **psychological measurement**. The only thing you can be sure of, for certain, is that the person responded that he or she did not value money to a great deal because that was what was circled on the page. If you are able to correlate that to lifestyle, expenditures, and other things money-related, you may be able to substantiate that measurement with behavioral observation. Self-report behavior can often be quite different from other behavior, but it is usually the "other behavior" that we consider more valid. As a scientist, your job is to figure out a way to tap into that and measure it properly (good luck, it will not be easy!).

Despite our caveats on the measurement of difficult phenomena, still, as mentioned, attempting to measure on a continuous scale rather than a dichotomous or polytomous one is probably usually the best first approach or attempt. You can also "bin" your data afterward if the results suggest that the underlying variable is more categorical than you once thought. For example, asking people how many car accidents they got into this month, while you may think it might contain a range of numbers, it is likely and more realistic that it will contain those having zero car accidents and those having one. That is, the true underlying variable that is even measurable is likely to be **binary** in this case. At the same time, some phenomena, at least at first glance, naturally implies discrete measurement. Whether you passed or failed a course in college is naturally discrete, though your grade is not necessarily so, as the measurement of "success" in the class can be broken down further into letters A, B, C, and so on, just as the letter grades can be broken down into actual measured scores (e.g. 70, 75, 80, etc.). The lumping into categories is probably not best here from a measurement perspective, and statistical analyses of such data would be best conducted on the actual measured scores. It is much easier to "bin" after the fact than initially. So, in an exploratory sense, you might first "try" for continuity, and then, if that fails, settle for dichotomous or polytomous coding.

8.2 The Generalized Linear Model

Based on our discussion thus far, it may seem at first glance that when working with a non-continuous response variable, one could never even conceive of implementing a model that comes anywhere near close to resembling the linear regression model of previous chapters. Since the regression model assumes continuity of the response, it would appear that we would require a completely different model to analyze response variables that are binary or polytomous. While there is truth to this, as we will see, the stepping stones to this new class of models will surprisingly have parallels to the original regression model. The **generalized linear model** (McCullagh & Nelder, 1990) **is a class of models that incorporates binary and polytomous models into its framework, and applies a transformation to the binary or polytomous variable**

such that it resembles a variable that is linear. Hence, the generalized linear model can be thought of as a way of "linearizing" a non-linear model such that it behaves more similarly to a linear model. In the case we will consider shortly, the **logistic regression model**, a special case of the **generalized linear model**, will effectuate a transformation on something called the **odds**, and generate a linear response from this transformation. From there, as we will see, the interpretation of the resulting coefficients will be similar to that of the traditional linear model. Hence, the name "generalized" linear model naturally encompasses the linear model as a special case, but, as we will see, also includes non-linear responses. The most popular of these special cases is the logistic model, which we now consider in some detail.

8.3 Logistic Regression for Binary Responses: A Special Subclass of the Generalized Linear Model

Binary logistic regression is one of the simpler and "easier to digest" of the generalized linear model class, and is also probably the most popular in the sciences. If you understand the basics of this model, understanding other models in the generalized linear model class should not be that difficult. Logistic regression works by transforming a binary response into a variable that is continuous and operates its analysis on this newly defined variable, making it much more similar to ordinary least-squares (OLS) regression than when the variable was binary.

The first step to understanding logistic regression is to define what is known as the **odds**:

$$\text{odds} = \left(\frac{p}{1-p}\right)$$

where p is the probability of an event occurring, while $1-p$ is the probability of an event not occurring. The odds are computed for a **mutually exclusive event**, meaning that the event occurs or does not occur on any given trial. One of these two possibilities must occur. And since by the rules of theoretical probability the probability of the total sample space must equal 1.0, it stands therefore that $p+(1-p)=1.0$. That is, the probability of the event occurring and not occurring on a single trial must sum to unity. You can think of odds as being "centered" at a value of 1.0, which in this case implies that the probability of the event occurring is equal to the probability of the event not occurring. Notice that this holds only if p is equal to 0.5, since

$$\text{odds} = \left(\frac{p}{1-p}\right) = \left(\frac{0.5}{1-0.5}\right) = 1.0$$

Hence, if the odds of an event occurring are 1 to 1, then it implies that the probability of that event is equal to 0.5. It is easy to see that as p increases, the odds in favor of the event are greater than 1.0. For example, for $p=0.6$, the odds in favor of the event are $0.6/0.4=1.5$ to 1. For $p=0.9$, the odds in favor of the event are 0.9/0.1 = 9 to 1. On the opposite end, as p decreases, the odds in favor of the event become less than 1.0.

The odds are useful here because our response variable is dichotomous and the categories are mutually exclusive. For instance, if we are trying to predict recovery from COVID-19 based on several predictors, we may be interested in the odds of recovering versus not recovering, or the odds of surviving versus not surviving. In both of these examples, the response variable has two possibilities, and the occurrence of one possibility precludes the other. That is, if you recover, then you cannot "not recover." If you survive, then this implies you did not die. This is where the concept of mutual exclusiveness fits into things here.

Now, the odds are one thing, but they are built on a binary principle of p vs. $1-p$. Regression analysis is not ideally suited to operate on a binary response variable. However, not all hope is lost. What if we were able to somehow transform the odds into something that is linear? This is exactly the approach we will take for the logistic regression. That is, our goal will be to transform the odds into a variable that we could perform a regression on, analogous to performing a least-squares regression on a continuous variable. The key to logistic regression then will be to transform the odds into something that is linear, which we will call the **log of the odds**. More specifically, we take the **natural log of the odds**:

$$\ln(\text{odds}) = \ln\left(\frac{p}{1-p}\right)$$

The natural log of the odds, given as **ln(odds)**, is also known as the **logit**. Referring to our previous example, for an odds of 1.0, the log of the odds is therefore equal to 0. Hence, a logit of 0 represents "even odds" and a probability of 0.5 for the even occurring. Notice that as p is greater than 0.5, the log of the odds will be greater than 0. For p less than 0.5, the log of the odds will be negative.

Hence, we have seen how we can go from probabilities to odds and from odds to the logged odds, or the logit, for short. The key point to logistic regression is in this transformation. Once we obtain the logit, we can interpret it as we would an ordinary regression coefficient, only now we are no longer interpreting the coefficient in the original units of the response variable. Instead, we are interpreting it in the units of the logit. For example, in a previous chapter we regressed verbal score onto quantitative score and obtained a regression coefficient of 0.56. This coefficient was computed on the original units of the variable, and hence our interpretation was that for a one-unit increase in quantitative, we could expect, on average, a 0.56-unit increase in verbal. Had the analysis instead been one of logistic regression, the 0.56 would be in units of the logit and our interpretation would have been for a one-unit increase in quantitative, we would expect, on average, a 0.56 unit increase in the logit of verbal. This is what we mean by the coefficient being computed on different units. Now in the case of verbal on quantitative, we would not be doing a logistic regression in the first place, and hence would never be interpreting the 0.56 figure in terms of a logit increase. Our example here is merely meant to compare how OLS regression compares to logistic in interpretation.

We can easily demonstrate the transformation of probabilities to odds and odds to logits directly in Python. We start with a probability of 0.5 and compute the odds:

```
odds = 0.5/(1-0.5)
odds
Out[102]: 1.0
```

Notice that, here, the odds are computed as p to $1-p$, where p in this case is equal to 0.5, and hence $1-p$ is equal to 0.5 also (given as `1-0.5` in the code). Not surprisingly, the odds are equal to 1.0, as we previously said they would be for a probability of 0.5. We now compute the log of the odds. For this, we will use **numpy**:

```
import numpy as np
np.log(odds)
Out[106]: 0.0
```

Python tells us the log of the odds are equal to 0, which again, makes perfect sense. Notice also that if the probability of an event occurring is equal to exactly 1.0, then the odds, computationally at least, are **undefined**, and Python will let us know such by the following:

```
odds = 1.0/(1.0-1.0)
ZeroDivisionError: float division by zero
```

The odds are undefined because **division by 0 is undefined**. Note carefully that this is different from the odds equaling 0. That is, "undefined" does not equate to equaling zero, and can be easily misunderstood. If you try dividing 0 by 2 on your pocket calculator, you will get 0. On the other hand, if you divide 2 by 0, you will get an error message, "E" informing you the division cannot be executed. That is, it is undefined. On a more theoretical level, one might say that as the probability of an event nears 1.0, the odds approach **infinity**. However, since infinity is more of a **concept** than an actual number, speaking of 1/0 as an actual number for the odds is very awkward. However, as $p \to 1$, the odds $\to \infty$.

If the probability of an event is equal to 0, then the odds are equal to zero:

```
odds = 0.0/(1.0-0.0)
odds
Out[109]: 0.0
```

If we can go from odds to logits, we should be able to go in the reverse direction as well, from logits to odds. However, how does one "undo" a logit? You may recall that the inverse operation of taking a logarithm is to **exponentiate**. For example, consider the logarithm of the number 8 to base 2:

$$\log_2(8) = 3$$

The log to base 2 of the number 8 is equal to 3. Why? Because 2 raised to the exponent 3 is equal to 8. In this case, 2 is the base and 3 is the exponent we raise to, to get 8. We can easily demonstrate this in Python:

```
2**3
Out[112]: 8
```

Notice that in Python, exponentiation is denoted by `**`. This is contrary to R software, where it is denoted by `^`. Notation for mathematical operations in software is unfortunately not standardized across programs so you must always be sure of what you are computing. You may think you are computing one thing, but in reality it may be something entirely different! Be slow and cautious in your work as misunderstandings regarding what the program is actually computing are relatively easy to come by. **Always double (and triple) check**.

Above, we exponentiated to base 2 simply as an example to demonstrate or remember how to compute a logarithm. In logistic regression, we will not be exponentiating to base 2, but instead to base e, which is the **exponential function**, where e is a constant equal to approximately 2.71828 (called "Euler's number"). We emphasize "approximately" here because e is an **irrational number**, and hence has no terminating or periodic (i.e. repeating) decimal. Hence, if you want to obtain the odds from the logit, then you should **exponentiate the logit**. Again, as a simple example, suppose the odds in favor of the event are 2 to 1. The log of the odds are therefore

```
np.log(2)
Out[113]: 0.6931471805599453
```

That is, the logit is 0.69. Now, to get the odds back, we exponentiate the logit, which means we raise e to the power of 0.69:

```
np.exp(0.6931471805599453)
Out[116]: 2
```

We can see that when we exponentiate the logit, we get the number 2, which is the original odds. Taking the log and exponentiating are reverse operations. We now have all the technical precursors in place to survey logistic regression in Python, which we do next.

8.4 Logistic Regression in Python

The data we will use comes from the accident and explosion of space shuttle **Challenger** in 1986. On January 28, 1986, the space shuttle lifted off from Cape Canaveral, Florida at 11:39 am, EST. Shortly into the flight, the shuttle exploded and all crew were lost in the tragedy. A follow-up investigation revealed that the o-rings on the shuttle's boosters that are designed to keep the fuel from seeping out of the booster, failed, allowing fuel to ignite and cause the disaster (Figure 8.1). Presumably, the o-rings expanded due to the temperature at which the shuttle was launched, which

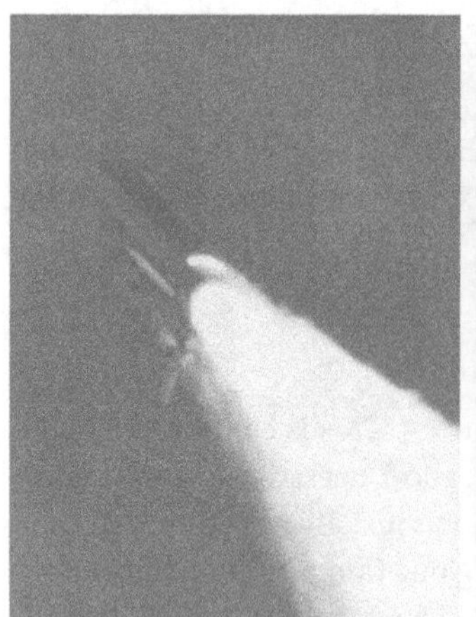

Figure 8.1 Challenger shuttle disaster of 1986. Challenger in flight with flame exiting the right booster (left). Explosion of the shuttle moments later (right).

was a frigid 31 degrees, much colder than any previous launch. The cold temperature apparently caused the o-rings to expand, allowing fuel to leak out onto the main booster, causing the explosion.

The following are the Challenger data that include data from prior launches. These are measurements as to whether or not there was any "blow-by" that was recorded as a failure. A blow-by is crudely defined as gases seeping beyond the seal of the booster. Hence, in the following, "1" is denoted as a **failure** (i.e. there is blow-by), while "0" is denoted as **no failure**. The temperature variable (`temp`) is in degrees Fahrenheit. We build the dataframe in Python (we print only the first six cases in the output):

```
import pandas as pd
data = {'oring' : [1, 1, 1, 1, 0, 0, 0, 0, 0, 0, 0, 0, 1, 1, 0, 0,
0, 1, 0, 0, 0, 0, 0],
'temp': [53, 57, 58, 63, 66, 67, 67, 67, 68, 69, 70, 70, 70, 70,
72, 73, 75, 75, 76, 76, 78, 79, 81]}
df_challenger = pd.DataFrame(data)
print(df_challenger)
    oring  temp
0       1    53
1       1    57
2       1    58
3       1    63
4       0    66
5       0    67
```

We first confirm the number of failures (1) and no failures (0) on the entire data set (not just the six data points listed above):

```
df_challenger['oring'].value_counts()
Out[124]:
0    16
1     7
Name: oring, dtype: int64
```

Hence, we can see from this that we have a total of 16 booster successes and 7 times the booster failed. That is, in this data set, 16 times there was no "event" of seepage or blow-by, and 7 times there was. A plot will allow us to visualize the distribution of counts. For this, we will use **seaborn**:

```
import seaborn as sns
sns.countplot(x='oring', data = df_challenger, palette='hls')
Out[126]: <matplotlib.axes._subplots.AxesSubplot at 0x1cc59710>
```

We can see the counts for o-ring failures (1) represented in this bar graph (below). We now run the logistic regression. Though there are a few ways of computing a logistic regression in Python, we will use **statsmodels** (sklearn can also be used):

```
import statsmodels.api as sm
```

We will add a constant to our model via `sm.add_constant()`, define our `y` vector as equal to the `oring` variable, and `X` as equal to `temp`:

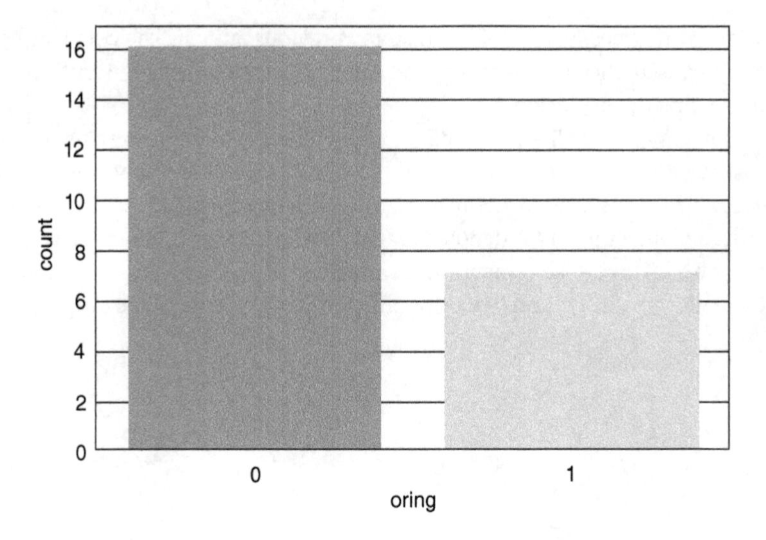


```
y = df_challenger['oring']
X = df_challenger['temp']
X_const = sm.add_constant(X)
```

We now specify our model as follows, and request model fit statistics through `model.fit()`:

```
model = sm.Logit(y, X_const)
results = model.fit()
Optimization terminated successfully.
         Current function value: 0.441635
         Iterations 7
print(results.summary())
                           Logit Regression Results
==============================================================================
Dep. Variable:                  oring   No. Observations:                   23
Model:                          Logit   Df Residuals:                       21
Method:                           MLE   Df Model:                            1
Date:                Sun, 20 Sep 2020   Pseudo R-squ.:                  0.2813
Time:                        12:14:39   Log-Likelihood:                -10.158
converged:                       True   LL-Null:                       -14.134
Covariance Type:            nonrobust   LLR p-value:                  0.004804
==============================================================================
                 coef    std err          z      P>|z|      [0.025      0.975]
------------------------------------------------------------------------------
const         15.0429      7.379      2.039      0.041       0.581      29.505
temp          -0.2322      0.108     -2.145      0.032      -0.444     -0.020
==============================================================================
```

The intercept of the regression equation is equal to 15.04, while the parameter estimate for `temp` is equal to –0.23. The model reports a pseudo R-squared value (McFadden's pseudo R-squared) of 0.2813, where "pseudo" here implies the statistic should not be interpreted one-to-one with a traditional R-squared value as one would have in an OLS regression. That is, it is not truly a "variance explained" measure, but nonetheless can be used as a general measure of effect size for the model. For details on pseudo R-squared statistics, see Cohen et al. (2002).

We can see that `temp` is statistically significant ($p = 0.032$). The model equation is thus:

$$\textbf{predicted logit=15.0429-0.2322}\left(\textbf{temp}\right)$$

This is the estimated equation for predicting the response variable, which in this case recall is the **logit**. Recall that the response variable here is the logged odds (i.e. the logit) and not the naturally-occurring original binary variable we started out with. Let us use the equation to predict some logits for different values of `temp`, which recall is "`X`" in our model set-up. First, we will enter our equation into Python:

```
predicted_logit = 15.0429 - 0.2322*X
```

For example, our first temp value in our data file is 53. The predicted logit is thus:

```
predicted_logit = 15.0429 - 0.2322(X)
                = 15.0429 - 0.2322(53)
                = 2.7363
```

That is, 2.7363 is what we would expect for a temp of 53. The other values follow:

```
predicted_logit
Out[229]:
        temp
0    2.736648
1    1.808432
2    1.576378
3    0.416108
4   -0.280054
5   -0.512108
6   -0.512108
7   -0.512108
8   -0.744162
9   -0.976217
10 -1.208271
11 -1.208271
12 -1.208271
13 -1.208271
14 -1.672379
15 -1.904433
16 -2.368541
17 -2.368541
18 -2.600595
19 -2.600595
20 -3.064703
```

```
21 -3.296757
22 -3.760865
```

Notice that we go from positive numbers toward zero, then increasingly into negative numbers of greater absolute magnitude. The reason for this is because our temp data follows a similar trajectory (from smaller values of temp to larger). Indeed, a Spearman rho should detect this **monotonicity** and yield a correlation in absolute magnitude of 1.0:

```
from scipy import stats
stats.spearmanr(predicted_logit, X)
Out[232]: SpearmanrResult(correlation=-1.0, pvalue=0.0)
```

We see that the obtained correlation is equal to −1.0. The *p*-value is irrelevant, as we simply wanted to demonstrate that as values of temp decrease, values of the logit increase (and vice versa). A plot of this relationship clearly reveals the perfect linear trend:

```
import matplotlib.pyplot as plt
import numpy as np
plt.plot(predicted_logit, X)
plt.xlabel("predicted logit")
plt.ylabel("temp")
plt.title("temp as a function of logit")

Out[38]: Text(0.5, 1.0, 'temp as a function of logit')
```

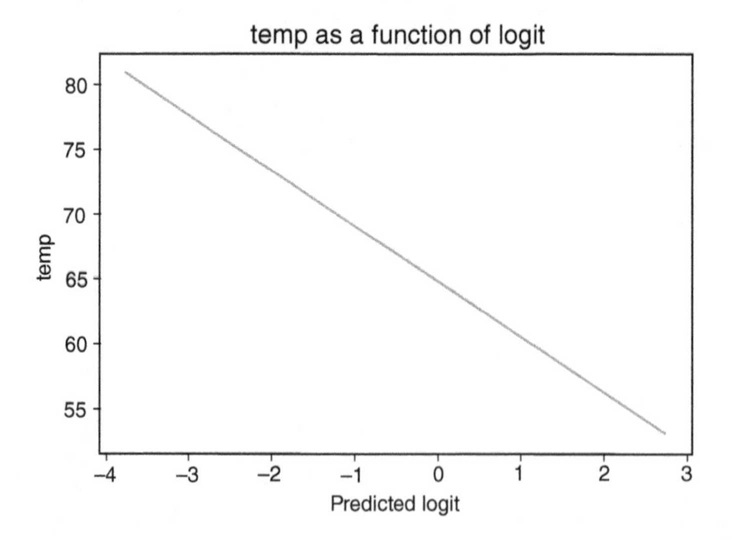

8.5 Multiple Logistic Regression

As when generalizing from simple linear regression to multiple, we can likewise generalize the simple logistic model to a multiple logistic regression. The data for this example comes from the **ISLR library** in R software, and consists of percentage returns for the S&P 500 stock index over the course of 1,250 days in the period beginning 2001 to the end of 2005. Let us first obtain the data and then describe its variables. If the data are not loading below, make sure Spyder, or whatever environment you are using, is searching the correct location for the file; for example, the `Smarket.csv` file for the following data I saved to my Desktop; hence, the top right of the Spyder interface looks like this:

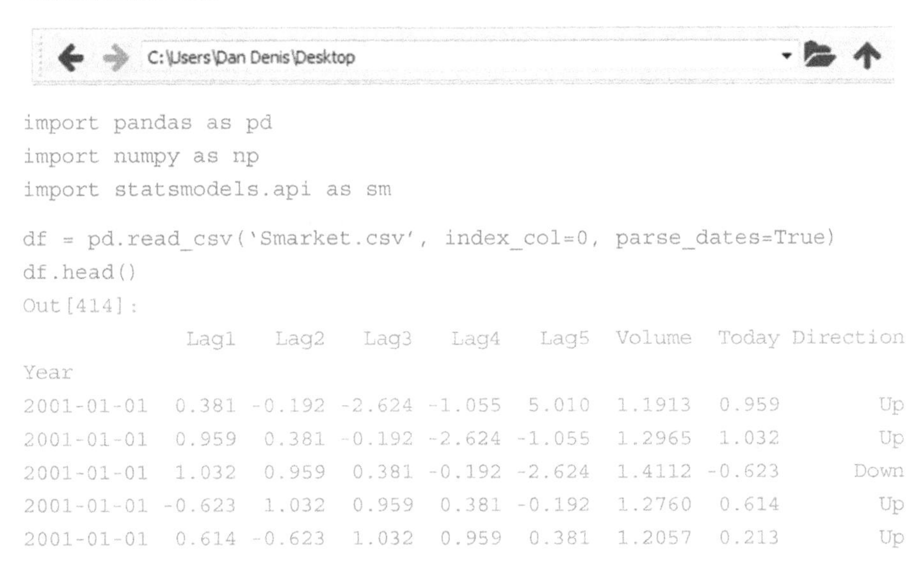


```
import pandas as pd
import numpy as np
import statsmodels.api as sm

df = pd.read_csv('Smarket.csv', index_col=0, parse_dates=True)
df.head()
Out[414]:
             Lag1   Lag2   Lag3   Lag4   Lag5  Volume  Today Direction
Year
2001-01-01  0.381 -0.192 -2.624 -1.055  5.010  1.1913  0.959        Up
2001-01-01  0.959  0.381 -0.192 -2.624 -1.055  1.2965  1.032        Up
2001-01-01  1.032  0.959  0.381 -0.192 -2.624  1.4112 -0.623      Down
2001-01-01 -0.623  1.032  0.959  0.381 -0.192  1.2760  0.614        Up
2001-01-01  0.614 -0.623  1.032  0.959  0.381  1.2057  0.213        Up
```

Here, `Lag1` through `Lag5` are the corresponding previous trading days. For each date, the percentage returns for each trading day are given. The volume (in billions) of the shares traded on the previous day of trading are also given. The variable `Today` is the percentage return on the given date, and `Direction` is the variable `Today`, but binary-coded into whether the market was up or down on the given day. Our goal will be to predict `Direction` based on the five lag variables as well as volume. Hence, our function statement is the following:

$$\textbf{Direction} = \textbf{Lag1} + \textbf{Lag2} + \textbf{Lag3} + \textbf{Lag4} + \textbf{Lag5} + \textbf{Volume}$$

We will again use **statsmodels** to perform the logistic regression:

```
import statsmodels.formula.api as smf
```

The following is the model we would like to run:

```
Direction ~ Lag1+Lag2+Lag3+Lag4+Lag5+Volume
```

To run the model, we use the `glm` function, which stands for **generalized linear model**, and specify the family as `binomial()`. To get the correct coefficients, we need to also adjust how Python is coding the binary response, and add an intercept:

```
change = np.where(df['Direction']=='Up', 1, 0)
model_constant = sm.add_constant(df[['Lag1', 'Lag2', 'Lag3',
'Lag4', 'Lag5', 'Volume']])
model = sm.GLM(change, model_constant, family=sm.families.
Binomial()).fit()
```

```
print(model.summary())
          Generalized Linear Model Regression Results
==============================================================================
Dep. Variable:                      y   No. Observations:                 1250
Model:                            GLM   Df Residuals:                     1243
Model Family:                Binomial   Df Model:                            6
Link Function:                  logit   Scale:                          1.0000
Method:                          IRLS   Log-Likelihood:                -863.79
Date:                Mon, 21 Sep 2020   Deviance:                       1727.6
Time:                        01:02:59   Pearson chi2:                 1.25e+03
No. Iterations:                     4
Covariance Type:            nonrobust
==============================================================================
                 coef    std err          z      P>|z|      [0.025      0.975]
------------------------------------------------------------------------------
const         -0.1260      0.241     -0.523      0.601      -0.598       0.346
Lag1          -0.0731      0.050     -1.457      0.145      -0.171       0.025
Lag2          -0.0423      0.050     -0.845      0.398      -0.140       0.056
Lag3           0.0111      0.050      0.222      0.824      -0.087       0.109
Lag4           0.0094      0.050      0.187      0.851      -0.089       0.107
Lag5           0.0103      0.050      0.208      0.835      -0.087       0.107
Volume         0.1354      0.158      0.855      0.392      -0.175       0.446
==============================================================================
```

We interpret the primary features of the output:

- The dependent variable is `Direction`, with categories `down` and `up`. We are modeling the category "`up`".
- The model family (`Binomial`) and link function (`logit`) confirm that we are working with a logistic regression.
- The total number of observations in the data file is equal to 1250.

Now, since this is a multiple logistic regression having many predictor variables, it is imperative to emphasize the interpretation of parameter estimates in the context of the model. For example, the coefficient for `Lag1` of –0.0731 is interpreted as follows:

> **For a one-unit increase in Lag1, we expect, on average, the logit on the response variable to decrease (because of the negative sign) by 0.0731 units given the context of the model, or, otherwise said, given the simultaneous inclusion of predictors Lag2 through to Lag5 as well as volume.**

Recall that just as was the case in least-squares regression, it would be incorrect to omit a statement about the "context of the model." That is, **the decrease of 0.0731 is not an expected decrease for Lag1**. Rather, it is an expected decrease for `Lag1` in the **context of also including Lags 2 through 5, as well as volume**. If you fail to mention the context of a model and simply cite a coefficient for a given predictor, you are misleading your audience and reporting incorrect information. **All models are context dependent. You must include the context of the model you ran when interpreting individual coefficients.**

The above are in units of the logit, or the "logged odds." We may prefer to interpret things in terms of **odds.** To get the odds for each predictor, recall that we are to exponentiate the logit. For example, for `Lag1`, the computation is

```
import numpy as np
np.exp(-0.0731)

Out[225]: 0.9295078745906393
```

The odds are thus 0.929 for `Lag1` holding all other variables in the model constant (or again, "given the context of the model"). Since an odds of 1.0 represents even odds, the number 0.9295 represents a **decrease**. That is, for a one-unit increase in `Lag1`, the likelihood of the event decreases (i.e. it is below a probability of 0.50 for our data). This is challenging to make sense of in terms of odds. Probabilities here will be much more intuitive. Fortunately, we can also convert this odds to a probability. We do this by dividing the odds by **1 + odds**:

```
prob = np.exp(-0.0731)/(1 + np.exp(-0.0731))
prob
Out[227]: 0.4817331335265174
```

As we can see, the probability is 0.4817 of being in the group labeled 1 (i.e. "1" corresponding to an upward direction in our data). The probability of 0.4817 makes good sense too, since the odds are just below 1.0. That is, had our computed probability come out much higher for some reason (e.g., 0.90), we would know that something was amiss, since the odds are actually lower than 1.0. It is always a good idea to conduct these kinds of informal **confirmatory checks** on your computations from time to time to make sure they agree with one another. If anything is amiss, "unreasonable" numbers sometimes appear and signal that a recheck of your work is in order. If something "doesn't seem right," it usually isn't (it is difficult enough ensuring things are correct even when they seem right!).

8.5.1 A Model with Only Lag1

As we emphasized earlier, it is incorrect to interpret parameter estimates without reference to the model. Why? Because if we did not include the other predictors, the parameter estimate for `Lag1` would change! Not convinced? We will include only `Lag1` to demonstrate this (we print only the model coefficients):

```
model_constant = sm.add_constant(df[['Lag1']])
model = sm.GLM(change, model_constant, family=sm.families.
Binomial()).fit()
print(model.summary())
==============================================================================
                 coef    std err          z      P>|z|      [0.025      0.975]
------------------------------------------------------------------------------
const          0.0740      0.057      1.306      0.191      -0.037      0.185
Lag1          -0.0702      0.050     -1.404      0.160      -0.168      0.028
==============================================================================
```

Note that the coefficient for `Lag1` has changed from its original value, from an estimate of –0.0731 in the original model to –0.0702 when used as the only predictor. The change in this case is slight, but the point is that **these coefficients do habitually change**. The *p*-value has changed from 0.145 to 0.160, as well as the confidence limits from the lower limit –0.171 to upper limit 0.025 in the full model to the lower limit –0.168 to the upper limit 0.028 in the one-predictor case. These changes can become quite impactful in other models. Again, whatever model you run, it should always be interpreted in the context of other predictors entered (or not entered) into the model. This is why **experimental designs** are so favored over correlational ones; we stand a better chance at controlling for these confounds. In correlational research, virtually almost anything and everything under the sun could be impacting your model. The problem of **omitted variables** in correlational designs is so relevant that many experimentalists will often not give correlational models much thought at all. There is a reason why experiments are usually considered the "gold standard" of evidence. Evidence that a COVID-19 vaccine works based on correlational evidence, for instance, is simply not good enough. Having said that, many times correlational designs are the best we can do because experimental control is either not do-able or entirely unethical. So long as their limitations are understood and appreciated, these models can still be very useful and should not be automatically disregarded.

Instead of obtaining output for the entire model as above, we can also request Python to print only information we immediately desire. For example, if you try `print(model.params)` and `print(model.pvalues)`, you will obtain the same information as earlier.

8.6 Further Directions

In this chapter we have surveyed the logistic regression model as a special case of the more generalized linear model. There are many more **link functions** that the generalized linear model can accommodate, such as a **Poisson link** or **negative binomial link**, among others. Fox (2016) is an excellent source for following up this chapter with a full treatment of the generalized linear model. A technique in machine learning and computer science, that of **support vector machines**, is yet another method whose purpose is to classify observations into categories, a kind of competitor to both logistic regression and discriminant analysis. For details, see James et al. (2013), or, for a much deeper treatment, see Izenman (2008).

Review Exercises

1. From a purely mathematical point of view, classic **logistic regression** features a binary response variable. From a scientific perspective, however, the decision of whether to consider a response variable binary or not is much more complicated. Why? Discuss how measurement issues are at the forefront of such decisions.
2. Why is it the case that true **continuity** is a **philosophical notion** and not a **scientific** one? Explain.

3. In a sentence or two, describe how **linear models** can be subsumed under the more **generalized linear model** framework. For example, how is a linear model also a generalized linear model?
4. If the **probability** of rain today is 0.80, and it must either rain or not rain, what are the **odds** of rain?
5. Justify that the odds are "**centered**" at a value of 1.0. That is, why is it that an odds of 1.0 indicates an even probability for or against an event from occurring?
6. Does the **odds** have an upper bound? Why or why not? Justify.
7. Is an **odds** less than zero possible? Justify by reference to the formula for the odds.
8. Can a **logit** be negative? Why or why not? Show, using the formula, whether this can or cannot be the case.
9. For what value of probability are the odds **undefined**? What does it mean to say they are undefined anyway?
10. Consider the **Heart Failure Clinical Records** data set in the **Machine Learning Repository**. Perform a logistic regression attempting to classify sex (woman or man) based on the age of the patient (in years) and serum sodium in the blood. Interpret the odds of heart failure given that you are a male relative to if you are a female in the context of serum sodium in the blood. What is the probability of heart failure given that you are a male compared to if you are a female?

9

Multivariate Analysis of Variance (MANOVA) and Discriminant Analysis

CHAPTER OBJECTIVES

- Learn how multivariate analysis of variance (MANOVA) is different from the analysis of variance (ANOVA) model.
- Why you should not necessarily perform MANOVA simply because you have several response variables.
- Understand why tests of significance in MANOVA are very different and more complex in configuration than in ANOVA, and why there is not only a single *F*-test as in ANOVA.
- Understand how the covariance matrix plays a fundamental role in the multivariate realm.
- Be able to compute and interpret an effect size measure for MANOVA.
- Why a multivariate model result does not necessarily break down into individual univariate findings.
- Learn what a discriminant function analysis (LDA) is, how it can be considered the "reverse" of MANOVA, and why it is sometimes used as a follow-up to MANOVA.
- Learn how to compute discriminant function scores and understand classification results based on derived discriminant functions.

In the models we have surveyed thus far, whether it is analysis of variance (ANOVA) or regression, both of these featured a **single dependent** or response variable. In both ANOVA and regression, the response variable was assumed to be a single continuous variable. Statistical models that analyze only a single dependent or response variable are generally known as **univariate models** for this reason. This is true regardless of how many independent variables are featured in the model. For example, even if an ANOVA model has three factors as independent variables, the model is still traditionally considered to be univariate since it only has a single response. Likewise, a multiple regression model may have five predictors, but is often still considered univariate since it only features a single response (though multiple regression is also often discussed under the heading of multivariate models due to its **multivariable** (i.e. several variables) nature).

Applied Univariate, Bivariate, and Multivariate Statistics Using Python: A Beginner's Guide to Advanced Data Analysis, First Edition. Daniel J. Denis.

Thanks to developments in statistics in the early 1900s, the so-called "boon" of mathematical statistics and inference with the works of R.A. Fisher, Karl Pearson, Harold Hotelling and others, techniques were developed that allowed scientists to evaluate statistical models featuring more than a single response variable simultaneously. But why would a researcher want to analyze a model having several responses in the same model? That is a very good question and one that should be considered very carefully before deciding that a multivariate model is suitable. Before introducing and discussing these models further, one would do well to issue a few caveats on the designation of "multivariate" and briefly explore just why a researcher would want to consider a multivariate model in the first place. Many times, researchers conduct multivariate models when they would be better off with univariate ones. We address these issues next.

9.1 Why Technically Most Univariate Models are Actually Multivariate

We have defined multivariate models as those in which more than a single response variable is analyzed at the same time. However, we should note that this specification for what is multivariate vs. univariate is more of a **nomenclature** and **convention** tradition than a true distinction between univariate from multivariate at a technical level. For example, suppose we choose to model verbal ability and quantitative simultaneously as a function of training program. Perhaps some individuals were trained in one program and others a different one and we would like to see if there are **mean vector differences** (we will discuss shortly what this means) between the two training programs on both variables considered simultaneously. Though this is, by definition, a multivariate model since it has two response variables, notice that if we "flipped" the model around, such that the training program is now the response and verbal and quantitative abilities the predictors, the model could then be amenable to **logistic regression** or, as we will see, **discriminant analysis**, in which case, there is only a single response variable. Hence, what was deemed multivariate now becomes univariate very quickly, yet at their technical levels, it is a fact that these models cannot be that different from one another. In fact, they are not, as we will later briefly discuss in more detail.

The above similarities and distinctions also hold true for regression analyses. Consider two response variables as a function of a predictor. Since there is more than a single response variable, the model would rightly and formally be considered a **multivariate regression**. However, if we flipped the model around such that the single predictor is now the response, and the two responses are now predictors, the model is simply a **multiple regression**. Though the models are not identical in terms of their technical details, they are, in reality, not that far off. Hence, realize that whether you are analyzing univariate models or multivariate models, the mathematical details underlying these models are often remarkably similar, even if different elements of each model are focused on in different applications. The theoretical or research drive behind each one may be quite distinct. Given their technical similarities, **your choice of model for your research needs to be driven primarily by substantive goals**.

9.2 Should I Be Running a Multivariate Model?

There is a long tradition in statistics, especially in fields that quite often do not conduct rigorous experiments with control groups, etc., to favor methods that are statistically very complex over simpler ones. Thousands of research papers employing the latest "cutting-edge" and very complex models can be found in journal articles and other publications, so much so that it is probably a reasonable bet to assume that journal editors and reviewers will be more "impressed" if you submit a complex model than they might be with a simpler one. In this sense, it is probably the case that multivariate modeling is perceived to "advance" science in some fields much further than much simpler models. But is this true? **Does running a complex multivariate model necessarily afford a study of greater scientific merit or bring us any closer to understanding reality?** Hence, instead of starting out with why you might want to run a multivariate model, let us instead begin with why you would **not want to run such a model**. The following are the **wrong** reasons for conducting a multivariate model:

- You have several response variables at your disposal and therefore, by a "rule of thumb" that you once learned from "cookbook" teaching or reading, believe that conducting a multivariate model is the right thing to do.
- You are doubtful of running a multivariate model, but nonetheless do so because you think it will make your dissertation or scientific publication appear more "rigorous" and "advanced."
- You believe that multivariate models are always better than univariate models regardless of the research context, since univariate models are "out of fashion" (i.e. "Nobody does ANOVA anymore, multivariate models are the 'hip' thing to do").

All of the above reasons and rationales for performing a multivariate model are 100% wrong and incorrect! First off, as a general comment, only rarely should you follow any "rules of thumb" in statistics. The decision trees that fill introductory statistics and research books that tell you which analysis is correct given this or that many response and predictor variables, should be used, if at all, as simply a **guide** to choosing the most appropriate model, and nothing more. **No decision tree of this sort can tell you which model you should be running. Only you can make that decision based on your research question and knowledge of the area under investigation (and, if needed, with the advice of a statistical consultant).**

Second, a complex multivariate statistical analysis does not imply your research is more "rigorous" than if you conducted simpler analyses. **A dissertation or research publication with a poor research design and poor measurement of variables will be just as poor if you subject it to a complex statistical analysis.** Aesthetically pleasing it may be statistically, on a scientific basis it will largely be a waste of time, and, worse yet, will serve to confuse your audience regarding whether you found or did not find something in your data. Never run a complex statistical model simply to make a fashion statement, not if you want to be a serious scientist, that is. On the other hand, if your goals are social and political such that you wish to make it seem like you are advancing science through sophisticated statistics, then by all means, fit the most complex model you can find and you may just fool enough to make them think you did something of scientific value. It is hoped, however, that **good science is your first priority**, and with this in mind, you should always choose your statistical model based on well thought out research questions, not based on technical complexity. As a scientist, your first priority is to contribute to the literature and to communicate a scientific finding, not appease or impress your audience with technical complexity. To some, you may succeed in pulling the wool over their eyes, but to others, they'll see it as it is, a **smokescreen** for a poorly run study with no experimental controls and poorly measured variables.

Finally, univariate models are never somehow "out of fashion" now that complexity rules the day in statistical modeling. For instance, in finding a cure for COVID-19, **experiments** are much more likely to use relatively simple univariate models. A 10-page paper on the cure for COVID-19 communicates more in terms of scientific utility than a 100-page paper on a poor research design flooded with multivariate and other advanced statistics. **The research design and quality of your study or experiment is always the first step to good science.** Now, from a statistical and mathematical point of view, multivariate models are indeed much more interesting and fun to work with. Our point is merely that from a scientific point of view, multivariate models do not necessarily lend more scientific credibility to your project. You are not doing pure mathematics, nor are you doing theoretical statistics. You are doing **science**, and Occam's razor and the principle of parsimony must always govern your work. **Simpler is usually better.**

The correct justification for performing a multivariate model, multivariate analysis of variance (MANOVA) in this case, is because your research question demands analyzing a linear combination of response variables at the same time. However, when might you want to do this? Let us consider a classic example, one in which a multivariate model is well-suited and appropriate. Suppose you have scores on verbal, analytical, and quant abilities on some psychometric test, such as the IQ measure we have alluded to in this book. It might make good research and theoretical sense to combine these variables together somehow, since a composite score of **verbal + analytic + quant** is probably representative in some sense of a more complete variable or construct, namely the variable you wish to analyze in this case, IQ. Hence, we can say that

$$\mathbf{IQ = VERBAL + ANALYTIC + QUANT}$$

This is what is known as a **linear combination** of variables. A linear combination is simply a sum of variables, each weighted by a scalar, which is a number. In mathematics, the general form for a linear combination is the following:

$$l_i = a_1x_1 + a_2x_2 + a_3x_3 + \dots + a_px_p$$

where a_1 through a_p are corresponding scalars that serve as weights for variables x_1 through to x_p. In our sum for IQ, it is implied that the weights are equal to "1" on each variable. That is, our linear combination, when unpacked and spelled out in more detail, is actually the following:

$$\mathbf{IQ = (1)VERBAL + (1)ANALYTIC + (1)QUANT}$$

Even with equal weights across all variables, the above is still defined as a linear combination, and in practice we could use it to generate the "variate" of IQ. That is, you could, if you wanted to, sum the scores so they give you a total IQ, then use this as your response variable as a function of training program. Indeed, many times this is done in clinical research and testing on psychometric instruments without direct awareness that what is being generated is a linear combination. It is usually simply referred to as **summing the scores on the scale**, but it is not recognized that there are implied weights of "1" before each variable. So, again, we could if we really wanted to, generate the above sum and use that as our response variable. The question is, however, whether that sum can be improved upon **to satisfy a purpose other than simply being a sum of scores.** This is where MANOVA comes in. The sum that you produced with implied weights of unity before each variable is not guaranteed to have any special properties. MANOVA (and discriminant analysis) will more "wisely" select these scalars. That is, MANOVA and discriminant analysis will select these scalars so as to optimize some function of the data. As usual, and as was the case for least-squares regression, it comes down to a problem of **optimization.**

In MANOVA, the simultaneous inference occurs on a sum of variables as above, only that now the linear combination generated will not necessarily consist of scalars all (or even any) equal to 1. That is, the linear combination will no longer be chosen "naively" as we have done by simply assuming scalars of 1. Instead, the linear combination will be chosen in a way that selects the scalars with some **optimization criteria** in mind. Recall that optimization is an area of mathematics that seeks to **maximize** or **minimize** the value of a function based on particular constraints unique to the problem at hand. In other words, it selects values that are "optimal" in some sense, where "some sense" means "subject to constraints." Techniques in calculus are often used to find such maximums or minimums. In MANOVA, optimization criteria is applied that will generate scalars for each variable, possibly all different, such that some criteria will be maximized. What is that criterion? The answer is found in the **discriminant function**. Hence, for an understanding of MANOVA, we need to first get a quick cursory glimpse of what discriminant analysis is all about, since behind the scenes of every MANOVA is a discriminant analysis. In what follows then, we quickly survey discriminant analysis before returning to our discussion of MANOVA. We will then later more thoroughly unpack the discriminant analysis model as a separate and unique statistical methodology.

9.3 The Discriminant Function

We have said that MANOVA generates a linear combination of response variables, each weighted by scalars, not necessarily all distinct, but usually so. What we need to understand now is how those scalars are selected. Using default values of "1" for each variable will not satisfy optimization requirements, but to know what scalars will, it is first important to understand which function we are trying to optimize in the first place. The key to understanding what makes MANOVA "tick" is to "flip it around" so that the multiple response variables are now predictors, and the independent variable is now the response variable. This will give us the **linear discriminant analysis model**. Recall our model thus far. We are hypothesizing IQ as a function of training program:

$$\textbf{IQ}\left(\textbf{verbal}+\textbf{analytic}+\textbf{quant}\right)\textbf{ as a function of Training Program }\left(\textbf{1, 2}\right)$$

To get the discriminant analysis, we flip this equation around:

Training Program (1, 2) as a function of IQ (verbal + analytic + quant)

In words, the **function statement** now reads that training program is a function of a linear combination of variables. In the MANOVA, we hypothesized the linear combination as a function of training program. Remarkably, these two ways of posing the question are near identical in terms of their technical underpinnings, though different details are focused on depending on which method we are studying. Hence, we are now in a position to address the question we asked previously, which is how to best select the scalars that will weight the linear sum of variables verbal + analytic + quant. Should we simply settle on scalars of "1" before each variable, or should we choose others? The answer we have come to so far is to choose scalars that will maximize some function. And it is in defining the discriminant function wherein lies the answer. **We will choose scalars that maximize the separation between groups on the binary response variable.** This linear combination, weighted by such scalars, is the same one used in the MANOVA, but we usually do not see it in our analysis. That is, it is usually not given in output or featured in the derivation of MANOVA. The discriminant analysis simply "unpacks" the linear combination so that we learn more about it. That is, we learn more about the linear combination that generated **mean vector differences** in the MANOVA. In MANOVA, mean vector differences are the focus, which, for our example, are the means on verbal, analytic, and quant as a **vector**, as a function of training program. Recall that, by definition, a vector typically has several numbers in it. For the MANOVA, the vector is one of means, which is why we call it a **mean vector**.

We will return to our discussion of discriminant analysis a bit later in the chapter. It is enough to appreciate at this point that the choice of scalars to generate the linear combination(s) in MANOVA will be determined by selecting a function or functions that maximally distinguish groups. These functions are called discriminant functions and underlie all MANOVA procedures. For now, however, we return to our discussion of MANOVA.

9.4 Multivariate Tests of Significance: Why They Are Different from the *F*-Ratio

We said earlier that testing multivariate models requires different tests of significance than when evaluating univariate models. The reason for this lies in the fact that in multivariate models, the simultaneous consideration of response variables requires us to model the covariance between responses. In univariate models, this covariance is not modeled since there is only a single response. When this covariance is modeled, as in multivariate models, it introduces a whole new world of complexity.

So, what does "modeling the covariance between responses" mean? To understand this, it is easiest to first introduce tests of significance for MANOVA and then unpack how each test treats this covariance. Contrary to ANOVA that featured only a single test of significance (the ***F*-test**), because of the potentially complex configuration of covariances in multivariate models, MANOVA features several different tests that all treat covariances a bit differently. Hence, gone are the days in which you can simply run an F-test, obtain statistical significance, and be on your way to interpreting the accompanying effect size. In the multivariate domain, when you claim a statistically significant effect, you now need to inform your audience or reader which precise test was used that was statistically significant, as different tests may report different results. We now survey the most common multivariate tests reported by software.

9.4.1 Wilks' Lambda

The first test, and undoubtedly historically the most popular, is that of **Wilks' lambda**, given by

$$\Lambda = \frac{|\mathbf{E}|}{|\mathbf{H}+\mathbf{E}|}$$

We can see that the test is defined by a ratio of $|\mathbf{E}|$ to $|\mathbf{H}+\mathbf{E}|$, where $|\ |$ in this case indicates not absolute value, but rather the **determinant** of the given matrix. The matrix in the numerator, $|\mathbf{E}|$, is a matrix of the sums of squares and cross-products that does not incorporate group differences on the independent variable. In this sense, $|\mathbf{E}|$ is analogous to **MS within** in univariate ANOVA. However, they are not one-to-one the same, since $|\mathbf{E}|$ incorporates cross-products (crudely, the aforementioned covariance idea) between responses, whereas of course MS within does not since there are no cross-products or covariance as a result of there being only a single response. $\mathbf{E}$ is typically called the "error" matrix in MANOVA and computing the determinant on $\mathbf{E}$ gives us an overall measure of the degree of "generalized" within-group variability in the data, again analogous in spirit to what MS within told us in ANOVA.

The matrix $\mathbf{H}$, like $\mathbf{E}$, contains sums of squares and cross-products, only that now, these sums of squares and cross-products contain between-group variation. Again, this is somewhat analogous to what is accomplished by **MS between** in ANOVA. It is often referred to as the **hypothesis matrix** for this reason, the hypothesis being that there are true population differences on the independent variable. And as is the case in ANOVA where **SS total = SS treatment + SS error**, MANOVA features an analogous identity, where $\mathbf{T} = \mathbf{H} + \mathbf{E}$. That is, in words, the total variation and cross-product variation represented by $\mathbf{T}$ can be broken down as a sum of the hypothesis matrix $\mathbf{H}$ and the error matrix $\mathbf{E}$.

So what does Λ accomplish? Notice that Wilks' is comparing, via a ratio, $|\mathbf{E}|$ to the total variation represented by $|\mathbf{H}+\mathbf{E}|$. Hence, the extent to which there is no between-group differences, $|\mathbf{E}|$ and $|\mathbf{H}+\mathbf{E}|$ will represent the same quantities, since under the case of no between-group variation, $\mathbf{H}$ will be equal to 0, and hence $|\mathbf{H}+\mathbf{E}| = |0+\mathbf{E}| = |\mathbf{E}|$. Therefore, we see that Wilks' is an **inverse criterion**, meaning that under the null hypothesis of no differences, $\Lambda = 1$, which is the maximum value Λ can attain. It is called an **inverse criterion** because contrary to F in univariate ANOVA, the null case is represented by a maximum value for Λ. In the univariate F-test, recall that larger values for F count as increasing evidence against the null hypothesis, and smaller values suggest that the null hypothesis is not false. For Wilks', when $\Lambda = 0$, it means that $|\mathbf{E}|$ must equal 0, and we have

$$\Lambda = \frac{|\mathbf{E}|}{|\mathbf{H}+\mathbf{E}|} = \frac{|0|}{|\mathbf{H}+0|} = \frac{0}{|\mathbf{H}|} = 0$$

Hence, the closer the value of Λ to 0, all else equal, the more evidence we accumulate against the null hypothesis. That is, the more evidence we have to reject the null hypothesis of equality of mean vectors.

9.4.2 Pillai's Trace

As mentioned, Wilks' lambda is only one of many multivariate tests. We now survey a second test, that of **Pillai's trace**, which focuses on different elements of the **H** to **E** components. Pillai's trace is given by

$$V^{(s)} = \text{tr}[(\mathbf{E}+\mathbf{H})^{-1}\mathbf{H}]$$

where tr is the trace of the quantity after it, that of $(\mathbf{E}+\mathbf{H})^{-1}\mathbf{H}$, and where we are taking the inverse of $\mathbf{E}+\mathbf{H}$ (recall the inverse of a matrix is denoted by $^{-1}$). But what is being assessed by $(\mathbf{E}+\mathbf{H})^{-1}\mathbf{H}$? Let us take a look at it more closely. Recall that the function of taking the **inverse** in matrix algebra is analogous (in a conceptual sense) to division in scalar algebra. Hence, the quantity $(\mathbf{E}+\mathbf{H})^{-1}\mathbf{H}$, if we were to write it in scalar algebra terms, could crudely be written as

$$\frac{\mathbf{H}}{\mathbf{E}+\mathbf{H}}$$

Thus, to understand Pillai's trace, we could say that it is comparing **H** to the total variation represented by $\mathbf{E}+\mathbf{H}$, then taking the trace of this "quotient." We put quotient in quotes here because it will not actually be a quotient when written in proper form. In proper form it is a **product**, the product of the inverse of $\mathbf{E}+\mathbf{H}$ and **H**. Hence, we can see that Pillai's is behaving a bit more analogous to traditional F in univariate ANOVA in this regard, where the hypothesis matrix is in the "numerator" (in terms of the scalar analogy) of the equation and the total variation is in the "denominator."

Pillai's can also be expressed with respect to **eigenvalues** λ_i via

$$V^{(s)} = \sum_{i=1}^{s} \frac{\lambda_i}{1+\lambda_i}$$

Note that in this formulation, each respective eigenvalue λ_i is being extracted and then compared via a ratio to the denominator $1+\lambda_i$. These ratios are then summed to provide the total value for Pillai's trace. Recall we introduced notions of **eigenvalues** and **eigenvectors** in earlier chapters. However, these elements are not intuitive and from a statistical point of view at least, they need to be grounded in a statistical application for them to have any real meaning. Hence, the interpretation of eigenvalues in the context of MANOVA will make much more sense when we survey **discriminant analysis** in more detail a bit later. Eigenvalues will make even more sense when we survey principal components later in the book as well, though those eigenvalues will not be the same eigenvalues extracted as in MANOVA or discriminant analysis.

9.4.3 Roy's Largest Root

A third multivariate test of significance is that of **Roy's largest root**, given by

$$\theta = \frac{\lambda_1}{1+\lambda_1}$$

where λ_1 is the largest of the eigenvalues extracted. Notice that Roy's only uses the largest of the eigenvalues, whereas Pillai's uses them all. This can easily be seen by comparing the two statistics side-by-side:

$$V^{(s)} = \sum_{i=1}^{s} \frac{\lambda_i}{1+\lambda_i} \quad \text{vs.} \quad \theta = \frac{\lambda_1}{1+\lambda_1}$$

Notice that both statistics are constructed the same way. We notice, however, that in Pillai's, we are summing across the ith eigenvalues. In Roy's, we are doing no such thing, and only feature a single eigenvalue, λ_1. **That eigenvalue is the largest of the eigenvalues extracted, and hence Roy's is focused on only the root accounting for the most information.** In the case where there is only a single eigenvalue extracted, Roy's will equal that of Pillai's, since the only eigenvalue must also simultaneously be the largest one. When would you want to use only the largest eigenvalue? The answer is when one eigenvalue dominates the size of the others, usually implying that the mean vectors lie in a single dimension (Rencher & Christensen, 2012), which, as we will see later, suggests that a **single discriminant function** is explaining most of the separation between groups. If remaining eigenvalues are sizeable and still important, then Pillai's would be preferable to compute over Roy's. All of this, again, will make more sense when we review discriminant analysis, since, as of now, it may be unclear to you why we are obtaining more than a single eigenvalue from a MANOVA problem. This cannot be adequately understood in the context of MANOVA. Discriminant analysis, however, will yield the answer.

9.4.4 Lawley-Hotelling's Trace

A fourth multivariate test is that of the **Lawley-Hotelling's trace:**

$$U^{(s)} = \mathbf{tr}(\mathbf{E}^{-1}\mathbf{H}) = \sum_{i=1}^{s} \lambda_i$$

Notice that $U^{(s)}$ is taking the trace (i.e. tr) of a product, the product in this case being $\mathbf{E}^{-1}\mathbf{H}$. Recall that $\mathbf{E}^{-1}\mathbf{H}$ means we are, in analogous scalar algebra, comparing $\mathbf{H}$ in the numerator to $\mathbf{E}$ in the denominator. That is, we are contrasting the hypothesis matrix to that of the error matrix, where the hypothesis matrix contains the group differences and the error matrix does not. In this sense, $U^{(s)}$ is computing something somewhat similar to Pillai's discussed earlier; however in Pillai's, we took the trace of $(\mathbf{E}+\mathbf{H})^{-1}$ multiplied by $\mathbf{H}$. $U^{(s)}$ is also equal to the sum of eigenvalues for the problem, $\sum_{i=1}^{s} \lambda_i$, which is again somewhat similar in spirit to Pillai's, but does not divide by $1+\lambda_i$ when performing the sum.

9.5 Which Multivariate Test to Use?

Though these multivariate statistics each assess something slightly different in the data as we have seen, in most cases, they will usually all hint at or strongly suggest the same decision on the multivariate null hypothesis. That is, though on a theoretical level they are different from one another (and their differences are important), in an applied sense, you usually will not have to concern yourself about these differences as they will typically all universally suggest a similar result when applied to real data. In cases where they do not (e.g. one test suggests you reject while another does not), **you are strongly encouraged to not ignore the situation**, but instead delve into your data deeper in an exploratory sense to disentangle why one test is yielding statistical significance and another is not. As we have seen, the sizes of each statistic depend greatly on the respective sizes and patterns of **H** and **E**, and if the tests do not agree in their decision, then investigating these matrices in more detail would be called for. This would be for two purposes. The first is to better understand the results you are obtaining, but also to potentially detect patterns in these matrices that may otherwise go unnoticed. In this sense, a failure to universally reject the multivariate null across all significance tests may help you discover something in your data that is meaningful on a **scientific** as well as **statistical** level. Why did one test reject the null while another did not? The answer may (or may not!) give you insight into the nature of your data. **Never just ignore the situation. Rather, do some additional exploratory analyses to find out why.**

This is as far as we take our discussion of multivariate tests here. For further details on these tests, including comparisons and contrasts between them, the reader is encouraged to consult Olson (1976) for a classic, excellent, and still very useful overview of all of them. More theoretical multivariate statistics texts will also discuss these tests in greater detail and provide the requisite technical context for understanding and appreciating them on a deeper level. Among these include Johnson and Wichern (2007) as well as Rencher and Christensen (2012).

9.6 Performing MANOVA in Python

We now demonstrate a MANOVA in Python. For this demonstration, we consider the aforementioned data, but only on quant, verbal, and train group. We first create the small data file:

```
import pandas as pd
data = {'quant': [5, 2, 6, 9, 8, 7, 9, 10, 10],
'verbal': [2, 1, 3, 7, 9, 8, 8, 10, 9],
'train': [1, 1, 1, 2, 2, 2, 3, 3, 3]}

df_manova = pd.DataFrame(data)
print(df_manova)
   quant  verbal  train
0      5       2      1
1      2       1      1
2      6       3      1
3      9       7      2
4      8       9      2
5      7       8      2
6      9       8      3
7     10      10      3
8     10       9      3
```

We confirm in this data that there are a total of nine cases, with three cases per train group. We now run the MANOVA using **statsmodels**. Before we run the MANOVA, we need to first label `train` as a **categorical variable** so it will be recognized in the model statement. As it stands, it is a **numeric variable**, as we can easily confirm:

```
print(df_manova.dtypes)
quant      int64
verbal     int64
train      int64
dtype: object
```

Indeed, we see from the above that all variables are being identified as numeric. For `quant` and `verbal` this is how we want them, as they are continuous variables in the MANOVA. However, for `train`, we want this to be a factor variable, so we convert it to a categorical one:

```
cols = ['train']
for col in cols:
    df_manova[col] = df_manova[col].astype('category')
print(df_manova.dtypes)
quant         int64
verbal        int64
train      category
dtype: object
```

We confirm above that `train` is now a categorical variable and we can now proceed with the MANOVA:

```
from statsmodels.multivariate.manova import MANOVA
maov = MANOVA.from_formula('quant + verbal ~ train', data = df_
manova)
print(maov.mv_test())

                  Multivariate linear model
==============================================================

--------------------------------------------------------------
         Intercept         Value  Num DF Den DF F Value Pr > F
--------------------------------------------------------------
            Wilks' lambda 0.1635 2.0000 5.0000 12.7885 0.0108
           Pillai's trace 0.8365 2.0000 5.0000 12.7885 0.0108
   Hotelling-Lawley trace 5.1154 2.0000 5.0000 12.7885 0.0108
      Roy's greatest root 5.1154 2.0000 5.0000 12.7885 0.0108
--------------------------------------------------------------

--------------------------------------------------------------
         train            Value  Num DF  Den DF F Value Pr > F
--------------------------------------------------------------
           Wilks' lambda  0.0561 4.0000 10.0000  8.0555 0.0036
          Pillai's trace  1.0737 4.0000 12.0000  3.4775 0.0417
  Hotelling-Lawley trace 14.5128 4.0000  5.1429 17.8112 0.0033
     Roy's greatest root 14.3516 2.0000  6.0000 43.0547 0.0003
==============================================================
```

The output for the intercept is of no use and we do not interpret it. We are more interested in the output for `train`. We can see that all tests of significance, from Wilks' lambda through to Roy's largest root, are statistically significant at least at $p < 0.05$, but what was our null hypothesis to begin with? Recall that the MANOVA analysis is not being done on means. Rather, it is being performed on **mean vectors**. That is, the null hypothesis for this problem is that mean vectors are equal across train groups:

$$H_0 : \boldsymbol{\mu}_1 = \boldsymbol{\mu}_2 = \boldsymbol{\mu}_3$$

where the boldface type $\boldsymbol{\mu}_1$ and so on are now **vectors**, not univariate means as in ANOVA. The alternative hypothesis is that somewhere amid these mean vectors is at least one pairwise difference. For our data then, we reject the null hypothesis of equality of mean vectors and conclude that at least one mean population vector on the `train` variable is different from another.

9.7 Effect Size for MANOVA

Having computed the significance tests for MANOVA and rejected the multivariate null, we would now like to compute an **effect size**. Recall the reason for computing effect sizes. Effect sizes inform us of the actual variance explained in our data and are not as influenced by sample size as are *p*-values. An effect size for Wilks' lambda can be computed as

$$\eta_{\mathrm{p}}^2 = 1 - \Lambda^{1/s}$$

where η_{p}^2 is **partial eta-squared,** Λ is the Wilks' value from the MANOVA, and s is the smaller of either the number of dependent variables or the degrees of freedom for the independent variable. Since for our data $\Lambda = 0.056095$ and s is equal to 2 since the minimum of 2 dependent variables and 2 degrees of freedom is equal to 2 (we will discuss shortly in discriminant analysis where this rule comes from), the computation we need is the following:

$$\eta_{\mathrm{p}}^2 = 1 - \Lambda^{1/s} = 1 - 0.056095^{1/2} = 1 - 0.23684383 = 0.76$$

Hence, the effect size associated with our multivariate result is equal to 0.76. Recall that in this case we are analyzing a linear combination and not individual variables. Hence, we are reporting the variance explained across a linear combination of variables, not a single variable as in ANOVA. There are some complexities that arise when using η_{p}^2 in a MANOVA context. For details, see Tabachnick and Fidell (2001, p. 339). This is as far as we discuss effect sizes in MANOVA. For a more thorough discussion of effect sizes in multivariate settings in general, see Rencher and Christensen (2012).

9.8 Linear Discriminant Function Analysis

Having surveyed the MANOVA model, we now consider in more detail **linear discriminant function analysis**. As mentioned earlier, on a technical level LDA ("linear discriminant analysis") can be considered as the "reverse" of MANOVA. That is, if you find a statistically significant effect in MANOVA, it suggests mean vector differences on a linear composite of response variables. The question asked by discriminant analysis is:

What is the nature of this one (or more) linear composite(s) that is (are) successful in providing group differentiation?

Here is a great way to think of what a discriminant function actually is, by way of analogy. Suppose you tell your friends that you are a master of being able to determine who is intelligent vs. who is not, simply by talking and watching them for a few minutes. So, you meet someone, and after an hour of talking with them, you draw the conclusion that they either belong in the intelligent group or the group not-so-intelligent. Suppose out of 100 trials with each time a different subject, you are correct every time. That is, you are able to correctly classify subjects into the correct IQ group. Your friends, of course, would ask you:

How are you so good at discriminating intelligent vs. non-intelligent people?

In other words, what they are asking is the nature of the **discriminant function** you are using to successfully discriminate intelligent vs. not-so-intelligent people. What might your discriminant function be made of? Let us consider some possibilities that you may be using, on an informal level, to determine group membership:

- **Quickness of speech** of the subject, in that people who speak faster seem to be "sharper," and so if they are speaking quickly and seem sharp, this variable contributes to them being classified into the "intelligent" group.
- **Use of vocabulary** by the subject; maybe you believe subjects who use a bit more advanced vocabulary belong in the intelligent group.
- **Degree of delay in responding in communication** by the subject, in that after you are finished speaking, the elapsed time until they start speaking might indicate their degree and speed of processing the information.

Of course, we are making up the aforementioned criteria for the convenience of our example. However, the point is that in your discriminating ability are likely factors or variables you consider mentally such that they help you group people into one category or another. This is your discriminant function for grouping individuals, and might be made up, informally, of the above variables, as one example.

The question then becomes, how much respective "weight" do you assign to each of the above variables? That is, do you deem quickness of speech much more "important" than use of vocabulary and delay in responding? If you do, then perhaps intuitively you are allowing that variable to carry more **impact** or **influence** in your discriminant function. In discriminant analysis, we will no longer rely on your "informal criteria," but rather will attempt to arrive at a discriminant function via much more precise and intelligent means. In other words, we will develop some **optimization criteria** for deriving the **best** discriminant function. This is the essence of what discriminant analysis is about, to obtain **scalars** (weights) such that when applied to the variables, best serve to discriminate between groups. That is, the selection of scalars maximizes some criterion, in a similar way that least-squares regression minimized some criterion (i.e. the sum of squared errors in prediction).

9.9 How Many Discriminant Functions Does One Require?

The question of how many discriminant functions should be kept from an analysis is different from the question of how many discriminant functions are generated. For a given problem, the number of discriminant functions generated will be equal to the **smaller of the number of predictor variables or one less the number of groups on the dependent variable**. Hence, if there are p populations on the dependent variable, then in general there will be $p-1$ discriminant functions generated. At first glance, you may think the number should equal p rather than $p-1$, but this intuition would be wrong. Why is this so? We can demonstrate why through our previous example of the mental discriminant function we used to distinguish among intelligent folks vs. not-so-intelligent people. Recall that your criteria consisted of three variables to do the discriminating, but there are only two groups to discriminate. Hence, you can think of those three variables as representing a new "variate," a new axis upon which the discriminating will take place. Now, how many variates will you need to discriminate between two groups? Only a single function will do the job, because that single dimension can "cut through" the two categories to provide separation, as shown in the following diagram:

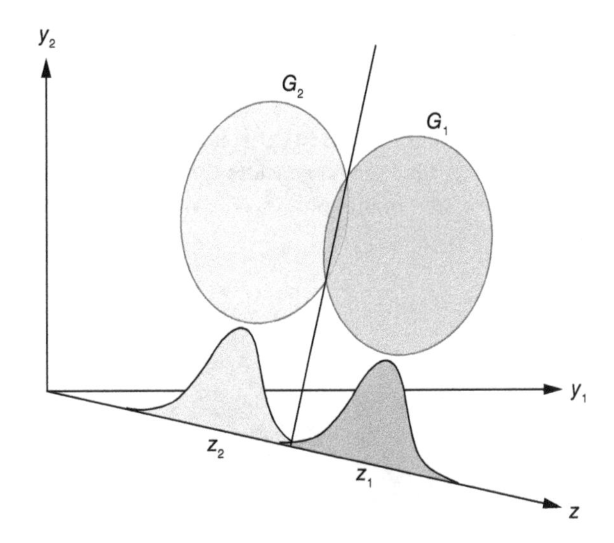

Notice that in the figure, the newly derived function is the line that is providing maximum separation between the groups. Only a single discriminant function is needed to separate two groups. In our analogy, you only need one **mental discriminator** (in this case) to distinguish between intelligent vs. not-so-intelligent people. Now, for three groups, the maximum number of functions is two, and what is more, one of those functions may do a better job at discriminating than the other. Note as well the overlap in the diagram. There is overlap because the function is not doing a perfect job. If the function is not doing a perfect job, then why use it? It is key to understand that while the discriminant function will rarely provide perfect separation between groups, it is not advertised to do so in the first place. Rather, it is designed to provide **maximum separation** between groups. This is key. But what does "maximum separation" mean? At first glance, it would seem to suggest "perfect" separation, but this is not the case, no more than a least-squares line promises to fit data perfectly or provide perfect prediction. The least-squares line is simply designed to minimize the sum of squared errors, or, equivalently, maximize R-squared (both amount to the same thing). It promises to do this better than any other line one could fit to the data, and so it is with the discriminant function. The discriminant function provides maximum separation even if the separation is far from perfect. What will determine the degree of separation achieved? The quality of the data and the true existence of mutually exclusive groups will determine how good the separation is. Beyond that, the discriminant function can only do the best it can, just as any other statistical method is only as good as the data on which it is fit. Again, as we have highlighted throughout this book, **statistical methods do not "save" data, they simply model it.** The extent to which a model fits well is up to your science, not the model. All the statistical method can guarantee is that it is doing the best it can. That is, all the statistical model or algorithm can do, in a very true sense, is usually to maximize or minimize a function. Despite what researchers want to believe of their statistics, statistical modeling usually cannot do much more than that. The rest is up to you and your research design.

9.10 Discriminant Analysis in Python: Binary Response

We demonstrate a couple of simple examples of LDA in Python. Our first example features two predictors on a binary response variable, that is, a response variable having only two categories. After this, we will consider the example of three levels on the response variable, that is, discriminant analysis on a **polytomous** response. We first build our data frame for the two-case:

```
data_discrim = {'y': [0, 0, 0, 0, 0, 1, 1, 1, 1, 1],
'x1': [4, 3, 3, 2, 2, 8, 7, 5, 3, 3],
'x2': [2, 1, 2, 2, 5, 3, 4, 5, 4, 2]}

df_discrim = pd.DataFrame(data_discrim)
df_discrim
Out[47]:
   y  x1  x2
0  0   4   2
1  0   3   1
2  0   3   2
3  0   2   2
4  0   2   5
5  1   8   3
6  1   7   4
7  1   5   5
8  1   3   4
9  1   3   2
```

For these data, we would like to discriminate groups 0 and 1 on y using predictors x_1 and x_2. This will call for a single discriminant function to be produced since there are only two categories on the response. Recall that even had we many more predictors, since there are only two response categories, we can only produce a single discriminant function. As an analogy, if I asked you to guess which hand behind my back contains the quarter, you require only a single "mental function" to make your choice; it is either in the right hand or the left hand, and hence whatever information (i.e. "linear combination" by analogy) you use to make this discrimination can be combined into a single function, in this case, a single linear combination. You only use one discriminating tool to make the decision, even if that tool is rather complex and made up of a linear sum of different variables.

Let us build our response and predictor sides. We will use `np.array()` for this:

```
y = np.array([0, 0, 0, 0, 0, 1, 1, 1, 1, 1])
X = np.array([[4, 2], [3, 1], [3, 2], [2, 2], [2, 5], [8, 3], [7, 4],
[5, 5], [3, 4], [3, 2]])
```

To demonstrate that Python will be unable to generate more than a single function, suppose we requested two functions via `n_components=2`:

```
import numpy as np
from sklearn.discriminant_analysis import LinearDiscriminantAnalysis
lda = LinearDiscriminantAnalysis(n_components=2)
lda.fit(X, y)
```

```
FutureWarning: In version 0.23, setting n_components > min(n_fea-
tures, n_classes - 1) will raise a ValueError. You should set n_
components to None (default), or a value smaller or equal to min(n_
features, n_classes - 1).
  warnings.warn(future_msg, FutureWarning)
```

The reason we received an error is that we should be requesting at most a single function (for reasons just discussed), as in the following correct implementation of the code where we specify `n_components=1`:

```
lda = LinearDiscriminantAnalysis(n_components=1)
model = lda.fit(X, y)
```

```
Out[257]:
LinearDiscriminantAnalysis(n_components=1, priors=None,
                           shrinkage=None,
                           solver='svd', store_covariance=False,
                           tol=0.0001)
```

The following are the discriminant function scores for the function:

```
scores = lda.transform(X)
scores
```

```
Out[259]:
array([[-0.43107609],
       [-1.35954768],
       [-0.92847159],
       [-1.42586708],
```

```
       [-0.1326388 ],
       [ 1.98958197],
       [ 1.92326257],
       [ 1.35954768],
       [-0.0663194 ],
       [-0.92847159]])
```

That is, the first case in our data via the model, obtained a discriminant score of −0.43, the second case −1.36 (rounded up), and so on. We can obtain the coefficients to the discriminant function using `lda.coef_`:

```
print(lda.scalings_)
[[0.49739549]
 [0.43107609]]
```

These are the raw **unstandardized** discriminant function coefficients. The constant for the function is equal to −3.283, and hence we can now use the function to generate discriminant scores:

```
y = -3.283 + 0.49739549(x1) + 0.43107609(x2)
```

For an example regarding how to use the function, our first observation in our data has a score of 4 on `x1` and 2 on `x2`. Let us compute that case's discriminant score:

```
y = -3.283 + 0.49739549(4) + 0.43107609(2)
y = -3.283 + 1.98958196 + 0.8621521
    = -0.43126586
```

Notice that the computation agrees (within rounding error) with the first score generated by Python earlier. We can duplicate all of the scores that Python produced automatically using `lda.transform(X)` earlier (output not shown).

The value of each discriminant function evaluated at group means can be computed:

```
m = np.dot(lda.means_ - lda.xbar_, lda.scalings_)
m

Out[265]:
array([[-0.85552025],
       [ 0.85552025]])
```

That is, the mean of the function at y = 0 is equal to −0.855 while the mean of the function at y = 1 is 0.856. Next, we will use the discriminant functions to predict group membership using `model.predict()`:

```
pred=model.predict(X)
pred
Out[195]: array([0, 0, 0, 0, 0, 1, 1, 1, 0, 0])
```

We can see that the model correctly predicts the first five cases (0, 0, 0, 0, 0), which recall were all in the designated "0" group. It also correctly classifies the first three cases of the five cases in group "1" (1, 1, 1, 0, 0). Notice it misclassifies the last two cases

(0, 0). Hence, cases 9 and 10 were misclassified. We can obtain a convenient summary of this via what is known as a **confusion matrix**:

```
from sklearn.metrics import confusion_matrix
from sklearn.metrics import accuracy_score

print(confusion_matrix(pred, y))

[[5 2]
 [0 3]]
```

Notice that if all classifications were correct, the "5" and "3" across the main diagonal would equal "5" and "5." The "2" in the upper right of the matrix corresponds to the two misclassified cases.

9.11 Another Example of Discriminant Analysis: Polytomous Classification

We demonstrate a second discriminant analysis, this time on the `train` data. First, we set up our data file:

```
train = np.array([1, 1, 1, 2, 2, 2, 3, 3, 3])
X = np.array([[5, 2], [2, 1], [6, 3], [9, 7], [8, 9], [7, 8], [9, 8],
[10, 10], [10, 9]])

X
Out[65]:
array([[ 5,  2],
       [ 2,  1],
       [ 6,  3],
       [ 9,  7],
       [ 8,  9],
       [ 7,  8],
       [ 9,  8],
       [10, 10],
       [10,  9]])

model = lda.fit(X, train)
model
Out[68]:
LinearDiscriminantAnalysis(n_components=None, priors=None,
                           shrinkage=None,
                           solver='svd', store_covariance=False,
                           tol=0.0001)
```

Since we have three groups on the response variable, Python will compute for us two functions. As before, to obtain the raw coefficients (which, incidentally, agree with SPSS's unstandardized coefficients if you also conduct the analysis using that software), we compute:

```
print(lda.scalings_)

[[ 0.02983363 0.83151527]
 [ 0.9794679 -0.59019908]]
```

As we did for the two-group case, we can obtain the discriminant scores for each function (which also agrees with SPSS):

```
lda.transform(X).shape
Out[75]: (9, 2)

lda.transform(X)

Out[76]:
array([[-4.31397269,  0.61732703],
       [-5.38294147, -1.28701971],
       [-3.30467116,  0.85864322],
       [ 0.70270131,  0.99239274],
       [ 2.63180348, -1.01952069],
       [ 1.62250195, -1.26083688],
       [ 1.68216921,  0.40219366],
       [ 3.67093863,  0.05331078],
       [ 2.69147074,  0.64350986]])
```

Finally, we can obtain the model prediction classifications (which, again, agrees with SPSS's discrim function):

```
pred = model.predict(X)
pred
Out[114]: array([1, 1, 1, 2, 2, 2, 2, 3, 3])
```

We see that only a single case was misclassified. This is reflected in the confusion matrix for the classification:

```
print(confusion_matrix(pred, train))
[[3 0 0]
 [0 3 1]
 [0 0 2]]
```

9.12 Bird's Eye View of MANOVA, ANOVA, Discriminant Analysis, and Regression: A Partial Conceptual Unification

In this chapter, we have only had space to basically skim the surface of all that makes up MANOVA and discriminant analysis. A deeper study will emphasize eigenvalues much more, the dimension reduction that takes place in each procedure, as well as the assumptions underlying each analysis, such as equality of population covariance matrices on levels of the categorical grouping variable. Entire chapters and books are written on these procedures, and hence beyond our basic cursory overview and simple demonstration in Python, additional sources should be consulted. Denis (2021) provides a deeper and richer overview, and Rencher and Christensen (2012) survey these techniques in much greater detail than in this chapter. Navigating the multivariate landscape can be challenging, and hence we close this chapter with an overview of how many models can be subsumed under multivariate models in general.

MANOVA, for whatever reason, has historically not received the "fame" that discriminant analysis has received, even though, as we have seen, it is essentially the "reverse" of discriminant analysis. Today, with **data science** and **machine learning** being quite popular, MANOVA seems to have taken a backseat to discriminant analysis, largely due to, I believe, the **substantive focus** of many of the problems in those areas of investigation. In machine learning, for instance, the focus is on **classification**, not on **mean vectors**, and less on the nature of the functions generated by the discriminant analysis. The emphasis is less on **theoretical construction** and **justification of dimensions** as it is on simply getting the classification "right" and optimizing prediction accuracy. "Black boxes" are more "tolerable" in these fields than in fields where identifying and naming dimensions is a priority. So long as you can predict the correct category, it does not necessarily matter as much what "ingredients" went into this prediction. This is what we mean by the "black box." If it works, it works, and that is what matters. At least this is the primary emphasis in some fields of quantitative practice.

In traditional statistical applied substantive practice, however, for good or for bad, scientists would nonetheless like to often "name" the dimensions emanating from the discriminant function analysis, in addition to learning how well they can predict and classify. **The theoretical nature of the discriminator is of great interest**. Hence, in many applied fields, knowing that one or two dimensions can successfully predict and classify is not enough; researchers would like to attempt to **identify such dimensions**, which is done by focusing on the coefficients (usually the standardized ones) resulting from the analysis. Coefficients that are greater in absolute value typically indicate that the given variable is more relevant or "important" (in some sense) to the discriminant function than coefficients with lesser absolute value.

As it pertains to mean vector differences in MANOVA, this also does not seem to be a priority of many in data science or machine learning as it was in the day when Wilks introduced lambda in 1932. However, for the applied researcher, the theory of a linear combination of variables as a function of one or more independent variables is still just as relevant today and many questions in the social and natural sciences require thinking of problems in exactly this way. The concept of "latent variables" is common in many of these sciences, and hence a focus on eigenvalues, multivariate significance tests, and all the rest of it becomes quite relevant when conducting such analyses. ANOVA and MANOVA are not somehow "out of vogue" or "outdated" methodologies. From a pedagogical point of view, understanding these methods paves the way to understanding relatively advanced statistical techniques such as multilevel modeling and certain very advanced psychometric methodologies.

The "gist" of the above comments is that simply because we are approaching a problem in one way or another does not imply a lack of underlying similarity among the approaches. This is precisely the case as it concerns statistical modeling. While each statistical approach has its own peculiarities, it is also true that they are more **unified** than they are **disparate**. And actually, we can identify one statistical method for which many other models can be considered "special cases." What is that model?

Canonical correlation. A **canonical correlation** model has several variables on each side of the equation,

$$y_1, y_2, y_3, \ldots, y_j = x_1, x_2, x_3, \ldots, x_k$$

where on the left-hand side are variables y_1 through y_j, while on the right-hand side are variables x_1 through x_k. All of these variables are assumed to be continuous. But what does a canonical correlation measure? **Canonical correlation assesses the linear relationship among linear combinations of y variables to x variables**, where the y variables and x variables, not surprisingly, are each weighted by **scalars** to maximize the bivariate correlation between linear combinations:

$$(a_1)y_1 + (a_2)y_2 + (a_3)y_3 + \cdots + (a_n)y_j = (b_1)x_1 + (b_2)x_2 + (b_3)x_3 + \cdots + (b_n)x_k$$

where a_1 through a_n are corresponding scalars for the left-side linear combination and b_1 through b_n are corresponding scalars for the right-side linear combination (usually different from those on the left-hand side). In canonical correlation, as with Pearson r, there are no dependent or independent variables. There are simply two sides of the equation as one would have in a bivariate Pearson r correlation. The scalars a_1 through a_n and b_1 through b_n are chosen to **maximize the bivariate correlation between linear combinations** in the same spirit that predicted values in regression were obtained in order to maximize the bivariate correlation between predicted and observed values in simple or multiple linear regressions. Canonical correlation underlies most of the statistical methods we have surveyed to date, even if output for each procedure does not report it. **Canonical correlation is a fundamental idea of multivariate analysis, which is to associate linear combinations between sets of variables (left side and right side).**

9.13 Models "Subsumed" Under the Canonical Correlation Framework

Now, consider how virtually all (well, many at least) of the models we have talked about thus far can be considered "special cases" of the wider canonical correlation framework, simply by **changing how variables are operationalized** for each given model. If we want a **multivariate multiple regression**, then we use the right-hand side to predict the left-hand side. Hence, we are regressing the linear combination $y_1, y_2, y_3, \ldots, y_j$ onto $x_1, x_2, x_3, \ldots, x_k$. The research question is to learn whether the linear combination $(a_1)y_1 + (a_2)y_2 + (a_3)y_3 + \cdots + (a_n)y_j$ can be predicted by the linear combination $(b_1)x_1 + (b_2)x_2 + (b_3)x_3 + \cdots + (b_n)x_k$. The scalars of each linear combination a_1 through $a_n$and b_1 through b_n(usually distinct) are chosen such as to maximize this predictive relationship, analogous to how scalars in multiple regression in the regression equation, $y_i = a + b_1x_{1i} + b_2x_{2i} + \cdots + b_px_{pi} + \varepsilon_i$, are chosen to maximize the correlation between "sides" of the multiple regression equation. The only difference is that in multivariate multiple regression, each side consists of a linear combination instead of as in multiple regression where the right-hand side is a linear combination and the left-hand side is simply y_i. Or, if you wish to make things even more general (i.e. a bird flying really high!), you can think of the left side y_i as a very **simple linear combination made up of only a single variable.** Even if the analogy is not perfect, the most general conceptual framework pays dividends when it comes to trying to place these models in context. That is, single variables can be conceived as very simple and elementary linear combinations where all other possible variables are scaled with values of zero. Multivariate analysis typically analyzes more complex linear combinations. When we were doing a simple *t*-test, for instance, we could, in this framework, conclude the response variable to be a very basic linear combination consisting of only a single variable.

What about good 'ol **ANOVA**? Recall in ANOVA that the left-hand side is a single continuous response variable, while the right-hand side is one (one-way ANOVA) or more (factorial ANOVA) independent variables, each scaled as a categorical factor variable. How can this be obtained from the wider canonical correlation model? Simply reduce the left-hand side to a single continuous variable and change the right-hand side to a categorical variable(s) with factor levels:

$$y_1, y_2, y_3, \ldots, y_j = x_1, x_2, x_3, \ldots, x_k \quad \textbf{Canonical model}$$

$$y_1 = x_1(1,2),\, x_2(1,2,3),\, x_3(1,2,3,4), \ldots, x_k(1,\ldots,n) \quad \textbf{ANOVA (one-way through factorial)}$$

where y_1 is the single continuous response, $x_1(1,2)$ is a categorical factor with two levels, $x_2(1,2,3)$ is a categorical variable with three levels, and so on, to represent factorial ANOVA. Hence, we see that **the ANOVA model can be conceptualized as a special case of the wider canonical model.** Formally, it will be different of course, but the above conceptualization is a powerful way to understand how ANOVA relates to the "bigger" more inclusive canonical model. All we need to do is change the coding on the right-hand side of the equation and simplify our "linear combination" on the left-hand side.

How about **MANOVA**? Add more response variables, while keeping the right-hand side of the equation the same as in the ANOVA:

$$y_1, y_2, \ldots, y_j = x_1(1,2), x_2(1,2,3), x_3(1,2,3,4), \ldots, x_k(1,\ldots,n)$$

How about **discriminant analysis**? Flip the MANOVA model around, and for binary discriminant analysis, reduce the number of variables to one, with two levels:

$$y_1, y_2, \ldots, y_j = x_1(1,2), x_2(1,2,3), x_3(1,2,3,4), \ldots, x_k(1,\ldots,n) \quad \textbf{MANOVA}$$

$$x_1(1,2) = y_1, y_2, \ldots, y_j \quad \textbf{DISCRIM}$$

Hence, we see that several of the models we have surveyed in this book can be conceptually (and to a remarkable extent, technically as well) subsumed under the wider canonical correlation model. In a deeper study of these relationships, it would be acknowledged that an **eigenvalue** in one-way ANOVA can be extracted, which will be equal to **SS between** to **SS error**, which is the exact same eigenvalue that is extracted when performing the corresponding discriminant analysis, where the response is the factor of the ANOVA and the predictor is the continuous response. That is, the **eigen analysis**, to a great extent, subsumes both analyses. There is good reason why earlier in the book it was said that eigen analysis is the underground framework to much of statistical models. Eigen analysis is, in a strong sense, what univariate and multivariate statistics are all about, at least at the most primitive technical level. Much of the rest is in the details of how we operationalize each model and the purpose for which we use it. Always strive to see what is in the deepest underground layer of statistical modeling and you will notice that it is usually a remarkably simple concept at work that unites. From "complex" models to sweet simplicity when you notice common denominators. **In statistics, science, and research, always seek common denominators, and you will learn that many things are quite similar or at minimum have similar "tones" in meaning.**

Music theory, for instance, at its core is quite simple, though you would never know it by how the piano player makes it sound complex! But when you understand what the piano player is doing and the common denominator to most jazz arrangements, for instance, you are better able to appreciate the underlying commonality. Unfortunately, you often need to experience a certain degree of complexity before you can "see" or appreciate the simplicity through the complexity. Likewise, if I told you from the start of the book that many statistical models can in a sense be considered special cases of canonical correlation, it would have made no sense. It probably makes more sense now because you have surveyed sufficient complexity earlier in the book to be in a position to appreciate the statement and "situate" the complexity into its proper context.

Hence, though statistical procedures are definitely not completely analogous to one another and do have important technical differences, we nonetheless can appreciate that different procedures have a similar underlying base. That is, the machinery that subsumes many "different" statistical methods is often quite similar at some level, and understanding a bit of this "underground" immediately **demystifies the different names we give to different statistical methodologies**. This is as far as we take these relations here in this book as the deeper comparisons are well beyond the scope of what we can do here, but if you study and explore these relations further, you will no longer perceive different names for many statistical methods as necessarily different "things" in entirety, thereby falling prey to semantics and linguistic distinctions. You will see them as special cases of a more **unified and general framework**, and a certain degree of "statistical maturity" will have been attained in consequence (Tatsuoka, 1971).

Review Exercises

1. What is the difference between a **univariate** and a **multivariate** statistical model? Explain.
2. Why are most models actually **multivariate** from a technical point of view even if they are not typically named as such?
3. Why is using a "**decision tree**" alone potentially unwise when selecting which statistical model to run? Why is it important to merge the **scientific objective** with the choice of statistical model?
4. How can running a **complex multivariate model** be virtually **useless** from a scientific point of view?
5. What is meant by a **linear combination** of variables, and how does this linear combination figure in the MANOVA?
6. In the example featured in the chapter, why is the **linear combination IQ = (1) VERBAL + (1)ANALYTIC + (1)QUANT** considered a "naïve" one? Why is this particular linear combination likely to not be the one chosen by MANOVA?

7. What is **discriminant analysis** and how is it related to **MANOVA**?
8. Why are **multivariate** tests of statistical significance necessarily more complex than **univariate** tests? What do multivariate tests incorporate that univariate tests do not?
9. Prove that **Wilks' lambda** has a minimum value of 0 and a maximum value of 1.0.
10. How is the **trace** involved in Pillai's trace? That is, why take the trace at all? Why is multiplying the inverse of $\mathbf{E} + \mathbf{H}$ and $\mathbf{H}$ not enough? Unpack Pillai's trace somewhat and explain.
11. When would interpreting **Roy's largest root** be most appropriate? Why?
12. Consider the **iris** data and run a **MANOVA** on a linear combination of iris features as a function of species. Interpret Wilk's lambda, and compute an effect size for the MANOVA. Is the result statistically significant, and if so, how much variance is accounted for?
13. On the same **iris** data as in Exercise 12, perform now a **discriminant function analysis** where species is the response variable and the set of features is the predictor space (select two species of your choice). Relate the MANOVA to the discriminant analysis results in as many ways as you can.
14. Conceptually justify the **number of discriminant functions** computable on a set of data. For example, if there are three levels on the response, why is the number of functions equal to 2? If there are two levels on the response, why is the number of functions equal to 1? Explain.

10

Principal Components Analysis

CHAPTER OBJECTIVES

- Learn what principal components analysis (PCA) is and why it can be useful as a multivariate methodology.
- Understand how PCA can be obtained from the basic eigenvalue–eigenvector problem.
- Appreciate the differences between PCA and exploratory factor analysis (EFA), and why they are not the same procedure and should not be treated as such.
- How to potentially make substantive sense out of components derived from PCA and when to recognize when a PCA may not have resulted in anything useful.
- How to conduct a basic PCA in Python and how to interpret results.

You will recall from our discussion of **discriminant analysis** that one of the objectives of the method, in addition to correctly classifying observations, was to reduce the **dimensionality** of the data. That is, the discriminant functions we obtained were derived from reducing the dimensionality of the original data set into one or more functions that could, essentially, "take their place" when trying to classify the data. In other words, instead of using several observed variables on a given number of dimensions, perhaps we could reduce the dimensionality into **linear composites** and then use those linear composites to discriminate between groups. The idea of dimension reduction was thus very important to discriminant analysis, and, as we will see, plays a central role in the methodology considered in the current chapter, that of principal components analysis (PCA).

10.1 What Is Principal Components Analysis?

Principal components analysis (PCA) is a **dimension-reduction** technique (Jolliffe, 2002). Sometimes it is also referred to as a **data-reduction** technique. Both designations are appropriate so long as one understands what each means. It is essential to learn what PCA can do just as much as it is important to understand what it cannot do

Applied Univariate, Bivariate, and Multivariate Statistics Using Python: A Beginner's Guide to Advanced Data Analysis, First Edition. Daniel J. Denis.

so that you are not using the procedure with a sense of hoping it will tell you something about your data and scientific hypotheses that it cannot provide you with. As emphasized throughout this book, understanding the **limitations** of a statistical procedure is just as, if not more important, as understanding what it can do. Hence, it is imperative that we survey PCA in a bit of detail here to understand its underlying mechanics, and, in a strong sense, better appreciate just how limited it can be in addressing scientific problems. An example will help in motivating its purpose.

Suppose a researcher has 1,000 variables and has these variables at his or her disposal. Each of these variables can be said to represent a **single dimension**. For now, the exact nature of these variables or what they are substantively is unimportant. We will simply assume for now that they are measurable on a **continuous** scale. Now, it stands that these 1,000 variables, as a group, account for a certain amount of variance. That is, the first variable, second variable, etc., all have variances associated with them, and hence when we consider the set of variables in entirety, the total number of variables accounts for a well-defined amount of variance; that is, a **totality** of variance. Hence, if a researcher were asked how much variance the set of variables accounts for, he or she would sum up the individual variances of each variable and obtain a sum total variance.

As a simple example, one that we will elaborate on in this chapter, consider the following two variables *x* and *y*, which we build in Python:

```
x = [.00, .90, 1.80, 2.60, 3.30, 4.40, 5.20, 6.10, 6.50, 7.40]
y = [5.90, 5.40, 4.40, 4.60, 3.50, 3.70, 2.80, 2.80, 2.40, 1.50]
pca_data = pd.DataFrame(x, y)
pca_data

Out[84]:
5.9  0.0
5.4  0.9
4.4  1.8
4.6  2.6
3.5  3.3
3.7  4.4
2.8  5.2
2.8  6.1
2.4  6.5
1.5  7.4
```

What is the variance of variable `x`? We can easily compute this in Python using the **statistics** package:

```
import statistics
statistics.variance(x)

Out[74]: 6.266222222222223
```

We see that the variance of `x` is equal to 6.27. Likewise, the variance of `y` is computable:

```
statistics.variance(y)
Out[75]: 1.9133333333333338
```

The **total variance** accounted for by variables x and y is therefore equal to the sum of the variances:

```
total_variance = statistics.variance(x) + statistics.variance(y)
total_variance

Out[77]: 8.179555555555556
```

We therefore see that the total variance accounted for by both variables is equal to 8.18 (rounded up). The question asked by PCA is a remarkably simple one, yet historically quite profound. It is the following:

Can the total variance in the two variables be re-expressed by a transformation?

This question is the only question answerable by PCA, so it is imperative that we examine what this question actually means and how it is answered by PCA. What does it mean to "re-express" variance and why would we want to do such a thing in the first place? The first step to understanding this is to appreciate that **PCA will not fundamentally alter or change the information present in your data.** It will simply **transform the information** on variables onto new axes and dimensions, similar in this respect to what occurred in discriminant analysis. Recall in discriminant analysis that the fact we obtained a number of discriminant functions from our analysis did not "change" the data. It simply **reorganized** it. Analogously, this is likewise what occurs in PCA as well, though the constraints imposed on the procedure will be different. Though PCA operates differently, both techniques seek to **reduce dimensionality**. Much of multivariate (and univariate, for that matter) analysis is about **dimension reduction.** Even ANOVA, "simple" univariate ANOVA, features dimension reduction, though it may not at first glance be apparent.

Consider the following diagram as a graphical depiction of what occurs in PCA, where in this case we only have two original variables:

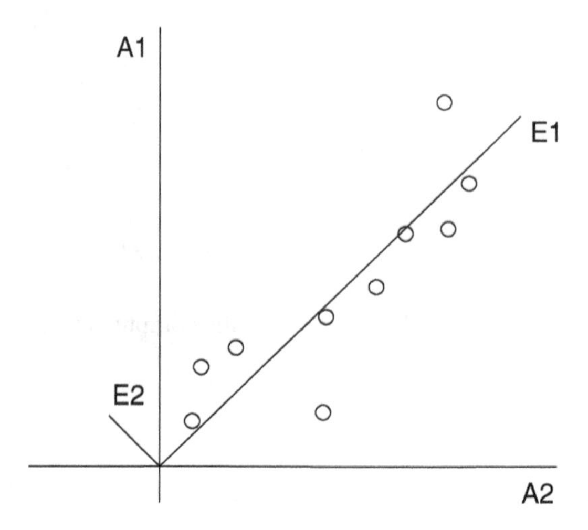

In the plot, A1 and A2 are the **original axes** of the variables. One interpretation of PCA is that it **rotates the original axes** to E1 and E2, where E1 is the first principal component and E2 is the second. Notice that E1 fits the data very well, while E2 does not. By "fits the data" in this sense, we no longer mean it minimizes the **vertical**

distances from data points to the line as in traditional least-squares. Rather, PCA seeks to minimize the **squared perpendicular distances** instead. Notice as well that the newly derived components are **perpendicular** to each other. That is, they are **orthogonal** to one another at 90-degree angles. This is one vital characteristic of PCA – it generates dimensions that are orthogonal to one another and hence each dimension presumably accounts for an individual proportion of variance in the original variables.

10.2 Principal Components as Eigen Decomposition

Though there are different ways of obtaining components in PCA, by far the most common way is to execute an **eigen decomposition** on the **covariance** or **correlation** matrix. Another way is **singular-value decomposition** (SVD). SVD can be used on general rectangular matrices, whereas eigen decomposition is restricted to square matrices (Mair, 2018). In this sense, PCA can be considered nothing more than obtaining the eigenvalues and eigenvectors of a covariance or correlation matrix. Though we have briefly surveyed this process of eigen decomposition earlier in the book, we briefly review it here as it is fundamental to PCA. It is imperative that you see how PCA can be likened to nothing more than an eigen decomposition, and hence, in this respect, is a mathematical **abstraction** first and foremost. What we mean by this is that PCA is an operation on mathematical objects and will not **necessarily** correspond to anything meaningful in **your** particular data.

For a square matrix **A**, it is well known that a scalar λ and vector **x** can be found such that the following equality is true:

$$\mathbf{Ax} = \lambda \mathbf{x}$$

What this means in English is that for a given matrix **A** multiplied by a vector **x**, this product represents the same quantity as taking that vector and multiplying it by a unique scalar λ. Now, that does not help much does it? Here is a better way of looking at it. The vector **x** is being **transformed** by the matrix **A**. This transformation is equivalent in nature to transforming **x** by simply multiplying it by that unique scalar λ. The historical significance of the above equality cannot be overstated. It literally serves as the foundation for many statistical techniques, especially those in multivariate analysis. It is present in univariate techniques such as ANOVA just as it is in more complex multivariate techniques, though it is only in multivariate tools where it is most apparent. The "special vector" is called an **eigenvector**, and the "special scalar" is called an **eigenvalue**. There will typically be as many eigenvectors extracted from **A** as there are rows and columns, and each eigenvector will be associated with its own eigenvalue. Eigenvectors, however, are only unique up to multiplication by a scalar. The eigenvectors, as we will see, are the coefficients to what will become the principal components, and the eigenvalue will be the variance explained by each component.

We can extract eigenvalues and eigenvectors in Python quite easily. First, we will construct the covariance matrix of our two variables, *x* and *y*:

```
data = np.array([x, y])
data
```

```
Out[85]:
array([[0.,  0.9, 1.8, 2.6, 3.3, 4.4, 5.2, 6.1, 6.5, 7.4],
       [5.9, 5.4, 4.4, 4.6, 3.5, 3.7, 2.8, 2.8, 2.4, 1.5]])

covMatrix = np.cov(data, bias = False)
covMatrix

Out[88]:
array([[ 6.26622222, -3.38111111],
       [-3.38111111,  1.91333333]])
```

We see from the above that the variance of the first variable is 6.26 while the variance of the second variable is 1.91. The covariance between variables is −3.38 and appears in the off-diagonal of the matrix. We can also obtain a **heatmap** of the covariance matrix using **seaborn**:

```
import seaborn as sn
sn.heatmap(covMatrix, annot=True, fmt='g')

Out[90]: <matplotlib.axes._subplots.AxesSubplot at 0x75b6438>
```

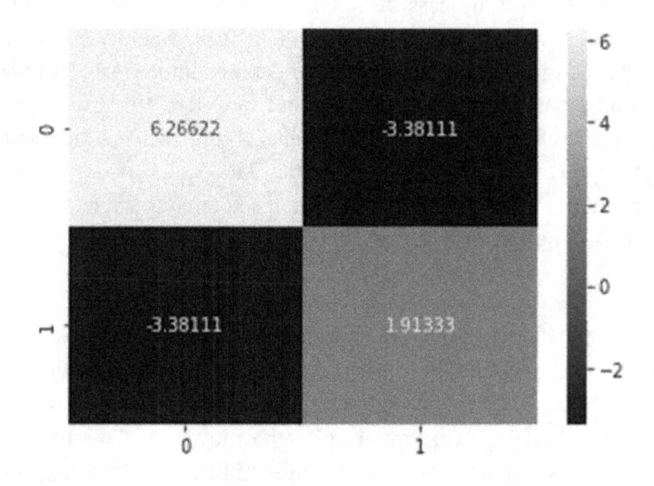

Now that we have successfully constructed the covariance matrix, it is a simple matter to extract eigenvalues and eigenvectors from it. For this, we can use numpy's **linalg** function, which stands for "linear algebra" since eigenvalues and eigenvectors are mathematical objects of linear algebra and are a central topic of that area of study:

```
import numpy.linalg as la
eigen = la.eig(covMatrix)

eigen
Out[94]:
(array([8.11082525, 0.06873031]), array([[ 0.87785621,
0.47892429],
       [-0.47892429,  0.87785621]]))
```

We see from the above that the two eigenvalues are equal to 8.11 and 0.06, with the first eigenvector equal to 0.878 and −0.478, while the second eigenvector having elements 0.479 and 0.878. What we have done with this simple eigen extraction can in fact be considered a rudimentary PCA on a covariance matrix. "Extraction" is probably not the best word here, since, as we will see in the next chapter, this could easily be confused with the process inherent in exploratory factor analysis (EFA) where a user typically designates in advance the number of factors to extract. In PCA, whether we keep a subset of the available components or not, they are still all computed when we perform the procedure. Had we wanted to be a bit more specific about the values we were "extracting," we could have requested and called them as follows:

```
eigenvalue, eigenvector = la.eig(covMatrix)
eigenvalue, eigenvector
```

```
Out[96]:
(array([8.11082525, 0.06873031]), array([[ 0.87785621,
0.47892429],
        [-0.47892429,  0.87785621]]))
```

This better indicates that we are extracting first the eigenvalues followed by the corresponding eigenvectors.

10.3 PCA on Correlation Matrix

Above we conducted a basic PCA on a covariance matrix. Had we wished to perform it on a **standardized covariance matrix**, also known as a **correlation matrix**, we would have computed:

```
import pandas as pd
data = {'x': [0, 0.9, 1.8, 2.6, 3.3, 4.4, 5.2, 6.1, 6.5, 7.4],
'y': [5.9, 5.4, 4.4, 4.6, 3.5, 3.7, 2.8, 2.8, 2.4, 1.5]}
df = pd.DataFrame(data,columns = ['x', 'y'])
corrMatrix = df.corr()
print(corrMatrix)
```

```
          x         y
x  1.000000 -0.976475
y -0.976475  1.000000
```

Why is the correlation perfectly **symmetric**? Notice that the off-diagonals of the correlation matrix are exactly the same. This is because what is contained in the main diagonal, from top left to bottom right, are correlations of the variables with themselves, and hence these correlations will always equal 1.0. In the off-diagonal are the actual correlations between the two variables and just as the covariances were the same in the covariance matrix, these correlations will also be identical. We now compute an eigen decomposition on the correlation matrix:

```
import numpy.linalg as la
eigen = la.eig(corrMatrix)
eigen
```

```
Out[83]:
(array([1.97647522, 0.02352478]), array([[ 0.70710678,
0.70710678],
       [-0.70710678,  0.70710678]]))
```

We see that the eigenvalues extracted are equal to 1.976 and 0.023 respectively for the two components. The first eigenvector is that of 0.707 and -0.707, while the second is 0.707 and 0.707. Notice that **the eigenvalues and eigenvectors extracted from the correlation matrix are not the same as those extracted from the covariance matrix**. Eigen analysis on correlation matrices will usually be different than on covariance matrices. We will revisit this important point later in the chapter. Which matrix you subject to the PCA can have important consequences.

10.4 Why Icebergs Are Not Good Analogies for PCA

The mathematics of PCA are what they are, and as we will see, they are relatively simple and straightforward (at least at the level we are surveying PCA). As we have seen, **PCA is nothing more than a transformation of information onto new dimensions**. That transformation is a technical one, a mathematically rigorous way of accounting for the same amount of variance in variables as by the original variables subjected to the analysis. But what does this really mean on a conceptual level? Not necessarily that much. Though it may be tempting to associate PCA (and its cousin, factor analysis) with "uncovering latent dimensions that underlie observed variables giving rise to them," as with any statistical method, we must be very careful not to project our own "substantive dreams" on the "innocent" mathematics. The mathematics are first and foremost **abstract**, though they may have historically been motivated (and often are) by substantive physical problems. **The point is, however, that the PCA does not know what you are thinking. It is only doing what it does, and that is a mathematical abstract transformation**. That is the only thing we can be sure of, for certain. All PCA does is perform a transformation that generates **orthogonal linear combinations of the original variables**. That's it! It does not do anything more that the scientist may have in mind. Does that mean it may not "mean" more in a particular substantive context? Absolutely not. But the point is that, by itself, PCA is quite "boring." It does not necessarily mean you are "uncovering" (in a physical sense) anything. Hence, when a scientist or researcher uses PCA with the goal of "uncovering" dimensionality, it is a matter of philosophical truth that PCA may not be able to do this for you in the way you may be thinking.

One way to think about PCA, and a lot of mathematics for that matter, is to consider the transformation as simply a **parallel transfer** of information. What we mean by "parallel" here is to suggest that in the transformation, **no new knowledge is *necessarily* being created; it is simply being moved from one form to another**. Though it may hint or suggest new knowledge, it is not a necessary implication of performing a PCA. A parallel transfer, in the way we are using the term, is simply "re-organizing" information from one format into another, such as illustrated in Figure 10.1.

Figure 10.1 Pumpkin with an eigenvector visible before the transformation (left). Pumpkin after the transformation with the eigenvector elongated by an eigenvalue (shown by the dotted arrow) associated with the eigenvector (right). The essential information in the pumpkin remains, but has been subjected to a linear transformation. Image used with permission.

In Figure 10.1, the right image features a pumpkin undergoing a transformation. The original shape is on the left side. The principal component vector, figuratively at least, is also drawn in. Notice that after the transformation, the vector is elongated (indicated by the dashed arrow), since it accounts for more variance in the transformed object than in the original. Though this analogy is not perfect, it is nonetheless useful for grasping what PCA accomplishes, which is nothing more than a **transformation of information onto new axes**, such as in the case of the pumpkin, or, Euclidean space more generally. The object of the pumpkin need not exist for the transformation to make sense. **PCA can be entirely an abstract operation on an abstract space with no recourse to physical representation**. Hence, when you perform a PCA on your variables, it does not necessarily mean the abstract system you are imposing on your physical objects will necessarily make the mathematical transformation "meaningful." It **may** make it meaningful, but it need not **necessarily**. This is very important to understand.

Despite our caveat above, some folks still like to interpret PCA using an iceberg analogy, where above the water are the observed variables and below it are the underlying components, as illustrated in the following:

This iceberg image is not an ideal representation of the PCA idea, however, since there is no reason why the iceberg below the surface of the water should have more information in it than the portion of the iceberg above water. The spirit of the analogy is that PCA will reveal the underlying components to the variables, but the iceberg analogy encourages us to see these underlying components as "hidden" and larger than the ice above the water. If you make the amount of information above and below the water line equal, then the analogy stands more of a chance at being a fair one. But even then, does the analogy hold? That is, why would the transformed information contain more "essence" (even if not more physical space) than the original information? The truth is that there is not necessarily any reason why this should be the case, because, as mentioned, it is simply a transformation of information onto new axes. However, in some applications of PCA, it may make sense to conceptualize it in this manner, but this is **only when the mathematics are merged with substantive theory and it makes sense to do so.** Otherwise, there is simply no basis to conclude on the mathematics alone that PCA or any other technique in statistics "unearths" underlying dimensions. It simply performs a rather anticlimactic and "boring" (I'm being flip, mathematics is far from boring) linear transformation. Anything more than that must come from substantive theory on which you, as a scientist, carry the burden of justifying what it might mean beyond that. The "merger" of statistics with science necessitates this point being very clear – **an analytical method or system cannot impose qualities onto a physical system**. It can only operate on an existing system if there is one to begin with. Hence, whether an underlying dimensionality has been "discovered" or "unearthed" will depend fully on whether you have the substantive material in your science (not in your statistics) to make that argument. Bottom line? A PCA may be nothing, or it may be something, depending on the quality of your data and the scientific inferences you can reasonably make. Otherwise, PCA is simply a mathematical abstraction and every scientist who uses the tool (or any statistical tool) needs to be aware of this.

10.5 PCA in Python

Though one can perform a basic PCA as we have done above using raw linear algebra and computing eigenvalues and eigenvectors, much more common of course in Python is to use a pre-programmed routine or function. We now demonstrate how we can obtain the same PCA as above, but through using `pca.fit()`. For this example, we will build our data file using an `array()` statement:

```
A = np.array([[5.9, 0.0], [5.4, 0.9], [4.4, 1.8], [4.6, 2.6], [3.5,
3.3], [3.7, 4.4], [2.8, 5.2], [2.8, 6.1], [2.4, 6.5], [1.5, 7.4]])

A
Out[97]:
array([[5.9, 0. ],
       [5.4, 0.9],
       [4.4, 1.8],
       [4.6, 2.6],
       [3.5, 3.3],
```

```
       [3.7, 4.4],
       [2.8, 5.2],
       [2.8, 6.1],
       [2.4, 6.5],
       [1.5, 7.4]])
```

Let us obtain a plot of the data:

```
import matplotlib.pyplot as plt
plt.plot(x, y)
```

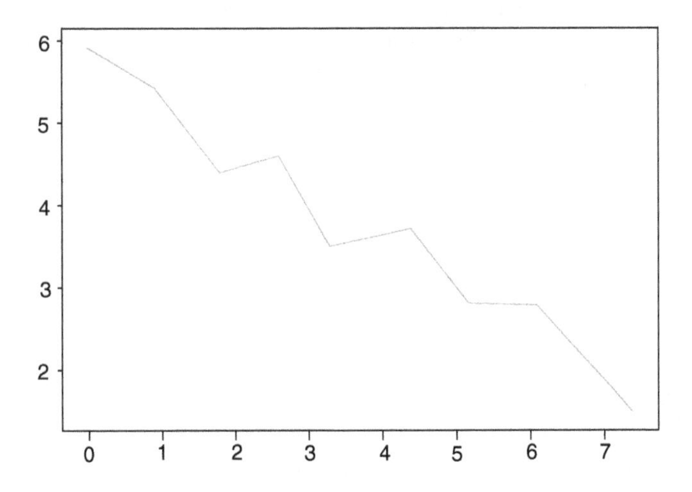

This does not tell us of the actual data points, so we may prefer the following option, where we can see the actual individual data more clearly:

```
plt.plot(x, y, 'o', color='black')
Out[107]: [<matplotlib.lines.Line2D at 0x1e66ce80>]
```

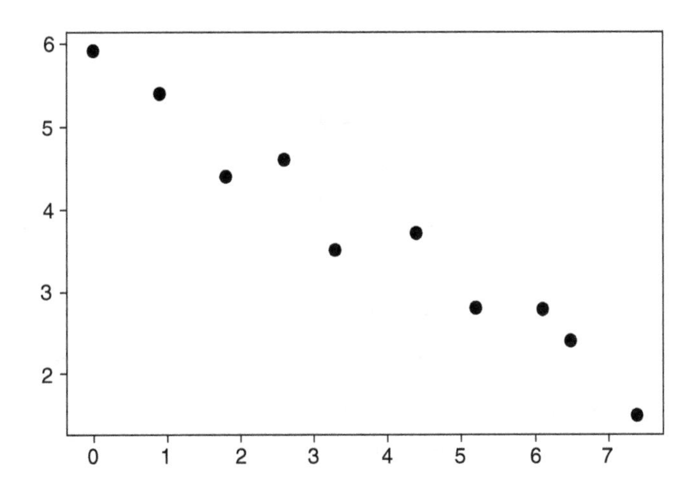

Recall that when we extracted eigenvalues and eigenvectors from our data earlier, there were two sets. This is because the problem had two variables. When we request a PCA in

Python, we can specify the number of components to "extract." We put "extract" in quotes here again to emphasize that in PCA, regardless of the number of components you specify to extract, the procedure is, under normal circumstances, always extracting as many components as there are variables. Hence, when we request the number of components to be less than the number of variables, all we are asking the software to do is to report fewer of the components. When we conduct factor analysis in the following chapter, we will see this is not the case. In factor analysis, **the number of factors that you choose to extract or retain will usually have a mathematical consequence on the structure of those and remaining factors.** This is a very important distinction.

With the above caveat in mind, we conduct a PCA in which two components are extracted, which is the maximum number for this analysis since there are only two variables:

```
from sklearn.decomposition import PCA
pca = PCA(2)

pca.fit(A)
Out[100]:
PCA(copy=True, iterated_power='auto', n_components=2, random_
state=None,
    svd_solver='auto', tol=0.0, whiten=False)

print(pca.components_)
[[ 0.47892429 -0.87785621]
 [-0.87785621 -0.47892429]]
```

Above are the eigenvectors associated with the components analysis. Notice that we have conducted the components analysis here directly on the data. That is, contrary to what we did earlier in the chapter, we did not need to first construct a covariance matrix. We see as well that the signs of the coefficients are a bit different from the original extraction of eigenvalues and eigenvectors. This is common across different software programs and is not a concern. We can obtain the eigenvalues for the PCA through `pca.explained_variance_`:

```
print(pca.explained_variance_)
[8.11082525 0.06873031]
```

As we saw earlier, the first eigenvalue, that of 8.11, is dominant among the two, suggesting that the first component explains about as much of the variance in the variables as both original variables. We can compute the proportion of variance explained by summing the variance and dividing each variance by that sum. That is, since the total sum is equal to 8.18, the proportion of variance accounted for by the two components is:

Component 1 Proportion of Variance Explained = 8.11 / 8.18 = 0.991

Component 2 Proportion of Variance Explained = 0.06 / 8.18 = 0.007

To emphasize our previous point, suppose we had, instead of requesting two components, requested only a single component:

```
from sklearn.decomposition import PCA
pca = PCA(1)
pca.fit(A)
```

```
Out[574]:
PCA(copy=True, iterated_power='auto', n_components=1, random_
state=None,
    svd_solver='auto', tol=0.0, whiten=False)

print(pca.components_)
[[ 0.47892429 -0.87785621]]

print(pca.explained_variance_)
[8.11082525]
```

Notice that the request for a single component rather than two does not change the structure of the first component in terms of its eigenvector; nor does it alter the proportion of variance explained. Again, we demonstrate this to contrast it with exploratory factor analysis of the following chapter, where the number of factors requested in the extraction will typically (for most versions of factor analysis) make a difference to both the vector structure of it, as well as the proportion of variance accounted for by that factor.

10.6 Loadings in PCA: Making Substantive Sense Out of an Abstract Mathematical Entity

We have seen how we can extract eigenvalues and eigenvectors from a covariance or correlation matrix. But what are these things? It needs to be emphasized at the outset that eigenvalues and eigenvectors are mathematical "realities." That is, their existence can be said to be purely abstract in that they naturally fall out of mathematical derivation. They characterize essential information inherent in a linear transformation. Now, to understand what this means requires a bit of maturity into understanding what a mathematical result actually is and why it may not be what you have already preconceived. Most people's perceptions of mathematics is far from the reality of what mathematics actually is. A very simple, if not trivial example will help in appreciating what an abstract mathematical consequence is, one that does not require even understanding the nature of the objects on which the reasoning is taking place. On an abstract level, a pure mathematician may not care that much about the true entity of the objects with which he or she deals. What a mathematician is often most interested is in how they **relate** to one another.

Imagine we have three objects, A, B, and C. What these objects are is not important. What is important is knowing they can be subjected to operations such as addition and subtraction. But again, the "nature" of A, for instance, is immaterial for our purposes (so far). Now, consider the following statements:

> **If C is greater than A, and B is in between A and C, but not equal to either of them, it stands therefore that C is greater than B.**

Wow! What we have just performed is **deductive logic** on relations between objects. Notice that the statement makes good sense even without knowing, understanding, or any **insight** into the objects on which the statements were made! We do not know what A, B, or C are, yet we know the statement holds true based on our own powers of reasoning that tells us it makes good sense. Nowhere in our query did we have to evaluate the **nature** of A, B, or C. It was good enough to simply study how they relate to one another.

Now, what on earth does any of this have to do with PCA? Everything! The above exemplifies how eigenvalues and eigenvectors "exist" distinct from us giving them any substantive or scientific meaning. That is, they are mathematical objects, neutral, and are a consequence of mathematical theorems and derivation. You might say they have no substantive meaning until we give them one. This, in a strong sense, is what PCA does. PCA gives meaning to eigenvalues and eigenvectors such that they may be of use to the scientist in making sense out of his or her data. But, fundamentally, eigenvalues and eigenvectors are, first and foremost, abstract mathematical objects.

So, what do these objects represent then? Eigenvalues and eigenvectors represent different values depending on the given statistical method. However, in PCA, as we have seen, the eigenvalues represent **variances**, while the elements of the corresponding eigenvectors represent **loadings** on the given components. That is, the eigenvalues correspond to the **variance of the components**. The components are the newly derived entities on which the original variables have been translated. The sum of components computed from the analysis explains just as much variance as the original variables, but the hope is that by transforming the original information, the scientist can explain most of the original variance by reference to the components only; hence the idea of variance consolidation or "data reduction" we emphasized earlier. To summarize:

Eigenvalues in PCA = variances of components.
Eigenvectors in PCA = loadings of components.

But just how can we use the loadings? We can potentially use the loadings to identify or "name" the given component. Loadings with greater size suggest the given component is "composed" more of that variable than others, but there are many difficulties with this. Let us explore a couple of them.

10.7 Naming Components Using Loadings: A Few Issues

Though we can use loadings to help "identify" the nature of a component substantively, a few caveats and warnings are in order. If these are not heeded or appreciated, the user runs the risk of concluding something from the data that simply is not justified. We consider these caveats in some detail:

- Whatever component is derived is based on the variables inputted into the analysis, and no other variables. Hence, the selection of variables for the PCA is the most important and crucial step of the entire components analysis. Of course, it may seem obvious that the components analysis is a function of the variables subjected to the analysis, but it is a point that can easily be lost on users. How is such unawareness evident? It is by drawing conclusions of the "physical nature" type that simply may not be warranted or justified for the area under investigation. For example, if a social scientist subjects 100 variables to a PCA, the PCA is not "uncovering" components or making a "discovery" of any sort. As we have said, all the PCA is accomplishing is a transformation. Hence, you guessed it, had more or less variables been subjected to the PCA than the original 100, the results and findings would likely change! That is, your PCA is only as good as the variables you subject to it. You should choose those variables very carefully, otherwise **your PCA may simply be misspecified**. As a scientist, spend 99% of your time selecting the correct candidate variables and 1% conducting the PCA! All loadings obtained in PCA are relative to all other variables subjected to the analysis.

- Related to the above point, social or psychological components may be more difficult to substantiate than components derived from an obvious physical process. For example, in digital images, components can be extracted that account for most of the variance in the image (Izenman, 2008). This is a guaranteed physical process, however. And though we can apply PCA to data collected on social data, it is more difficult to make the argument that such extracted components represent physical "building blocks" as they do in things like the digital image. Indeed, early motivations behind PCA were in representing physical and biological objects. Descriptions of such systems may be more "convincing" than claiming that IQ has a three-component structure, for instance. As a researcher, especially one doing social science, you should be aware of these distinctions. Do not simply subject your data to PCA and draw conclusions of the type a physical scientist would draw unless you have a very strong argument that your components are a "breakdown" of a physical or similar process as that for a physical object.
- Components derived from the correlation matrix will usually not be the same as those obtained from the covariance matrix. Hence, component extraction and retention can be quite subjective since they depend on the original matrix subjected to the analysis. This is related to the above point, since it tells us that "**existence**" of components underlying variables may be the wrong way to think of PCA. All PCA is doing (for the millionth time!) is a **transformation**, and so the details of that transformation must be clearly understood if one is to understand what has actually been obtained in the PCA. As we will see when we survey exploratory factor analysis in the following chapter, this concern is even of more significance in that procedure. The bottom line recommendation here is to not immerse yourself into issues of "existence" (or at least be very cautious about them) when performing either PCA or factor analysis, as they will not necessarily arise in the context you are working in. These procedures are not akin to an archeological dig where you eventually have a good sense of certainty in what has been found and can touch and feel it.
- PCA results may not be comparable across studies. It would be a mistake to conclude that you found three components while another study found four without unpacking **all of the details inherent in and a part of each study**. Even if assuming the same measures were used, they were surely used on different populations, and hence the details of each population should be advanced if any comparison is to be made. And as mentioned, even the inclusion of an additional variable can "throw off" your component structure, and hence comparing studies where different variables were used (even if that difference is slight) can have significant consequences to each analysis.

In using PCA (or any other statistical method for that matter), the user needs to be intimately familiar with what can vs. cannot be concluded from the procedure or algorithm. For example, PCA does not "discover" components. Rather, it merely transforms a high-dimensional "structure" onto another. Hence, an argument that a component "exists" is usually a weak one, unless one is working with physical variables (or equivalent substantively defensible variables) on which such a conclusion can be justified by reference back to the physical objects. Concluding that a component structure underlies something like intelligence, for instance, will usually require a bit more work and substantive evidence than in the purely physical (and simpler, in this manner of speaking) case.

10.8 Principal Components Analysis on USA Arrests Data

We now demonstrate a full components analysis on a more complex and realistic data set than the previous one on two variables. Our data contains statistics on the number of arrests per 100,000 residents in all states in the USA in the year 1973. The variables in our data are **murder**, **assault**, **urbanpop**, and **rape**, and the total number of observations is 50. The murder, assault, and rape variables are the number of such crimes per 100,000 cases, while urbanpop is simply the percent of the population that is urban. The data file is in `.csv` format and we use `pandas` to read the file, printing only the first 10 cases:

```
data = pd.read_csv('usarrests.csv')
data.head(10)

Out[583]:
   Unnamed: 0  Murder  Assault  UrbanPop  Rape
0     Alabama    13.2      236        58  21.2
1      Alaska    10.0      263        48  44.5
2     Arizona     8.1      294        80  31.0
3    Arkansas     8.8      190        50  19.5
4  California     9.0      276        91  40.6
5    Colorado     7.9      204        78  38.7
6 Connecticut     3.3      110        77  11.1
7    Delaware     5.9      238        72  15.8
8     Florida    15.4      335        80  31.9
9     Georgia    17.4      211        60  25.8
```

We select only the quantitative variables for analysis. That is, we will not include the state variable, as it is nominal in nature. Incidentally, when selecting variables from a dataframe, if you misspell the variable, it will likely come out as `NaN` as follows:

```
df = pd.DataFrame(data, columns=['Murder', 'Assault', 'Urbanpop',
'Rape'])
df

Out[351]:
   Murder  Assault  Urbanpop  Rape
0    13.2      236       NaN  21.2
```

```
1      10.0       263         NaN  44.5
2       8.1       294         NaN  31.0
3       8.8       190         NaN  19.5
4       9.0       276         NaN  40.6
```

Our error above was requesting `Urbanpop` instead of `UrbanPop` (i.e. with the capital P). When we request it correctly, we obtain the correct data layout:

```
df = pd.DataFrame(data, columns=['Murder', 'Assault', 'UrbanPop',
'Rape'])
df

Out[353]:
   Murder  Assault  UrbanPop  Rape
0    13.2      236        58  21.2
1    10.0      263        48  44.5
2     8.1      294        80  31.0
3     8.8      190        50  19.5
4     9.0      276        91  40.6
5     7.9      204        78  38.7
```

Now that our data file is set up correctly, let's obtain some means of the variables before we conduct the PCA. For this, we will use `np.mean()`:

```
import numpy as np
np.mean(df)

Out[355]:
Murder        7.788
Assault     170.760
UrbanPop     65.540
Rape         21.232
dtype: float64
```

The variances of the variables are given by `np.var()`:

```
np.var(df)

Out[356]:
Murder        18.591056
Assault     6806.262400
UrbanPop     205.328400
Rape          85.974576
dtype: float64
```

As we will see later when we also compute the variances using pandas, `np.var()` uses "n" in the denominator for the sample variance and not "$n-1$" as does pandas. That is, the computation of the variance above for murder of 18.591056 is computed as

$$S^2 = \frac{\sum_{i=1}^{n}(y_i - \bar{y})^2}{n}$$

rather than by

$$s^2 = \frac{\sum_{i=1}^{n} (y_i - \overline{y})^2}{n-1}$$

The difference between the two calculations will usually not be that great, especially for large sample size, but typically, since we usually desire an unbiased estimator, the second choice with $n-1$ is what is reported by most software. To get the unbiased estimate in numpy (i.e. with $n-1$), you can code `np.var(df, ddof=1)`, where `ddof=0` is the default with simply n in the denominator. Regardless of what denominator we use, we nonetheless notice that the variances are quite different from one another. Hence, if we run the components analysis on the raw data, the assault variable with its quite large variance might dominate the solution. The way around this is to **standardize** the data. We do so using `StandardScaler()`:

```
from sklearn.preprocessing import StandardScaler
scaler = StandardScaler()
scaler.fit(df)

Out[357]: StandardScaler(copy=True, with_mean=True, with_std=True)

scaled_data = scaler.transform(df)
scaled_data

Out[358]:
array([[ 1.25517927,  0.79078716, -0.52619514, -0.00345116],
       [ 0.51301858,  1.11805959, -1.22406668,  2.50942392],
       [ 0.07236067,  1.49381682,  1.00912225,  1.05346626],
       [ 0.23470832,  0.23321191, -1.08449238, -0.18679398],
       [ 0.28109336,  1.2756352 ,  1.77678094,  2.08881393],
       [ 0.02597562,  0.40290872,  0.86954794,  1.88390137],
```

These data are now standardized, but we notice that Python has removed the headers of the variables. We can easily get the headers back onto our data file via the following:

```
scaled_data_df = pd.DataFrame(scaled_data, columns = df.columns)
scaled_data_df

Out[363]:
     Murder   Assault  UrbanPop      Rape
0  1.255179  0.790787 -0.526195 -0.003451
1  0.513019  1.118060 -1.224067  2.509424
2  0.072361  1.493817  1.009122  1.053466
3  0.234708  0.233212 -1.084492 -0.186794
4  0.281093  1.275635  1.776781  2.088814
5  0.025976  0.402909  0.869548  1.883901
```

Now, if the variables have been standardized as we requested, each variable should now have a mean of 0 and variance of 1. Let's confirm this by obtaining first the means, using once more `np.mean()`:

```
np.mean(scaled_data_df)

Out[578]:
Murder      -8.437695e-17
Assault      1.298961e-16
UrbanPop    -4.263256e-16
Rape         8.326673e-16
dtype: float64
```

We can see that within rounding error, the mean of each variable is now equal to 0 (recall that `-8.437695e-17` means to move the decimal point 17 positions to the left, hence making it an extremely small number). Since the variables have been standardized, the variance of each variable should equal 1, as we can easily confirm via `np.var()` of our scaled data:

```
np.var(scaled_data_df)
Out[579]:
Murder      1.0
Assault     1.0
UrbanPop    1.0
Rape        1.0
dtype: float64
```

Based on the above then, we have confirmation that Python has indeed correctly standardized the data. Let's proceed now to run the PCA on the scaled data. We will request a total of four components, which is the maximum number of components possible for this data since there are only four variables. We specify the number of components via the statement `n_components=4`:

```
from sklearn.decomposition import PCA
pca = PCA(n_components=4)
pca.fit(scaled_data_df)

Out[364]:
PCA(copy=True, iterated_power='auto', n_components=4, random_
state=None,
    svd_solver='auto', tol=0.0, whiten=False)

pca.components_
Out[366]:
array([[ 0.53589947,  0.58318363,  0.27819087,  0.54343209],
       [ 0.41818087,  0.1879856 , -0.87280619, -0.16731864],
       [-0.34123273, -0.26814843, -0.37801579,  0.81777791],
       [ 0.6492278 , -0.74340748,  0.13387773,  0.08902432]])
```

In the above, **each component is a row, not a column**. For example, the first component has loadings of 0.53589947, 0.58318363, 0.27819087, and 0.54343209, and so on for the remaining components. The first component is thus

$$\mathbf{0.53(Murder) + 0.58(Assault) + 0.27(UrbanPop) + 0.54(Rape)}$$

The second component is

$$\mathbf{0.41(Murder) + 0.18(Assault) - 0.87(UrbanPop) - 0.16(Rape)}$$

and so on for the last two components. So, what does each component represent? Notice that the first component seems "composed of" primarily murder, assault, and rape. The coefficient for UrbanPop is a bit less with a loading of only 0.27, hence maybe the first component represents a combination of crimes. The second component is a bit more heavily weighted on UrbanPop (though the weight for murder is also creeping up a bit too). This component is much more difficult to make sense of. And this fact is a good thing, at least for demonstration and pedagogical purposes. It exemplifies (again) that "naming components" can be very elusive. While components analysis has done its job, it does not mean the resulting components will "make sense" on a substantive level. They make sense **mathematically**, yes, but they may not make sense **substantively**. This last point cannot be emphasized enough if you are to have a clear-headed understanding of what components analysis accomplishes vs. what it does not.

We can easily obtain the proportion of variance explained using `pca.explained_variance_ratio`:

```
print(pca.explained_variance_ratio_)
[0.62006039 0.24744129 0.0891408  0.04335752]
```

Note that the sum of the proportions equals 1.0. The sum of the variances for each component will also sum to 4. This is as a result of performing the PCA on the **standardized data** rather than the **raw data**. Had we performed it on the raw data, then the sum of variances should have equaled the **sum of the variances of the raw variables** inputted into the analysis. Let us verify this. We perform the PCA on the raw data:

```
from sklearn.decomposition import PCA
pca = PCA(n_components=4)
pca.fit(df)

Out[368]:
PCA(copy=True, iterated_power='auto', n_components=4, random_
state=None,
    svd_solver='auto', tol=0.0, whiten=False)

pca.components_

Out[369]:
array([[ 0.04170432,  0.99522128,  0.04633575,  0.0751555 ],
       [ 0.04482166,  0.05876003, -0.97685748, -0.20071807],
       [ 0.07989066, -0.06756974, -0.20054629,  0.97408059],
       [ 0.99492173, -0.0389383,   0.05816914, -0.07232502]])
```

Next, we compute the actual variances of each component:

```
print(pca.explained_variance_)
[7.01111485e+03 2.01992366e+02 4.21126508e+01 6.16424618e+00]
```

We then sum these variances:

$$7{,}011.11+201.99+42.11+6.16=7{,}261.37$$

The total variance of the components is thus equal to 7,261.37. If we sum the variances we computed earlier using `np.var()`, that sum will equal 7,116.15. However, if PCA is working correctly, the sum of variances of the original variables should be equal to that of the components. But 7,261.38 is clearly different from 7,116.15. This is where the denominator issue discussed earlier matters. When we use pandas' `df.var()`, we obtain instead:

```
df.var()
Out[395]:
Murder          18.970465
Assault       6945.165714
UrbanPop       209.518776
Rape            87.729159
dtype: float64
```

The sum of these variances is equal to 7,261.38, which agrees with the sum of the variances of the components just above. Again, this is different from the variance computed by numpy, because pandas is using the "$n-1$" denominator by default, whereas `np.var()` used only n in the denominator (unless we specify `np.var(df, ddof=1)`). We can also quite easily obtain the standard deviations of the original variables, as they are merely the square roots of the above variances:

```
df.std()

Out[404]:
Murder         4.355510
Assault       83.337661
UrbanPop      14.474763
Rape           9.366385
dtype: float64
```

Summing these squared standard deviations is also equal to the sum of the variances for the components (verify this for yourself).

10.9 Plotting the Components

A convenient way to obtain plots of the components solution is to use the **pca** library in Python or the **yellowbrick** package. We use both below. Because these packages do not come already installed with the Anaconda distribution, we will need to install them via **PIP** (refer to our previous discussion in Chapter 2 on how to do this in Anaconda):

```
pip install pca
from pca import pca
```

```
pip install yellowbrick
import yellowbrick
```

We then, as before, run the model obtaining all four components:

```
model = pca(n_components=4)
```

We will now first generate plots on the standardized data. To generate these plots, we use the scaled data with no headers (i.e. recall that we had put the headers back on earlier; do not use that dataframe, use the one with no headers instead):

```
X = scaled_data
results = model.fit_transform(X)
```

Requesting the results object will provide much information (e.g. to obtain outlier details, try `print(results['outliers'])`, most of which we do not show here since our objective is simply to produce a plot (for details, see the PCA library). We do, however, print the loadings that result:

```
results
Out[305]:
{'loadings':1          2          3          4
 PC1  0.535899  0.583184  0.278191  0.543432
 PC2  0.418181  0.187986 -0.872806 -0.167319
 PC3 -0.341233 -0.268148 -0.378016  0.817778
 PC4  0.649228 -0.743407  0.133878  0.089024
```

We first produce a generic plot before generating a biplot. The first plot is a scatterplot of the first two component scores (left) using the pca package, while the second plot (right) is a biplot using yellowbrick:

```
fig, ax = model.scatter()


fig, ax = model.biplot(n_feat=4)
```

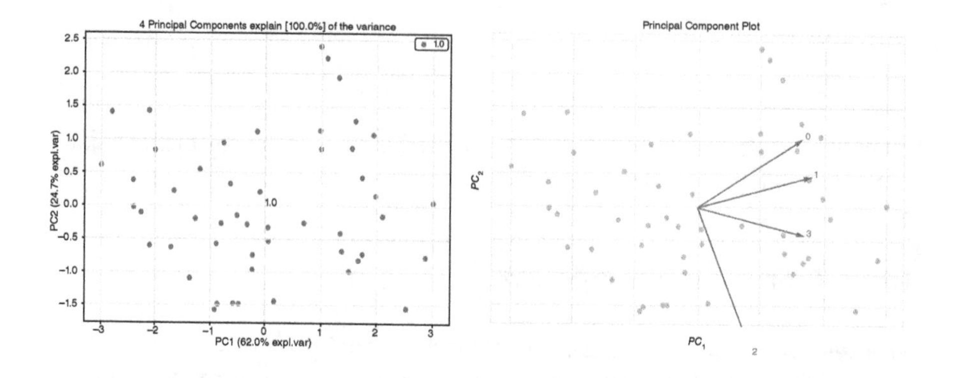

In the biplot (right), we can see that the first PC, scaled along the *x*-axis, has variables 0, 1, and 3 (corresponding to murder, assault, and rape, respectively) that are further to the right of the plot than the third variable ("2" in the plot, since the variables are numbered 0, 1, 2, 3), corresponding to urbanpop. In agreement with the component loadings in the above table, these variables are most "important" on the first component. Now, consider the plot from the vantage point of the second component (PC2). From this vantage point, we can see that the third variable ("2" in the plot, urbanpop) dominates the component, while the other three variables are much less in magnitude. Again, this agrees with what we are seeing in the loadings output. Hence, the biplot allows us to visualize variables across components in terms of their sizes on each component.

The following plots the variances explained by each of the components. As expected, the first component accounts for the majority of the variance in the variables, while the fourth and last component accounts for the least. The sum total of components extracted account for 100% of the explained variance. This is not surprising, since, as mentioned, PCA maintains the total variance in the original variables. It simply (and perhaps rather anticlimactically) re-expresses it into orthogonal dimensions. Thus, if the total variance did not sum to 100%, it would be suggestive of a problem with the components analysis:

```
fig, ax = model.plot()
```

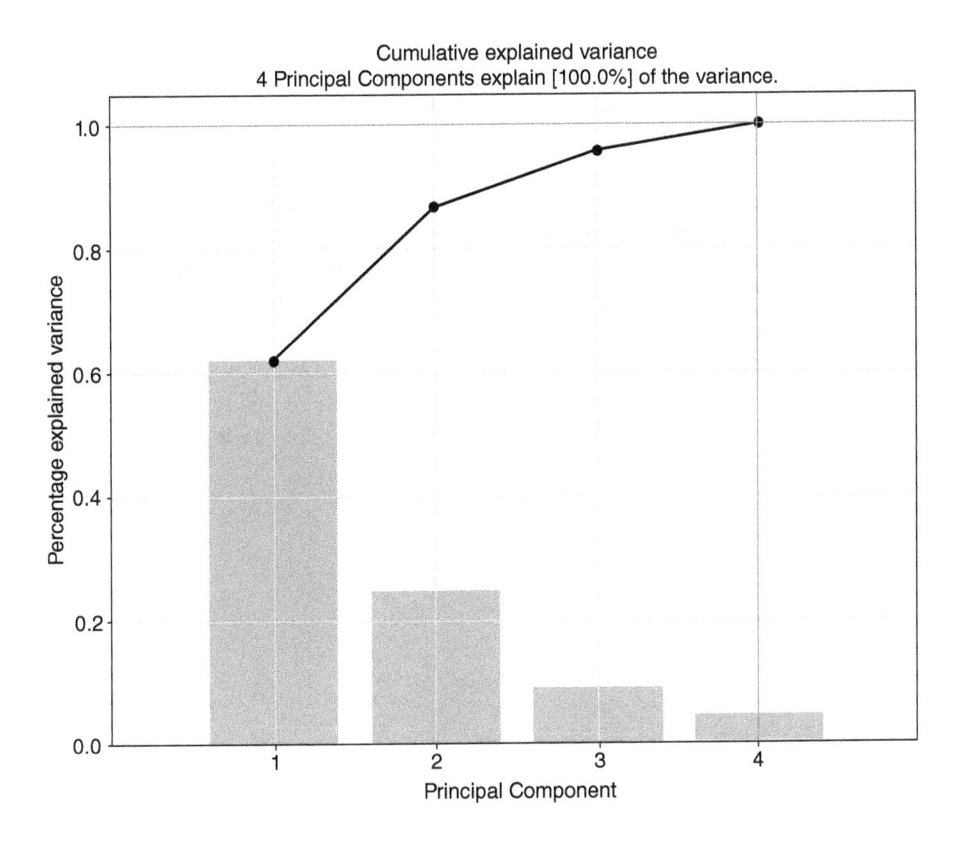

We can also generate a **3D plot** (not shown), which, as a result of the new dimension, we can plot the first three components rather than the first two (`fig, ax = model.scatter3d()`) as well as the corresponding biplot (`fig, ax = model.biplot3d(n_feat=4)`).

Review Exercises

1. Discuss the purpose of the **PCA**. What might motivate a researcher to consider such a method for his or her variables?
2. Why is it incorrect to say that PCA **"uncovers underlying components"** unless one actually knows what this statement means? What is the danger of interpretation in not understanding the technical base of PCA yet still using this phraseology?
3. Your colleague says to you, "PCA issues a **transformation** on data and nothing more." Do you agree with your colleague? Why or why not?
4. Why is it true that the **total amount of variance** in the variables before conducting a PCA is the same found in the components after conducting one? Explain.
5. Critically evaluate the use of **icebergs** as an analogy to what PCA accomplishes. Why might such an analogy be problematic?
6. Discuss PCA in terms of an **eigenvalue–eigenvector** problem. What is actually taking place when computing these mathematical entities? Describe the process.
7. Is running a PCA on a **covariance matrix** the same as that of running one on a **correlation matrix**? Why or why not?
8. Consider the **Breast Cancer Wisconsin (Original)** data set on several attributes related to breast cancer, including clump thickness, uniformity of cell size, up to and including mitoses. Consider all 10 of these variables, all measured on a continuous scale. Perform a PCA on this data to determine whether fewer than 10 dimensions can account for the majority of the variance in the original variables, and, if so, attempt to interpret these components. What proportion of the total variance do these components account for?
9. Confirm for your analysis in exercise 8 that the total variance of all components equates to the total variance of the original variables. Obtain a plot that summarizes the results of the PCA.
10. Consider the idea of extracting components from a digital image. Suppose that underlying a digital image made up of 100 variables are five components. How might this use of PCA differ from using it to extract components from social variables? Is it different at all? Explain.

11

Exploratory Factor Analysis

CHAPTER OBJECTIVES

- Learn the nature of exploratory factor analysis (EFA) and why, as a scientist, you may wish to perform one.
- Distinguish between EFA and PCA, and understand why they should not be likened to one another and are quite distinct statistical methods.
- Appreciate the fact that EFA has several philosophical pitfalls that can make interpreting findings from the approach relatively subjective.
- Understand that solutions in factor analysis are not invariant to the number of factors selected for extraction.
- Understand the nature of factor rotation and why rotating factors is not "cheating" or "fudging data" so to speak.
- How to perform a basic EFA in Python and interpret results.

As was the case for principal components analysis (PCA) of the previous chapter, **exploratory factor analysis** (EFA) can be seen as a **dimension reduction** technique in which a researcher begins with several observed variables and wishes to "uncover," in a manner of speaking, dimensions that underlie the observed data. We place "uncover" in quotes here for the good reason that we must be cautious in how we interpret this word in relation to factor analysis, as we also had to be cautious in our discussions of PCA of the previous chapter. As we will see throughout this chapter, factor analysis does not "find" factors in one's data in an objective concrete sense. In fact, we will see, whether or not factor analysis results in a viable solution often depends on very subjective decisions made by the researcher. Hence, "finding factors" in factor analysis is quite unlike that of the archaeological dig we referred to in which an explorer discovers remnants of past civilizations. In factor analysis, **what you "find" often depends on assumptions and inputs you a priori decide for the analysis**. This point cannot be overstated.

By the end of the chapter, it is hoped that you will appreciate both the benefits but also the drawbacks to using and interpreting an EFA. Without understanding factor analysis, just as with any statistical method, you are more likely to misuse and

Applied Univariate, Bivariate, and Multivariate Statistics Using Python: A Beginner's Guide to Advanced Data Analysis, First Edition. Daniel J. Denis.

misinterpret its findings. Though this is also true for methods such as ANOVA and regression, the situation is even more dire with factor analysis, as we will discuss. It is exceptionally difficult for a statistical consultant to recommend that you perform factor analysis unless they are intimately familiar with your research area. That is, when it comes to factor analysis, **substantive knowledge of the domain of interest is paramount**. Do you believe **latent variables** give rise to the observed variables you have measured? If so, then EFA may be an option. Be forewarned that factor analysis is an immense and very technical subject and hence our treatment here is very cursory. For a detailed account of all of its complexities, consult Mulaik (2009).

Before surveying EFA in some detail, it is key at the outset that you do not equate factor analysis with the method of the previous chapter, that of PCA. **EFA, while sharing some similarities with PCA, is not an equivalent method**, and there are very important differences between the two procedures. A failure to understand the differences between these two procedures can lead to a misuse and misinterpretation of either or both.

11.1 The Common Factor Analysis Model

As mentioned, factor analysis is a huge topic spanning many technical articles, chapters, and books. Of all the statistical methods featured in this book, factor analysis, for good or bad, is probably the one that has the most **complexity** associated with it. Why is this the case? It is so because there are so many options and inputs to factor analysis that to truly become an efficient expert at it, you should have at least some understanding of these many issues before using the technique. In this book, we survey only one approach to factor analysis, which is undoubtedly the most common. This is referred to as the **common factor analysis model**, and is given by the following:

$$\mathbf{x} = \mu + \Lambda\mathbf{f} + \varepsilon$$

where $\mathbf{x}$ is a vector of **observed random variables**. These are variables the researcher has collected and would like to subject to the factor analysis (you might call them the "candidate" variables). The number of variables in this vector can range from very few to a very large number, depending on the complexity of the research problem. To use a previous example, suppose a researcher's observed variables include quantitative, verbal, and analytical abilities as measured on some psychometric test. These are regarded as "observed" variables because we are able to obtain numerical measurements on them, and hence they are considered readily available to the researcher. In the language of factor analysis, they can be "seen" by the score on the psychometric test. In factor analysis language, this is different from an **unobserved** or **"latent"** variable, which is a variable that is regarded as not readily observable and hence must be "inferred" in a manner of speaking from other more observable variables. In the example we have used throughout the book, perhaps quantitative, verbal, and analytical, taken as a set, are indicators of an underlying construct or more complete variable, perhaps that of IQ. The primary rationale for performing a factor analysis is because the analyst hypothesizes or believes that underlying such observed variables is a **latent structure** that cannot be seen but rather is "responsible" for generating the correlations among the observed variables. Hence, if a researcher does not have this hypothesis in mind, then performing factor analysis may be ill-advised. Factor analysis focuses on the covariances and correlations among variables. If the researcher simply wants to extract orthogonal sources of variance, and hence simply reduce dimensionality without the assumption of underlying latency giving rise to observed variables, then the principal components of the previous chapter would usually be the preferred analytical strategy.

Returning to the factor analysis model, μ is a vector of means for the random variables contained in $\mathbf{x}$, and Λ is a matrix of factor loadings applied to the factors in $\mathbf{f}$. Hence, $\mathbf{f}$ in this case is a vector of **unobservable factors** that are weighted by the loadings in Λ. This idea should sound very familiar, because it is somewhat analogous to regression analysis. But how is factor analysis different from regression? In regression analysis, recall that instead of $\mathbf{f}$, an unobservable vector, the regression model instead contained $\mathbf{x}$, that is, **observable variables.** Each of these variables were, analogous to Λ, weighted by β, the vector of regression weights. Hence, we see the primary distinction between factor analysis and regression analysis. In regression analysis, nowhere in the model was there hypothesized to be a latent term, other than the error term ε, which can always be considered latent (i.e. errors are usually considered "unobservable"). Otherwise, all predictor variables are considered known and measurable without error in the classic regression model. In the factor analysis model, however, we are hypothesizing observed variables in $\mathbf{x}$ as a function in part of unobservable variables called "factors" in $\mathbf{f}$.

As mentioned, the factor analysis model contains a vector of errors, ε. These are generally known as **specific factors** in factor analysis and account for the variation unaccounted for by the common factors in the model. That is, anything that cannot be explained by $\mathbf{f}$ winds up in the error term. This is somewhat analogous to regression analysis where what cannot be accounted for by predictors ends up in the error term. In a more advanced study of factor analysis, some theorists will further differentiate specific factors into further components of variance, but doing so is beyond the scope of this book. For details, see Mulaik (2009).

Hence, we can see that the factor analysis model is, in a sense, closely analogous to the regression model, but with the caveat that other than the error term, no other unobservable variables appear in the regression model. In both cases, however, both models attempt to account for a systematic component of the data vs. an unsystematic one. But factor analysis does so by theorizing **unobservable latent dimensions**, and hence right from the start is more controversial than the traditional regression model.

11.2 Factor Analysis as a Reproduction of the Covariance Matrix

Though we have seen that the factor analysis model can be correctly interpreted as observed variables being a function of unobserved variables, mathematically it can be shown that when the factor analysis model fits the data well, the model itself implies a **reconstruction of the covariance (or correlation) matrix**. Though it is well beyond the scope of this book to justify why this is the case (see Denis, 2021, for details), it turns out that the factor model implies that the covariance matrix of observed variables can be reconstructed by the following:

$$\Sigma = \Lambda\Lambda' + \Psi$$

where $\mathbf{\Sigma}$ is the covariance of observed variables, and $\mathbf{\Lambda}$ are the factor loadings we just discussed and surveyed. The matrix $\mathbf{\Psi}$ is one of specific variances. Hence, what the equation says is that the covariance matrix can be written as a function of "squared loadings," where $\mathbf{\Lambda}\mathbf{\Lambda}'$ is the "squared" (i.e. we are multiplying by the transpose and not technically squaring, but it is helpful to think of it as squaring in this case) component represented by the loading matrix multiplied by its transpose, in addition to the matrix of specific variances. If the factor model is a good fit to the data, then it follows that the covariance matrix should likewise be reproducible by $\mathbf{\Lambda}\mathbf{\Lambda}' + \mathbf{\Psi}$. The sum of such "squared" factor loadings for a variable across the factor solution is what is known as a **communality**, and, as the name suggests, tells us the degree to which the variance in the variable is "explained" by the factor solution. The balance of the communality, that is, the variance **unexplained** by the factor solution for each variable, is the **specific variance** associated with that given variable. These quantities will make more sense later when we see them estimated in our software examples. Notice that it is the variance in the observed variable we are seeking to explain by the factor solution, and not the other way around. In the factor model, the observed variable is theorized to be a function of latent variables. Given all this, we can also interpret the goal of factor analysis in this slightly more technical yet simplified way; **factor analysis attempts to estimate loadings in such a way that they maximize the reproduction of the covariance matrix.** In the case of a correlation matrix, which recall is simply a standardized covariance matrix, the goal is the same. Regardless of the matrix, factor analysis is, "behind the scenes," trying to "reproduce" it.

This idea of "reproducing" a function of the variables may sound familiar. We can again relate it to regression. Recall that the goal of least-squares regression is to minimize the sum of squared errors in prediction. What this translates into as well is maximizing the correlation between observed and predicted values on the response variable. Hence, when pondering the question of how well a regression model fits, one asks the question, **"How well does it reproduce a function of the observed variables?"** The extent that it does is the extent to which the model is well-fitting. This is a similar idea as in factor analysis, where we are hoping the model regenerates the observed covariance matrix of variables. Recall that models, by their nature, are in the business of hypothesizing a structure to observed data. More generally, this is precisely what mathematical modeling (of which statistical and "decision sciences" modeling is a special case) is all about, that of describing empirical data by an abstract mathematical equation or algorithm of sorts.

11.3 Observed vs. Latent Variables: Philosophical Considerations

Having introduced this idea of "latent" or "unobservable" variables, the insightful reader at this point should not be anywhere close to at ease or satisfied with the definitions given earlier. In fact, he or she should be quite troubled by them! Being critical and unsatisfied with trivial definitions is a first sign of intellectual depth and curiosity, which is a great thing! The question of whether an observed vs. latent distinction is "real" is definitely a good one, and we discuss the matter now in an attempt to resolve it, even if imperfectly.

At first glance, the idea of latent variables even "existing" may seem silly. That is, what does it mean to have a variable that "gives rise" in this sense to other more observable variables? Indeed, the idea of hidden variables or variables that lie below the surface (maybe "limen"?) is philosophically problematic, but then again, so is the idea of "observed" variables just as problematic. Yes, even so-called "observed" variables are problematic too. How so? Allow me to explain using a simple example of a variable none of us would, at first, deem philosophically debatable or troublesome, that of **weight**. Is weight observed or latent? Most people would say it is observable, since one can step onto a bathroom scale, receive a measurement, and learn of one's current weight. However, even in this simple procedure, there are numerous psychometric issues that arise that at first thought you may not have considered.

First, weight is a **construct**, no matter how much we would like to think it is not (just follow me on this for a second). Weight is an "idea" that we believe underlies numerous other more observable events such as the amount of pressure being applied to the scale, the instrumentation involved in constructing the physical weight scale, and all of the other components to the measurement of weight. Hence, the variable "weight" does not exist separately from our inference of it based on all of these other components that contribute to the construct. That is, nowhere in nature is there "natural" weight. Like any variable, it is a **human construction**, and we choose to designate it as a sum component of numerous more observable events or procedures. Hence, the measurement of weight involves a lot more than just "weight." To be blunt, the thing you step on in the morning (you know, to ruin your day) is also involved in this thing called "weight." **The way we measure weight has a lot to do with what weight is.**

Second, when you step on the bathroom scale, what you are receiving as a measurement of your weight is not your actual weight. Rather, it is an **estimate** of your actual **true weight**. Your true weight will forever be unknown. Why? Because there is understood to be **measurement error** in the estimation of your true weight. Stepping on the scale does not reveal the true score for weight, it simply serves as an estimate of what the true score may be. Psychometrics 101! This is also true for all variables you may consider, not only weight, but virtually anything. Hence, "true weight" is virtually **unknowable**. However, if your measurement instrument is a good one, then you will get a good approximation to the true weight. Though measurement error may be reduced, it is never, at least on a theoretical level, eliminated. Hence, in this regard, "true weight" is a "hidden" variable. Now, surely launching **SpaceX** into space does not involve such measurement error, does it? The answer is yes it does! Good scientists and engineers use instruments that minimize measurement error, but it can never be avoided, at least not on a theoretical level, which means that even if you can avoid it in most practical situations, there will always be the possibility that a new measurement will contain such an error. In events like space shuttle launches, measurement error is extremely small, but it exists nonetheless. Hence, we must concur that **measurement error exists in all variables, from physical to social ones. In some variables it exists more than others.**

Thirdly, and related to the point just made above, it needs to be recognized that between an object and our seeking to know it is a **measurement procedure**. The measurement procedure is usually made up of some instrument or other process by which we seek to assign a value to the object, or at least our estimate of it. Again, this is true for whatever object we consider, even ones where you would never think such issues exist! For instance, the measurement of heart rate is even an example of this. The medical doctor wishes to know something called heart rate and uses an instrument of measurement to learn of this information. What is the patient's true heart rate? That value is subject to a **measurement process**, and the measurement process is subject to generating measurement error in the thing that is being measured. Hence, the stethoscope may provide one reading of heart rate, while another measurement process may provide another (at least in theory). In theory, we hope they agree, but they may not. The key point is between anything we seek to "know," there is a need to **measure** it. By way of analogy, you are using a measurement process to read the pages of this book. Your eyes, combined with the rest of your sensory organization are capturing a representation of what is on the page. How accurately is it measuring? You may think even if your eyes are "20/20" you are capturing the true value, but you would be wrong! You are capturing the value afforded by your current instrument of measurement. We must allow for the possibility that, in the future, what is 20/20 measurement today becomes 18/20 measurement tomorrow. What you are seeing on the page is your current "best shot" at capturing what is there. Hence, in line with our discussion, what is on the page is a **latent** object you are trying to detect.

The analogy need not stop here. We can be even more general. The world is filled with other constructs or **processes** we wish to capture. Consider the tennis great Rafael Nadal, who is well known as the best clay court tennis player of all time. Regardless, every time he steps out onto the court, he is **learning how to play clay court tennis**. In our language, he is attempting to accurately measure a process, an ideal that is currently unknown to him, similar in some ways to your bathroom scale attempting to measure something unknowable. The key idea is to realize and understand that virtually all processes, variables, and otherwise "things" are unknowable in this regard, and in this sense, are latent, and not "observable" as they may at first glance appear.

We can safely conclude then that **all variables are latent, but some are more easily and precisely measured than others.** That is, in any measurement or process, we are seeking to know the true variable or "thing" under inspection. In some cases, such as weight, that approximation one may argue is quite close and requires a relatively simple measurement instrument. In other cases, such as IQ, the measurement is likely to be more complex and may require several other variables in order to "**triangulate**" on the variable that is being sought. At least, that is the argument. Of course, what defines correct measurement is rather arbitrary and not necessarily related to measurement error. That is, there is nothing preventing us from designating IQ as simply the number of finger taps on a desk one can accomplish in a one-minute time interval. Foolish and simple as such a method would be to capture or assess such a complex construct, it remains that how we choose to measure something is not only a function of the degree of measurement error that may exist, but is first and foremost an issue of **validity**. Hence, **validity in measurement** is even more important. Is measuring beats-per-minute of a heart a valid measurement of heart-rate? As a society, we generally agree that it is. Of course, this example is extreme (we are at the point of forcing these philosophical notions to their limits), but in essence is no different in nature than our agreement (or disagreement) as to what intelligence is (or is not). Ideally, we first agree that we are measuring the correct thing and then we attempt to measure that thing with as little measurement error as possible using the **very best instruments currently available in line with how we operationalize the construct**. But we must also allow for the possibility that future instruments will reveal something different! Hence, in all of these cases, spanning the full spectrum, one may consider at an operational level for all variables to be **latent**, not observed. When we designate a variable as "observed" in factor analysis, we are simply considering it more readily available and as an indicator of a more latent or hidden process. Conclusion? Latent variables are those less readily available or observable than so-called "observed" variables. However, **all variables, every one, are in essence latent variables when subjected to proper scrutiny and philosophical (i.e. "intelligent" and "critical") unwrapping.**

11.4 So, Why is Factor Analysis Controversial? The Philosophical Pitfalls of Factor Analysis

Having briefly surveyed the factor analysis model, it may seem as though our initial statement of it being potentially controversial is unfounded. After all, it looks like a pretty innocent model. So, what could be the problem? Why the controversy? The problem is that regardless of the solution obtained in factor analysis, **factor loadings are not uniquely defined**. That is, the loadings obtained in Λ when the factor model is estimated are not unique. The idea of "unique" in this capacity has different meanings, but for our purposes, it implies that the loadings obtained are not the only ones that could have been obtained for the factor analysis model to successfully regenerate the covariance or correlation matrix. For applied purposes, this has two major consequences for the user of factor analysis:

- **Estimated factor loadings for a particular factor are typically contingent upon how many other factors are estimated along with it.** That is, a given factor loading in an EFA solution is not a unique solution. The solution will depend on how many other factors are extracted along with that factor. This does not occur in PCA. In PCA, whether you extract one component or five, for example, the loading for the given component is **stable** and does not change.
- Given the above, the determination of the number of factors in data is a complex decision that many argue is a function just as much of "art" as it is of science. For example, if a factor analyst extracts a five-factor solution and finds evidence for two factors, that solution might change if the researcher re-analyzes the data extracting only two factors. Hence, the determination of how many factors "exist" in the data can be quite arbitrary in EFA. Again, this is not the case in PCA. In PCA, weights and loadings are stable across the number of components extracted or interpreted.
- A third, but not unrelated issue to the one mentioned just above, is that the loadings in EFA can be subjected to an **orthogonal rotation**, but thankfully still reproduce the covariance matrix. Hence, in this sense, we can say the loadings are unique, but only up to an orthogonal rotation. That is, a researcher can rotate the solution in EFA and still regenerate the same covariance or correlation matrix. Again, this would appear to make the **determination of the weights seem quite arbitrary**.

These issues should not be easily dismissed by the researcher as he or she continues on with performing a factor analysis on his or her variables. Neither do they suggest that factor analysis should not be used as a tool to help make sense of one's data. What these caveats do suggest, however, is that researchers should interpret factor-analytic solutions with caution and realize that the solution they obtained is not the same as the rigorous "**discovery**" as some would make it seem. That is, the researcher did not "find" a given factor solution in the sense that there is a concrete **existence** of the given number of factors. Rather, using the factor-analytic model, and all the assumptions along with it, the researcher obtained what he or she believes is a feasible and useful solution to help make sense of the data and make progress in the scientific field. That is all that has been found, nothing more. Again, to use the analogy of an archaeological dig, "finding" factors is not the same as digging into the ground and discovering an ancient skeleton relic. It is not even close to that! Researchers who perform factor analysis and those who consume its results need to be aware of this. Factors are not "discovered" in the sense that a vaccine for COVID-19 is invented, for instance. What is "found" in factor analysis is much more transient and subject to several inputs and assumptions. Truth be told, the identification of the COVID-19 virus under a microscope also depends on instrumentation, inputs, and assumptions, but that becomes a bit more of an academic discussion and distinction than a real one. At some point, we have to work with the instrumentation we have and assume what we are seeing "exists." However, on a philosophical level (i.e. a "thinking about what we are doing" level, which is all "philosophical" means here), the "reality" of the virus under the microscope is subject to those same influences. Biologists usually do not have to confront such issues, but psychologists, if they are to do their work correctly and with care, regularly do (or at least, should). **Psychometrics** is a very exciting and challenging field. **The instrument of measurement is just as important as what is being measured.**

11.5 Exploratory Factor Analysis in Python

We now demonstrate an EFA on the same data on which we performed a PCA in the previous chapter. Recall the dataframe we built for the PCA. We rebuild it here:

```
import pandas as pd
import numpy as np

pca_data = np.array([[0, 5.90], [.90, 5.40], [1.80, 4.40], [2.60,
4.60], [3.30, 3.50], [4.40, 3.70], [5.20, 2.80], [6.10, 2.80],
[6.50, 2.40], [7.40, 1.50]])

pca_data
Out[57]:
array([[0. , 5.9],
       [0.9, 5.4],
       [1.8, 4.4],
       [2.6, 4.6],
       [3.3, 3.5],
```

```
[4.4, 3.7],
[5.2, 2.8],
[6.1, 2.8],
[6.5, 2.4],
[7.4, 1.5]])
```

Recall that in PCA of the previous chapter, we sought a transformation such that the total variance in the variables is preserved. In EFA, we do not seek such a transformation, but rather wish to **fit the factor model** discussed earlier to the data, and in the process **estimate loadings** and **specific variances**.

Before we proceed with the factor analysis, we can conduct **Bartlett's test of sphericity** to evaluate the null hypothesis that the correlation matrix is an **identity matrix**. If it is an identity matrix, then it makes little sense in performing the factor analysis, since it would indicate that we have zero pairwise correlations between variables (i.e. values of "0" on the off-diagonal positions of the matrix):

```
pip install factor_analyzer
import factor_analyzer
from factor_analyzer.factor_analyzer import calculate_bartlett_
sphericity
chi_square_value,p_value=calculate_bartlett_sphericity(pca_data)
chi_square_value, p_value

Out[69]: (23.012894934627802, 8.370237810650897e-07)
```

The value of 23.01 is an approximate chi-square, evaluated on a single degree of freedom. The *p*-value accompanying the test is extremely small, indicating that we can reject the null hypothesis of an identity matrix and infer the alternative hypothesis, that the correlation matrix is not an identity matrix, which gives us reason to proceed with the factor analysis. Even if the null hypothesis was not rejected here, it is certainly of no harm in trying the factor analysis anyhow. However, results may be dismal given there is little pairwise correlation among variables.

We can also obtain the **Kaiser-Meyer-Olkin (KMO) measure of sampling adequacy**, which is a measure of the likely utility for conducting a factor analysis in the sense of providing an estimate of the proportion of variance in variables that might be due to underlying factors (Denis, 2020). Values of 0.70 are preferred, though even lower values (such as the 0.5 we obtain below) should not necessarily discount conducting the factor analysis. Both Bartlett's and KMO should be used as guides only, not dogmatic criteria that should "halt" the factor analysis from taking place. We can compute the KMO in Python as follows:

```
from factor_analyzer.factor_analyzer import calculate_kmo
kmo_all,kmo_model=calculate_kmo(pca_data)

kmo_model
Out[71]: 0.5000000000000001
```

Now, let's run the EFA on the data, requesting a single factor solution. We use the default method called **minimum residual** (`minres`) (see Comrey, 1962, for details) and code `n_factors=1` to request a single factor solution:

```
pip install FactorAnalyzer
from factor_analyzer import FactorAnalyzer
fa = FactorAnalyzer(rotation=None, method = 'minres', n_factors=1)
fa.fit(pca_data)
Out[225]:
FactorAnalyzer(bounds=(0.005, 1), impute='median', is_corr_
               matrix=False,
               method='minres', n_factors=1, rotation=None,
               rotation_kwargs={},
               use_smc=True)
```

To get the factor loadings, we request `fa.loadings_`:

```
fa.loadings_
Out[226]:
array([[-0.98816761],
       [ 0.98816761]])
```

With two variables, the solution is not very meaningful, but if we were to interpret the above loadings at all, we could say that each variable loads very high onto the factor. This is hardly surprising, since the Pearson correlation coefficient between the variables is very high as well, equal to –0.976. The **communalities** for each variable are shown. Each is equal to the squared loadings for each variable. That is, the first communality is computed as (–0.98816761)(–0.98816761) while the second, (0.98816761)(0.98816761):

```
fa.get_communalities()
Out[227]: array([0.97647522, 0.97647522])
```

As in our example on PCA, this factor analysis on two variables is rather contrived. You would rarely ever conduct a factor analysis on only two variables. Typically, factor analyses are conducted on data sets in which the number of variables is much larger. We ended the previous chapter with an example of PCA on the USA arrests data and we will do the same here with an EFA on the same data.

11.6 Exploratory Factor Analysis on USA Arrests Data

Recall the USA arrests data from the previous chapter. We print the first 10 cases using `data.head(10)`, where "10" indicates the top 10 cases:

```
data = pd.read_csv('usarrests.csv')
data.head(10)

Out[639]:
   Unnamed: 0  Murder  Assault  UrbanPop  Rape
0     Alabama    13.2      236        58  21.2
1      Alaska    10.0      263        48  44.5
2     Arizona     8.1      294        80  31.0
3    Arkansas     8.8      190        50  19.5
```

```
4    California     9.0       276         91  40.6
5      Colorado     7.9       204         78  38.7
6   Connecticut     3.3       110         77  11.1
7      Delaware     5.9       238         72  15.8
8       Florida    15.4       335         80  31.9
9       Georgia    17.4       211         60  25.8
```

Instead of a PCA as we did in the previous chapter, let's run a factor analysis on the variables murder, assault, urbanPop, and rape. To perform the factor analysis, we will use Python's **FactorAnalyzer** library:

```
pip install factor_analyzer
import pandas as pd
from factor_analyzer import FactorAnalyzer
```

As we did when we analyzed the PCA data, we first designate the variables on which we will subject to the factor analysis, as well as import `StandardScaler()`, which will allow us to standardize the data:

```
df = pd.DataFrame(data, columns=['Murder', 'Assault', 'UrbanPop', 'Rape'])
```

We see here that we have designated our dataframe having variables murder, assault, urbanPop, and rape. Let's now standardize our dataframe:

```
from sklearn.preprocessing import StandardScaler
scaler = StandardScaler()
scaler.fit(df)
```

```
Out[649]: StandardScaler(copy=True, with_mean=True, with_std=True)
```

We create the standardized data using `scaler.transform()`:

```
scaled_data = scaler.transform(df)
scaled_data


Out[650]:
array([[ 1.25517927,  0.79078716, -0.52619514, -0.00345116],
       [ 0.51301858,  1.11805959, -1.22406668,  2.50942392],
       [ 0.07236067,  1.49381682,  1.00912225,  1.05346626],
       [ 0.23470832,  0.23321191, -1.08449238, -0.18679398],
       [ 0.28109336,  1.2756352,   1.77678094,  2.08881393],
       [ 0.02597562,  0.40290872,  0.86954794,  1.88390137],
       [-1.04088037, -0.73648418,  0.79976079, -1.09272319],
       [-0.43787481,  0.81502956,  0.45082502, -0.58583422],
```

We first try the factor analysis without rotation (by specifying `rotation = None`) and using `fa.fit()` to run the factor analysis. For demonstration, we will attempt to extract four factors:

```
fa = FactorAnalyzer()
fa.set_params(n_factors=4, rotation=None)
fa.fit(scaled_data)
```

```
Out[654]:
FactorAnalyzer(bounds=(0.005, 1), impute='median', is_corr_
               matrix=False,
               method='minres', n_factors=4, rotation=None,
               rotation_kwargs={},
               use_smc=True)
```

We request the loadings via `fa.loadings_`:

```
fa.loadings_
Out[406]:
array([[0.84370394, -0.37474146, -0.07321271, 0.],
       [0.91937036, -0.10367227,  0.17283121, 0.],
       [0.33167926,  0.54884826,  0.06269644, 0.],
       [0.78479779,  0.29235871, -0.15025673, 0.]])
```

Now, recall that we requested the factor analysis to run on four variables, and hence, by default, it was only able to generate a maximum of three factors. To the contrary, recall that the equivalent PCA was, by definition, able to generate four components. Again, this is one big difference between PCA and EFA. In EFA, **commonality** among variables is a priority, which means for most versions of factor analysis, the technique will typically not generate as many factors as there are variables subjected to it (see Johnson and Wichern, 2007, p. 488, for an exception, which is a version of factor analysis basing itself directly on principal components). Each column in the above output is a presumed factor, and each number under it corresponds to the loading for the given variables of murder, assault, urbanpop, and rape. Hence, the first factor seems to be made up primarily of murder, assault, and rape, having loadings of 0.84, 0.91, and 0.78, respectively. The second and third factors are not defined very well having relatively small loadings (though an argument could be made for a second factor).

We can get the original (i.e., pre-EFA) eigenvalues from the correlation matrix, which will mirror those obtained in PCA:

```
ev, v = fa.get_eigenvalues()
ev

Out[375]: array([2.48024158, 0.98976515, 0.35656318, 0.17343009])
```

Next, we obtain the **communalities** for each variable across the factor solution:

```
fa.get_communalities()
Out[656]: array([0.85762759, 0.88586043, 0.41517638, 0.72395827])
```

There are four communalities because there are four observed variables (not the fact that we are extracting three factors). We now summarize the computation of the first communality, that of 0.85762759:

```
(0.84370394)**2 + (-0.37474146)**2 + (-0.07321271)**2
Out[661]: 0.8576276011199994
```

Note that the sum of squared loadings across the factor solution is equal to the computed communality. The specific variance is thus equal to 1 – 0.85762759:

```
1-0.85762759
Out[662]: 0.14237241
```

The remaining communalities are computed in an analogous fashion. Let's now try a **varimax rotation**, where again we will obtain a three-factor solution. The idea of rotation here will be to rotate the loadings such that factors remain orthogonal. Varimax will attempt to maximize the variance in loadings for a given factor, which translates to boosting high loadings even higher and minimizing smaller loadings. Since the original axes of the factor analysis are in truth arbitrary, rotation in factor analysis is allowable, and, in a strong sense, the new axes have just as much "truth" to them as the original axes. What will determine ultimate justification regarding whether the rotation was successful or useful is whether it helps to make better sense of the factor solution, which should be informed by researcher judgment. We perform the rotation:

```
fa_varimax = FactorAnalyzer(rotation='varimax')
fa_varimax.fit(scaled_data)

Out[429]:
FactorAnalyzer(bounds=(0.005, 1), impute='median', is_corr_
               matrix=False,
               method='minres', n_factors=3, rotation='varimax',
               rotation_kwargs={}, use_smc=True)

fa_varimax.loadings_

Out[666]:
array([[0.91516486, 0.02762848, 0.13905951],
       [0.88036791, 0.32020407,-0.09100617],
       [0.05909459, 0.64142576,-0.0160376],
       [0.56292624, 0.59517677, 0.22986287]])
```

We can see that the varimax rotation had the effect of more or less making urbanpop (the third variable) pretty much irrelevant to the first factor, decreasing the loading from 0.33 in the original solution to only 0.05 in the rotation. However, the overall structure for the first factor remains pretty much intact, in that it is still made up of murder, assault, and rape. For the second factor, urbanpop and rape are relatively dominant in size. In both solutions, no case can be made for a third factor, as in both solutions, unrotated and rotated, the loadings are quite small. Hence, depending on one's theoretical stance or what one hoped to see from the solution, a case can be made for either a one-factor or two-factor solution (though the second factor does not look nearly as convincing as the first).

The communalities follow. We can see that urbanpop is the least relevant variable to the factor solution having a communality of only 0.41, while the other variables' communalities are relatively strong. That is, urbanpop is not explained as well by the factor solution as the other variables subjected to the analysis.

```
fa_varimax.get_communalities()
Out[668]: array([0.85762759, 0.88586043, 0.41517638, 0.72395827])
```

Review Exercises

1. What is **exploratory factor analysis (EFA)**? Why might a researcher want to perform one?
2. Why is **EFA** not equivalent to **PCA**? Explain two reasons why the two approaches are not the same and are used for different purposes.
3. Review and discuss the **common factor analysis model**. How is the model similar yet different from that of the regression model?
4. What does the **error term** indicate in the common factor analysis model?
5. What does it mean to say **factor analysis** is nothing more than an attempt to reproduce a covariance (or correlation) matrix? Why is this definition important to understand? What does this definition emphasize that other definitions may not?
6. Distinguish between an **observed** vs. **latent** variable. How are they different? Similar?
7. Discuss the nature of a **construct** in science. What is a construct exactly?
8. How does **measurement error** figure so prominently in science? Why can it be said to exist in virtually all variables?
9. What does it mean to say **factor loadings** are not **unique**, and why is this a major technical as well as philosophical pitfall to factor analysis?
10. Consider the **Breast Cancer Wisconsin** data from the **Machine Learning Repository**. Recall from the previous chapter that you conducted a PCA on this data. Attempt an **EFA** of the same variables and summarize your findings. How do they compare to the PCA you ran? Perform the EFA with and then without **varimax rotation**.

12

Cluster Analysis

CHAPTER OBJECTIVES

- Understand the essentials of how cluster analysis works.
- Why cluster analysis cannot define the nature of a cluster.
- Appreciate why seeking a cluster solution may be an elusive pursuit and how there may not be any substantive clusters present in your data.
- Why the determination of the existence of clusters is as much of a scientific decision as it is statistical and can often be subjective.
- How to conceptually and methodologically relate and distinguish cluster analysis from analysis of variance and discriminant analysis.
- How cluster analysis is based on a variety of distance measures and why selecting a distance measure is an important step in cluster analysis.
- Distinguish between k-means and hierarchical cluster analysis.
- How to interpret a dendrogram when performing hierarchical clustering.

Cluster analysis is a multivariable technique that seeks to uncover or identify **natural groupings** of objects for which on each object a number of measurements were taken (Everitt et al., 2001). In the machine learning and statistical learning fields, cluster analysis is usually housed in the domain of **unsupervised learning techniques**. "**Unsupervised**" here generally means that the user wishes to explore the data (or "allow the data to speak for itself") without any preconceived notions regarding the number of clusters one may find. As we will see, however, techniques such as **k-means** cluster analysis often require the user to specify in advance the number of clusters he or she seeks to uncover, and hence, in a sense, the technique is often still **supervised** to some extent, especially when used in this way. As was the case for principal components analysis, however, it is the spirit of the word that matters. That is, if you are doing cluster analysis, you are in a strong sense trying to see what the data have to say without imposing too much of your own a priori theoretical constraints. If you are doing **discriminant analysis** or analysis of variance (ANOVA) or multivariate analysis of variance (MANOVA), on the other hand, you presumably already have a grouping structure in mind. ANOVA, MANOVA, discriminant analysis, and similar techniques? **Supervised.** Principal components, factor analysis, and cluster analysis?

Applied Univariate, Bivariate, and Multivariate Statistics Using Python: A Beginner's Guide to Advanced Data Analysis, First Edition. Daniel J. Denis.

Unsupervised. We could write a book critically dissecting these labels and evaluate why they can be problematic and incomplete, but let's get on instead with the purpose of the chapter (i.e. the point is to use them as convenient guidelines only, not rigid ones). The distinction, in a sense, is similar to that between **exploratory** vs. **confirmatory** factor analysis. No technique is truly "exploratory." The use of it may be exploratory or unsupervised, but the technique itself is, again, built on abstract mathematics. Always remember that the mathematics are "innocent." It is the use of the technique that typically drives its substantive purpose and whether it is exploratory or confirmatory, unsupervised or supervised.

The variables in cluster analysis are typically **continuous** in nature, though work has also been done in developing techniques to cluster non-continuous variables such as dichotomous or polytomous ones. In this chapter, we consider only the case of continuous variables. But what is a cluster? As we will see, and strange as it may seem, cluster analysis is unable to answer this question! That's correct, cluster analysis is unable to answer the question regarding what is a cluster (Hennig, 2015; Izenman, 2008). However, it does make an attempt to define the nature of a cluster by operationalizing it based on a variety (i.e. not just **one**) of **distance** measures. That is, the very definition of a cluster depends, from a technical sense at least, on how we define it **mathematically**. Otherwise, it has no precise meaning and is simply a word. Recall our discussion earlier in the book about what "prediction" means in regression. Unless you define it precisely mathematically, it is simply a word, and if the researcher using it is not familiar with how it is being defined mathematically, then it becomes a nonsense statement. We will survey a few of these distance measures in this chapter as they are vital to understand how something called a "cluster" can even exist in the first place.

In the age of the COVID-19 pandemic, obtaining an intuitive understanding of cluster analysis is not difficult. Indeed, the study of COVID-19 across the world has largely been a study of clusters in one way or another. One might say that the job of epidemiologists (those who study infectious diseases) over the course of the COVID-19 pandemic has been one of identifying **clusters of disease** and then hypothesizing what may be the commonality uniting one cluster vs. another. As an example, consider the COVID-19 map of Montana by county, as of October 2020 (Figure 12.1).

For these data, the clusters are predetermined and defined by county. The total number of cases (in terms of a shade) is given within each county. An epidemiologist looking at this map may ask why some counties have more cases in them than others. From this perspective, the counties can be considered as clusters. Perhaps density of population is a contributing variable or maybe popularity of mask usage is another, or maybe it is the number of individuals being tested that is the dominant variable, or the accessibility to testing services. The list goes on and on regarding what could be the potential reasons for the number of cases in some counties versus others. Note carefully that for these data, the variable of "county" is a **categorical variable**, akin to what we would have in ANOVA as an independent variable. We will return to this point later.

Now, to truly understand cluster analysis, imagine for a moment that we remove the county lines in the map. That is, for a moment pretend they do not exist. Erase them in your mind such that all you see are the different shades of density across the map. Under these circumstances, we might still ask a similar question as before, but this time, assume that as of yet we have no awareness of what might be the clustering variable. That is, we ask:

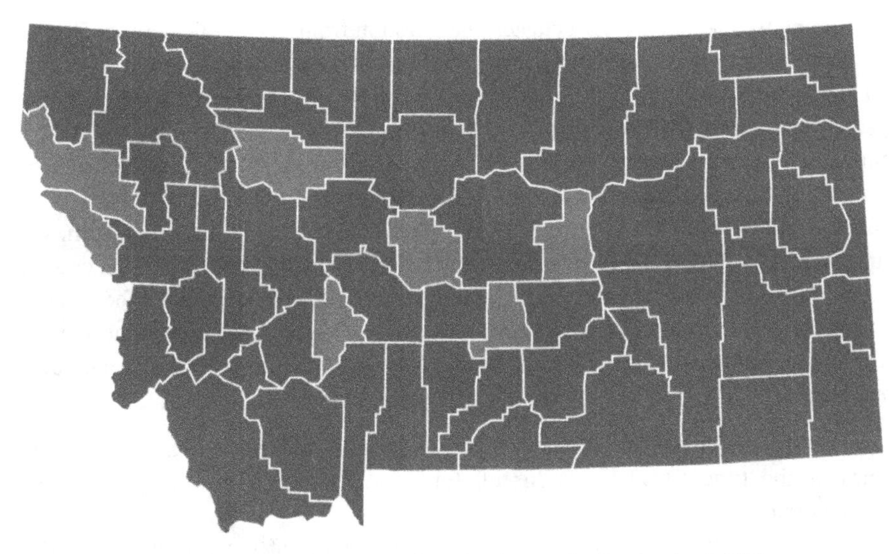

Figure 12.1 COVID-19 Map of Montana Counties during the Pandemic of 2020.

What is responsible for the presence of apparent clusters?

Why do some areas contain more cases than others? Now if we then analyzed the data further and gathered membership information on cases, we might begin to theorize **county membership** as a cluster membership variable. Notice how we have worked backwards here, where now we are pretending as though we never knew the cluster membership variable **a priori**. In the absence of the cluster variable, our job would be to identify or "uncover" (in a sense) the potential underlying variable that is somehow "responsible" for the grouping of COVID-19 cases. Our theory, of course, might be county-level membership.

However, we are just getting started. Even if we correctly identified county as a **potential** cluster variable, it in no way suggests that it is the **only** cluster variable or even the most important. For example, consider the county of **Gallatin county**, circled in the map:

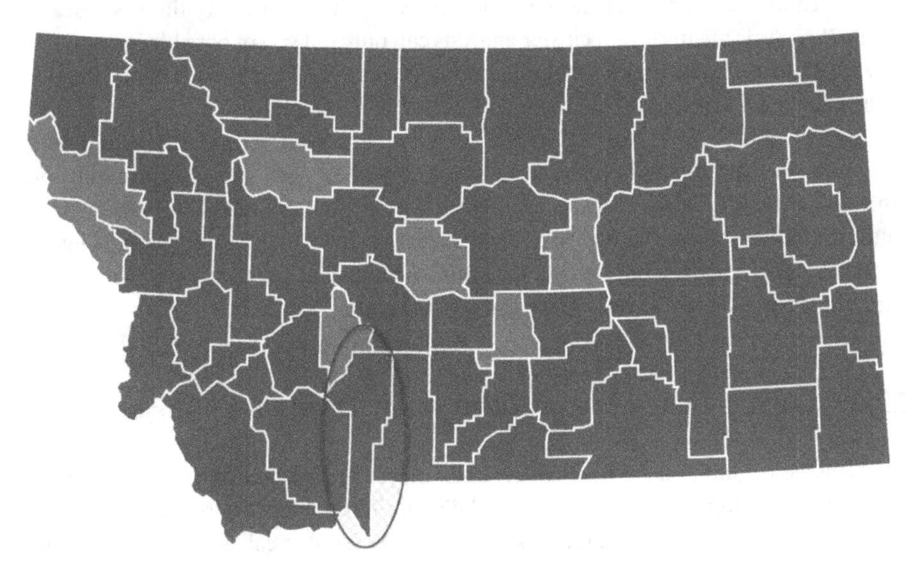

Assuming we identify this as a cluster, the question from a scientific point of view is why does this county contain such density? Individuals share the same county, yes, but that does not **explain** why more cases are in that county versus others. We merely know by some measure **quantifying proximity** that these cases are more "similar" within the county than between counties. Again, notice the parallel to ANOVA here, as the idea of "within" vs. "between" variation is central to cluster analysis as it is to ANOVA.

Back to our explanation as to why we might be observing a cluster for Gallatin. We might hypothesize that the number of COVID-19 cases is high because of the ease of transmission of the disease. Hence, imagine if the epidemiologist stopped there. That is, suppose we stopped on this explanation. Attractive as the cluster explanation may be, it may still not be getting at the root or causative factor behind the cluster membership. Here are some potential alternative explanations:

- The county may have low population density, but has had numerous close gatherings at the time COVID was in full force, and hence it encouraged ease of transmission.
- The county may feature individuals who travelled out of the county frequently, potentially contracting the disease from neighboring counties and bringing it "home" to their county.
- The county may be one that collectively resisted wearing face masks or coverings during the outbreak, contributing to ease of transmission.
- The county may consist mainly of elderly individuals, which evidence presumably showed at the time of this writing more easily contracted the disease compared to younger persons.
- The county may feature individuals with pre-existing conditions, possibly making the contraction of the disease easier.

Again, the theories the epidemiologist may advance are endless. The point of this discussion here is to emphasize that even if clusters are found and identified, learning **why** the cluster membership exists on a substantive level can be tremendously difficult, and, in some cases, virtually impossible. As we will see, cluster analysis itself is not able to answer the most important question of cluster analysis. The best cluster analysis can do is identify clusters **numerically**. It cannot identify them **substantively**. What this means is that cluster analysis can only tell you of possible groups that may exist in your data, but cannot on its own tell you why these groups might exist or their underlying **etiology**. Only the scientist can do that. Sounds limiting, doesn't it? For the case of COVID-19, we are likely to arrive at an answer. However, for unknown data with no hypothesized cluster structure, the pursuit may prove elusive and the ethical scientist should be aware of this possibility upfront and report the absence of a solution if that is what the algorithm, combined with the science, suggests. Cluster analysis is an elegant mathematical procedure, but substantively is often a "shot in the dark." Sometimes it works out, other times, it does not.

12.1 Cluster Analysis vs. ANOVA vs. Discriminant Analysis

As discussed throughout the book, many if not most statistical methods have technical similarities. Having surveyed ANOVA and discriminant analysis already, we are in a good position to compare them, at least on a methodological level, to the cluster analysis framework of the current chapter. Notice that in cluster analysis, **we are unaware of group membership**. This is the exact reason why we are performing the cluster analysis. In contrast, in ANOVA, MANOVA, and discriminant analysis, group membership has already been defined on one or more variables. Referring to the case of COVID once more, suppose that based on incoming data we began to hypothesize that COVID rates were higher among men than women. This becomes our grouping variable. Framed in the context of cluster analysis, we have two clusters of observations, one of males and one of females. To visualize this, consider the following scatter of cases:

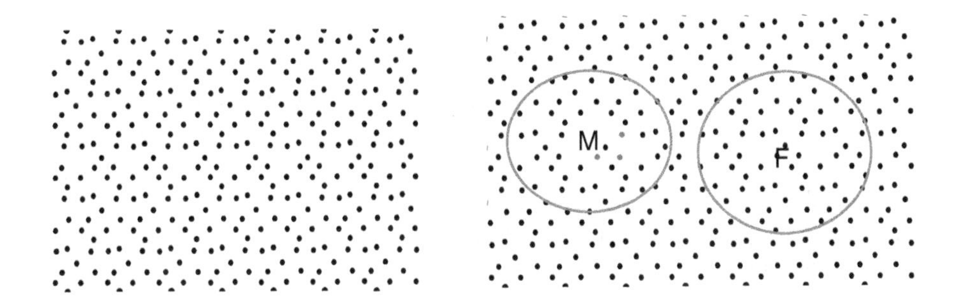

In this example, once we identify the males vs. females (M vs. F), two clusters appear to emerge (right). Hence, it may suggest that the rate of acquiring COVID-19 is different for males than it is for females. Now, had we hypothesized this might be the case beforehand and wanted to test our theory, then gender (male vs. female) would have made a suitable **independent variable**. Notice that perceived from this vantage point, cluster analysis is a more "primitive" technique from a scientific point of view (but only from a scientific point of view, not a technical point of view) when compared to ANOVA. In ANOVA, the researcher presumably has a theory as to why groups may be different, which is expressed as a factor with levels on the independent variable. How about a discriminant analysis? Since a discriminant analysis is essentially an ANOVA or MANOVA flipped around, group membership is now a response variable because we are attempting to predict the classification into one group vs. the other. Now, this is not meant to imply the procedures all have an identical technical base or are derived in exactly the same way, but it is to suggest they have similar methodological underpinnings. If you perceive these techniques from the vantage point of which to select for a given purpose, these distinctions and similarities can be helpful. ANOVA or discriminant analysis are not used as much in the spirit of "shooting in the dark" as is cluster analysis.

12.2 How Cluster Analysis Defines "Proximity"

We have said that cluster analysis seeks to generate groupings for which cases are most **similar**. How does it go about this? It is obvious that to liken observations by a measure of **proximity**, this measure must be well-defined and precise. That is, we have to mathematically define it. It turns out there are several ways of measuring proximity. These measures are often presented as measures of **distance**, or how closely or separated two or more objects are from one another. However, whichever you choose to call them makes little difference. Though there are many numerous measures, we consider only a few of them here. Our first is by far the most popular, that of Euclidean distance.

12.2.1 Euclidean Distance

The **Euclidean distance** between two p-dimensional objects, $\mathbf{x} = \lfloor x_1, x_2, \ldots, x_p \rfloor$ and $\mathbf{y} = \lfloor y_1, y_2, \ldots, y_p \rfloor$ is given by

$$\begin{aligned} d(\mathbf{x}, \mathbf{y}) &= \sqrt{(x_1 - y_1)^2 + (x_2 - y_2)^2 + \cdots + (x_p - y_p)^2} \\ &= \sqrt{(\mathbf{x} - \mathbf{y})'(\mathbf{x} - \mathbf{y})} \end{aligned}$$

where $d(\mathbf{x}, \mathbf{y})$ is the distance between the two vectors, $\mathbf{x}$ and $\mathbf{y}$. When "unpacked," this translates to what we see on the right-hand side of the equation, the corresponding distances between observations $(x_1 - y_1)^2, (x_2 - y_2)^2, \ldots, (x_p - y_p)^2$. We can easily illustrate Euclidean distance on two vectors $\mathbf{y}$ and $\mathbf{x}$ with a simple example. Consider two vectors, $\mathbf{x} = [2, 4]$ and $\mathbf{y} = [4, 7]$. The Euclidean distance between the two vectors is computed as follows:

$$\begin{aligned} d(\mathbf{x}, \mathbf{y}) &= \sqrt{(x_1 - y_1)^2 + (x_2 - y_2)^2 + \cdots + (x_p - y_p)^2} \\ &= \sqrt{(2 - 4)^2 + (4 - 7)^2} \\ &= 3.61 \end{aligned}$$

Euclidean distances, or their squares (i.e. squared Euclidean distances) are quite popular in cluster analysis. However, other distances can be defined as well. For example, the **Minkowski metric** is given by

$$d(\mathbf{x}, \mathbf{y}) = \left[\sum_{i=1}^{p} | x_i - y_i |^m \right]^{1/m}$$

We can see that m is a variable that operates in the exponent. When $m = 2$, the Minkowski metric reduces to a simple Euclidean distance, as can be easily seen:

$$\begin{aligned} d(\mathbf{x}, \mathbf{y}) &= \left[\sum_{i=1}^{p} | x_i - y_i |^m \right]^{1/m} \\ &= \left[\sum_{i=1}^{p} | x_i - y_i |^2 \right]^{1/2} \\ &= \sqrt{(x_1 - y_1)^2 + (x_2 - y_2)^2 + \cdots + (x_p - y_p)^2} \end{aligned}$$

A third measure is the **city-block distance**, also known as the **Manhattan distance**, defined by

$$d(\mathbf{x}, \mathbf{y}) = \sum_{i=1}^{p} | x_i - y_i |$$

In the above, $| x_i - y_i |$ is the **absolute distance** between observations. As mentioned, the choice of which distance to use is often subjective and different distance choices may yield different cluster solutions. Often, though not always, if a cluster solution is strong in a data set, then a variety of distance measure selections should yield more or less a consistency in the cluster discovery. There is absolutely nothing wrong with a researcher trying out different distance metrics to learn what cluster solutions they provide. Such is not "cheating," it is simply operationalizing the measure of proximity differently to see what comes about. This is the exploratory nature of cluster analysis, and trying out different distance measures is a good idea. There is no type I error rate to worry about by trying different distances in the search and potential establishment of a cluster solution. But if you find evidence for a cluster solution using some distances but not others, then that may require some explanation or discussion, which may not prove to be easy. It may require some further digging into your data to figure out why this difference is occurring. Is one distance measure emphasizing absolute distances whereas the other is not? Are there outliers in your solution that are disproportionately affecting one measure but not the other? All of these are good questions when exploring solutions with cluster analysis. It is an exercise in **dating mining**, so mine your data! In this sense, all cluster solutions are rather **subjective,** even if they are based on mathematically rigorous distance measures. "Are there clusters in my data?" "Well, it depends ..." is a good place to start, even if you do find evidence for clusters. As we said at the start of this chapter, the existence of clusters is only as good as they are defined mathematically. Otherwise, from the point of view of cluster analysis, they have no existence.

12.3 K-Means Clustering Algorithm

As mentioned, there are a variety of clustering methods. In the **k-means clustering algorithm**, a researcher first designates the number of clusters he or she is seeking to find in the data (which again, does not sound very "unsupervised"). That is, from the start, the researcher has some idea of the number he or she would like to find. Remarkably, k-means usually guarantees the user will find a cluster solution. However, the specifying of how many clusters to find is entirely subjective and must be merged with scientific considerations for the solution to usually make any sense. If the researcher believes there are two clusters underlying the data, for instance, then this is the number that is set for the algorithm to find. Similar to factor analysis, the user is free to specify the number of clusters analogous to searching for a specific number of factors. As with virtually all statistical methods, **you have to start somewhere.**

While k-means cluster analysis may seem quite modern, its origins are actually quite dated. James MacQueen (1967) best described the technique in an early paper:

> The main purpose of this paper is to describe a process for partitioning an N-dimensional population into k sets on the basis of a sample. The process, which is called "k-means", appears to give partitions which are reasonably efficient in the sense of within-class variance ... State [sic] informally, the k-means procedure consists of simply starting with k groups each of which consists of a single random point, and thereafter adding each new point to the group whose mean the

> new point is nearest. After a point is added to a group, the mean of that group is adjusted in order to take account of the new point. Thus at each stage the *k*-means are, in fact, the means of the groups they represent (hence the term *k*-means).
> (pp. 281, 283)

To paraphrase MacQueen, k-means cluster analysis joins or "fuses" observations that are most similar, and does so iteratively until a **minimization criterion** has been achieved. There can be no overlap in k-means, which means that if a case is in one cluster, it cannot be in another. The cluster means that are updated at each step along the way are typically called **cluster centroids**. The idea of imposing a minimization criterion on the cluster solution is an example of other techniques studied in this book where a minimization or maximization criterion is imposed. Recall again that in least-squares regression, the objective was to fit a line such that the sum of squared errors is kept to a **minimum** value. Likewise, the objective in discriminant analysis was to find a function that **maximizes** group differences on the response. The goal of principal components analysis was to derive linear combinations of observed variables with **maximum** variance. Hence, we see again that so much of statistical modeling is a problem of optimization, that of **maximizing** or **minimizing** some quantity subject to constraints. It is rarely, despite what the "data deluge" zeitgeist implies, simply "allowing the data to speak for themselves." Even if very rudimentary, a mathematical structure of some type, even if minimal, is typically still imposed on virtually any data that is subjected to statistical interpretation. "What do the data say?" The correct response is, "Well, the data say this under these conditions, operations, and constraints, and possibly something else under others." **That is the correct answer! It is usually the only answer!** The correct response is never "The data say what they say, period." This point cannot be emphasized enough. If you do not understand this point, you will likely make faulty scientific conclusions when interpreting data. You will be failing to appreciate and recognize the proverbial microscope through which you are observing your findings. You will be able to conduct this or that analysis as you exhibit extraordinary software skills, but you will fail to appreciate that, again by way of analogy, **the microscope is every bit as important as what is being visualized under it**. Be sure you understand this point! When a researcher says, "I found evidence for clusters," that should be the beginning of your questions into the tool used, not the end.

12.4 To Standardize or Not?

As we will see shortly when we demonstrate cluster analysis in Python, the user has the option to leave the data as raw or to **standardize** the data, meaning to transform it to ***z*-scores** with a mean of 0 and variance of 1 for each variable. The decision to standardize or not is not an automatic one and the user or interpreter of cluster analysis should not assume that the decision is inconsequential for the cluster analysis. So how should you decide? Typically, if the variables are not measured on the same units and have wildly different variances, standardizing is probably a good option. However, if they are on the same units with similar variances, then you may opt not to. The best option, of course, is to compare solutions with and then without standardization; then you can see for yourself how your analysis behaves under each condition. Generally, however, the above guideline (regarding variances) can be used as a decision as to whether or not to standardize.

12.5 Cluster Analysis in Python

We demonstrate a simple k-means cluster analysis in Python. One advantage of clustering the iris data is that we are already well aware of one potential cluster solution, that of **species**. However, for our purposes, we act as though we do not know this cluster solution or that we are seeking out a different one altogether. If we were deliberately attempting to use iris features to predict species, we would be in the realm of a supervised learning technique such as discriminant analysis or logistic regression, or MANOVA. However, since we are going to "let the data speak for themselves" as it were, we are proceeding as if we do not know a good clustering variable. Let's first import some of the libraries and tools we will need:

```
from sklearn.cluster import KMeans
import pandas as pd
import matplotlib.pyplot as plt
import numpy as np
```

Next, we import the **iris** data from sklearn datasets, after first importing the **seaborn** package:

```
import seaborn as sns
iris = sns.load_dataset("iris")
print(iris.head())

   sepal_length  sepal_width  petal_length  petal_width species
0           5.1          3.5           1.4          0.2  setosa
1           4.9          3.0           1.4          0.2  setosa
2           4.7          3.2           1.3          0.2  setosa
3           4.6          3.1           1.5          0.2  setosa
4           5.0          3.6           1.4          0.2  setosa
```

We only want the first four variables, however, so we select them from the dataframe (i.e. we do not want to include species in our cluster analysis, it is categorical, and does not belong):

```
df = iris[['sepal_length', 'sepal_width', 'petal_length', 'petal_
width']]
df.head(3)
Out[710]:
   sepal_length  sepal_width  petal_length  petal_width
0           5.1          3.5           1.4          0.2
1           4.9          3.0           1.4          0.2
2           4.7          3.2           1.3          0.2
```

Now that we have the variables we want, let's perform the k-means cluster analysis, requesting three clusters. The number of clusters is designated by `n_clusters=3` in this case. We then use `.fit()` to obtain the solution:

```
kmeans = KMeans(n_clusters=3)
kmeans.fit(df)
kmeans.predict(df)
```

```
Out[721]:
array([0, 0, 0, 0, 0, 0, 0, 0, 0, 0, 0, 0, 0, 0, 0, 0, 0, 0, 0, 0, 0, 0,
       0, 0, 0, 0, 0, 0, 0, 0, 0, 0, 0, 0, 0, 0, 0, 0, 0, 0, 0, 0, 0, 0,
       0, 0, 0, 0, 0, 0, 1, 1, 2, 1, 1, 1, 1, 1, 1, 1, 1, 1, 1, 1, 1, 1,
       1, 1, 1, 1, 1, 1, 1, 1, 1, 1, 1, 2, 1, 1, 1, 1, 1, 1, 1, 1, 1, 1,
       1, 1, 1, 1, 1, 1, 1, 1, 1, 1, 1, 1, 2, 1, 2, 2, 2, 2, 1, 2, 2, 2,
       2, 2, 2, 1, 1, 2, 2, 2, 2, 1, 2, 1, 2, 1, 2, 2, 1, 1, 2, 2, 2, 2,
       2, 1, 2, 2, 2, 2, 1, 2, 2, 2, 1, 2, 2, 2, 1, 2, 2, 1])
```

Using `kmeans.predict()` above allows us to obtain the predicted group membership for each case. For instance, the first case was predicted into the cluster = 0, the second case as well, and so on. If you count up the numbers from left to right above, you will see that the 51st case is classified into cluster = 1, and so on for the remaining cases. Let's now obtain the centroids of the cluster analysis. We can obtain this using `cluster_centers_`:

```
centroids = kmeans.cluster_centers_
centroids
```

```
Out[727]:
array([[5.006     , 3.428     , 1.462     , 0.246     ],
       [5.9016129 , 2.7483871 , 4.39354839, 1.43387097],
       [6.85      , 3.07368421, 5.74210526, 2.07105263]])
```

We now generate a plot showing the separation achieved across petal length and petal width (left) and then across sepal length and sepal width (right):

```
kpredict = kmeans.predict(df)
plt.scatter(iris['petal_length'], iris['petal_width'], c =
kpredict, cmap = 'cool')
Out[153]: <matplotlib.collections.PathCollection at 0x1db04358>
```

```
plt.scatter(iris['sepal_length'], iris['sepal_width'], c =
kpredict, cmap = 'cool')
Out[154]: <matplotlib.collections.PathCollection at 0x21b5eac8>
```

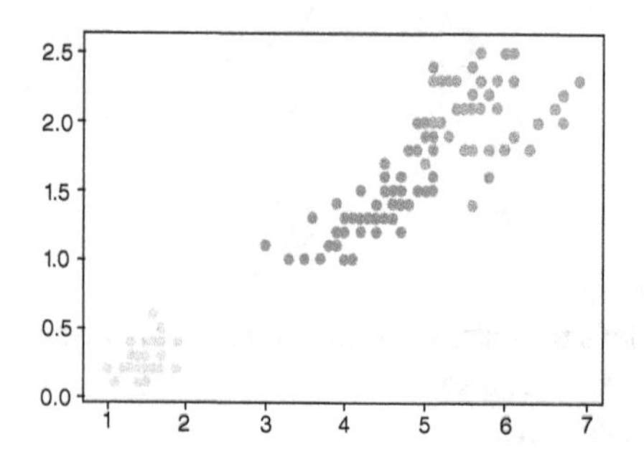

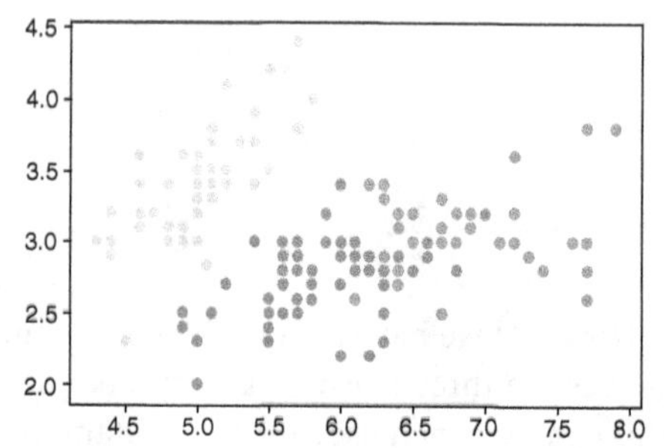

As we can see from the plots, there appears to be pretty good separation from two of the species from that of a third, especially across petal length and petal width. How do we know species is a suitable clustering variable? Simply because we have the variable already in our data file. That is, we already have group information on our cases, **we know they are "classifiable" by species**. This is why earlier we were able to look at the first 50 cases and realize that they all belong to the species setosa. Had we not had the species variable, or were naïve about what a suitable clustering variable could be for these data, we would have had to do more exploratory work to discover a suitable clustering variable or give up and acknowledge that the clustering solution "makes no sense" to us as of yet. Be sure this is clear, as it is **fundamental to the nature of clustering**. It is the most important point. The clustering algorithm did not "know" species was a suitable clustering variable since we did not arm the algorithm with this information. It simply performed a cluster analysis based on the k-means algorithm. It did so "unsupervised." It operated mathematically according to a blind algorithm. It is us, as users of cluster analysis, that "validated" the cluster solution against a variable we already knew existed in our data file, that of species. Had we not had the species variable, we would be scratching our heads wondering why those first 50 cases were classified into the first cluster, and so on for the remaining clusters. To obtain the identification case information for each cluster in the three-cluster solution, you can code the following (Larose and Larose, 2019), then compute some descriptive information on each cluster using `cluster1.describe()`, `cluster2.describe()`, etc.:

```
cluster = kmeans.labels_
cluster1 = iris.loc[cluster == 0]
cluster2 = iris.loc[cluster == 1]
cluster3 = iris.loc[cluster == 3]
```

When you try the above for the first cluster (too many numbers to list here), you will see they are all of the species "setosa." And when we compute `cluster1.describe()`, we note the means featured earlier when we obtained the centroids:

```
cluster1.describe()
Out[166]:
       sepal_length  sepal_width  petal_length  petal_width
count      50.00000    50.000000     50.000000    50.000000
mean        5.00600     3.428000      1.462000     0.246000
std         0.35249     0.379064      0.173664     0.105386
min         4.30000     2.300000      1.000000     0.100000
25%         4.80000     3.200000      1.400000     0.200000
50%         5.00000     3.400000      1.500000     0.200000
75%         5.20000     3.675000      1.575000     0.300000
max         5.80000     4.400000      1.900000     0.600000
```

This is why the current cluster analysis on the iris data is "easy" from a substantive or scientific point of view, in that we already have a good "theory" as to what might be "responsible" for the clusters, that of species. In practice, clustering is rarely that easy and you may have to sit for a while, or even years (in theory) with a solution until you find a scientific explanation for it. The variable "responsible" for the clustering you are witnessing may even, in theory, be "unknowable" as of yet, and may require other advancements in human history to actually account for it. Hence, the best clustering variable may actually be **latent** (analogous to a latent factor from the previous chapter), at least for now.

For instance, in the COVID-19 pandemic, why some people got the illness and others did not could theoretically remain unknown for a long time, perhaps even until a new gene is identified (at least in theory). Obtaining clusters is easy, those who have the illness vs. those who do not. But what differentiates the groups? That is the cluster solution you are seeking, or, in discriminant analysis terms, the response variable you are wanting to identify, or, in ANOVA terms (or *t*-test in this specific two-group situation), the levels of the independent variable you are seeking to operationalize. If you know in advance, or have a strong **theory** about what the clustering variable might be, you will have little reason to seek out a cluster solution. Instead, you will try a **discriminant analysis** or **M(ANOVA)** to evaluate your "group" or "cluster" theory. **Notice how unified, from a conceptual standpoint, all of these statistical methods really are**. That is, if you approach them from a scientific perspective, rather than a technical one, you gain a good appreciation for how these techniques are related to the theory-building interests of the scientist. This is precisely why techniques such as principal components and cluster analysis are often regarded as more "exploratory" or "unsupervised" when compared to the more "confirmatory" or "supervised" techniques of ANOVA, discriminant analysis, logistic regression, etc.

Hence, the central point here is the fact that a clustering algorithm that "found" clusters does not **necessarily** imply they exist apart from the abstract distance measures that created them. Mathematics is an abstract system and its results need not necessarily accord with our current theories or explanations of nature. This is a theme that has repeated itself throughout this book and hence is true of cluster analysis as well. Always seek to separate the mathematics from the material on which you are applying the mathematics to, at least at first. Then, as you would in putting your hands together and seeing how well you can overlay them exactly finger by finger, entertain how closely they might fit. If they fit well, you might have something of scientific value. If they do not, then you may be back to square one. Eliminating a proverbial square, however, is still progress (though sadly, often not publishable). We now survey the second primary approach to clustering, that of hierarchical or "agglomerative" clustering.

12.6 Hierarchical Clustering

In k-means clustering, we specified in advance the number of clusters we hypothesized from the data or were seeking to find. The k-means algorithm then worked iteratively to obtain a cluster solution. In a second general approach to clustering, that of **hierarchical clustering** (also often more generally going by the name of "**agglomerative**"), the researcher does not specify directly the number of clusters he or she is seeking to uncover from the data. In hierarchical clustering, the researcher allows the algorithm to parse things out and potentially uncover a cluster solution. In this sense, then, hierarchical clustering may be perceived as even more "exploratory" or "unsupervised" of a method in that it is truly exploring the data without any preconceived theory as to even the number of clusters – that is, at least no theory that the researcher is advancing overtly.

There are different methods of joining clusters in hierarchical clustering. These include **single linkage**, **complete linkage**, and **average linkage** methods. Single linkage (or "nearest-neighbor") features merging cases with the smallest distance, then merging remaining cases with the smallest distance from the newly created cluster. The algorithm proceeds in this way at each step of the hierarchy, making successive mergers along the way. Results can be depicted in what is known as a **dendrogram**, which is a tree-like structure that reveals a history of the mergers made at each step. An example of a simple dendrogram is shown:

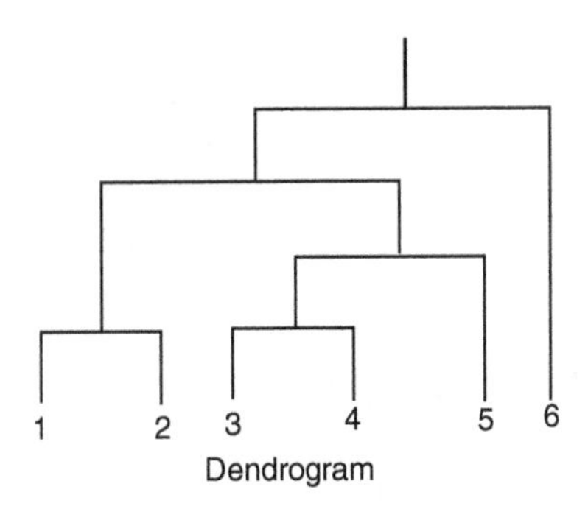

Dendrogram

The dendrogram should be read from the bottom-up. We see that cases 1 and 2 were merged, as well as 3 and 4. Cases 3 and 4, forming a new cluster, were merged to case 5. The final cluster formation was between case 6 and the cluster generated by the left-hand side of the dendrogram. While in simple linkage the smallest distance is a priority for a merger, in complete linkage the largest of the dissimilarities are calculated and the smallest of these are merged at each step. In average linkage, as the name suggests, it is the average of the dissimilarities at each step that is taken as a measure of distance. For more details on these distance measures, see Johnson and Wichern (2007).

Incidentally, as recommended throughout this book, if your data is not that large, it is well worth your time to study the dendrogram linkage by linkage, fusion by fusion to better understand why two points were fused. For instance, given that cases 1 and 2 were merged, the next question is, why? How about cracking open your data file to look at those two cases? Are they similar in some sense? If they were both dogs out of all the animals you are seeking to cluster, then you may of course have a hint. In this way, good statistical analysis is encouraging you to **get to know your data better**. That's the point! These types of merger-by-merger inspections can be very exciting on data that is well-calibrated and precise. "Why are those two animals clustering and why did they then cluster to a third? Wow, the third animal is a cat, while the other two were dogs. What's going on?" Asking questions like this is the objective of science, and it can get pretty exciting!

So, which linkage to use? Like many decisions in cluster analysis, the decision is ultimately **subjective.** Yes, rigorous applied statistics and science can come down to subjective decisions! Those who specialize in cluster analysis often recommend trying out a few different measures to learn if they might converge on the same or similar solution. However, the issue of whether clusters truly "exist" in your data must be coupled with scientific expertise, otherwise it is simply an exercise in mathematical partitioning. Does this sound at all familiar? It might, since we concluded something similar about ANOVA. Well, to be exact, R.A. Fisher, its founder, actually said it. Recall that he said ANOVA was a procedure of "arranging the arithmetic" and nothing more. Paraphrased, it meant that you can perform an ANOVA on any data and it will partition variability into that of **between** vs. **within**. Whether that partition makes any scientific sense for your data, however, is not a question that can be answered by ANOVA! It must be answered by your **science**. In applied statistics, as opposed to theoretical statistics, it must always be **science first, statistics second**. And so it is with cluster analysis as well. Again, cluster analysis, like many exploratory techniques, is nothing more than an attempt to partition the **signal from the noise**. And though statistical techniques are mathematically precise, their application is often quite messy. The morale is to not have unrealistic expectations of the tool. It usually cannot, on its own, answer the deeper questions you have as a scientist. Only good science can do that, by whatever statistical method is used in the pursuit. The "promise" is not with the tool. It is with the scientist.

12.7 Hierarchical Clustering in Python

We now run a hierarchical cluster analysis in Python. As when we ran k-means, we do not include the species variable. When we ran the k-means, recall that we defined a new data set without species by identifying the variables we wished to include:

```
df = iris[['sepal_length', 'sepal_width', 'petal_length',
'petal_width']]
```

We could have just as easily have dropped "species" from the iris data as follows, instead of calling on which variables we wanted to include. In the following, the data set is still iris, but it is now without species. We execute this via `iris.drop()`:

```
iris.drop(['species'], axis=1, inplace = True)
iris

Out[241]:
```

	sepal_length	sepal_width	petal_length	petal_width
0	5.1	3.5	1.4	0.2
1	4.9	3.0	1.4	0.2
2	4.7	3.2	1.3	0.2
3	4.6	3.1	1.5	0.2
4	5.0	3.6	1.4	0.2
5	5.4	3.9	1.7	0.4

Before conducting the cluster analysis, let's first obtain some **pairwise plots** of the sepal and petal features through **seaborn**, specifically `pairplot()`:

```
import seaborn as sns
sns.pairplot(iris)

Out[242]: <seaborn.axisgrid.PairGrid at 0x871e7b8>
```

We can see from these figures that some groups within the bivariate plots seem to differentiate themselves. For example, in the plot sepal width by petal length, notice that

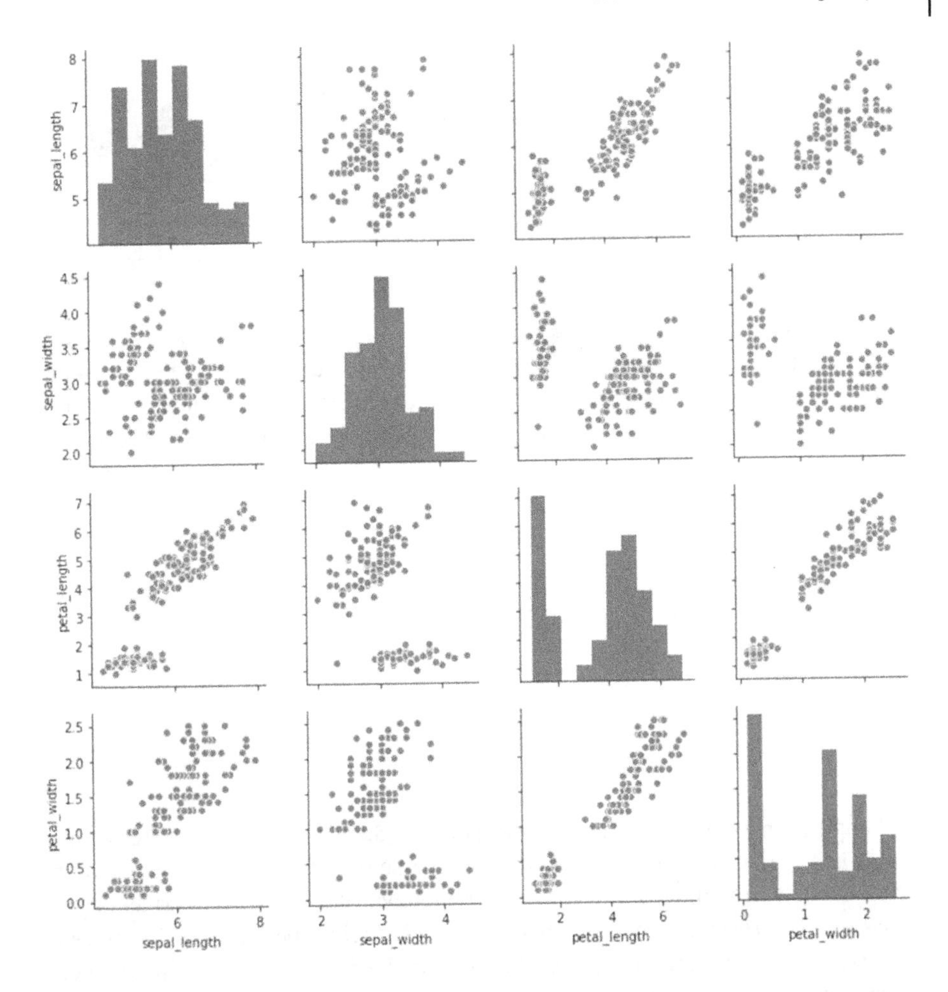

the scatter is not bivariate uniform, but rather seems to contain a degree of separation. We see similar separation in the bivariate plot corresponding to petal length and petal width. Although these plots do not indicate what the final cluster solution will look like in the end, they are nonetheless informative as an exploratory portrayal of the data before conducting the analysis.

Let's now conduct the cluster analysis requesting a total of three clusters, using **Euclidean** distance with single linkage:

```
from sklearn.cluster import AgglomerativeClustering
groups = AgglomerativeClustering(n_clusters=3,
affinity='euclidean', linkage='single')
groups.fit_predict(iris)
```

By using `fit_predict()` above, as in the case of k-means earlier, we obtain the predicted cluster grouping for each case below. Because we named our cluster object "groups," we could have also requested the cluster membership classification results using `print(groups.labels_)`. For example, the first case is predicted to go into cluster 1, the second into cluster 1, and so on. Notice that the first 50 cases (in bold) belong to the first cluster.

```
Out[502]:
array([1, 1, 1, 1, 1, 1, 1, 1, 1, 1, 1, 1, 1, 1, 1, 1, 1, 1, 1, 1, 1, 1,
       1, 1, 1, 1, 1, 1, 1, 1, 1, 1, 1, 1, 1, 1, 1, 1, 1, 1, 1, 1, 1, 1,
       1, 1, 1, 1, 1, 1, 0, 0, 0, 0, 0, 0, 0, 0, 0, 0, 0, 0, 0, 0, 0, 0,
       0, 0, 0, 0, 0, 0, 0, 0, 0, 0, 0, 0, 0, 0, 0, 0, 0, 0, 0, 0, 0, 0,
       0, 0, 0, 0, 0, 0, 0, 0, 0, 0, 0, 0, 0, 0, 0, 0, 0, 0, 0, 0, 0, 0,
       0, 0, 0, 0, 0, 0, 0, 2, 0, 0, 0, 0, 0, 0, 0, 0, 0, 0, 0, 0, 0, 2,
       0, 0, 0, 0, 0, 0, 0, 0, 0, 0, 0, 0, 0, 0, 0, 0, 0, 0], dtype=int64)
```

Now, what are those first 50 cases? That is, what might be "responsible" or "associable" (yet not necessarily "causal") for clustering those 50 cases? Well, let's look at our data set to see if we can find any clues. What were the first 50 cases? The first six cases are listed (of the original iris data set, not the one in which we removed the species variable):

```
iris
Out[266]:
```

	sepal_length	sepal_width	petal_length	petal_width	species
0	5.1	3.5	1.4	0.2	setosa
1	4.9	3.0	1.4	0.2	setosa
2	4.7	3.2	1.3	0.2	setosa
3	4.6	3.1	1.5	0.2	setosa
4	5.0	3.6	1.4	0.2	setosa
5	5.4	3.9	1.7	0.4	setosa

Though we have printed only the first six cases (due to keeping things reasonable spacewise), we see that these are all of the species setosa. If you look at the rest of the list of iris data in Python, you will see the remainder of the first 50 cases are also of species setosa. Hence, as we tentatively concluded when performing the k-means analysis, one interesting clustering idea is that the first cluster is that of setosa! Monumental! Now, does that necessarily define the cluster? No. "What?" I said no! The cluster need not be defined by setosa for certain. **It need not be defined by setosa at all. True cluster membership could, in theory, be something for which setosa is merely a correlate.** Ah, correlational research, and why your instructor keeps telling you to do experiments! However, for our purposes, identifying the cluster as setosa is simply too convenient, and we will go with that. The point is, however, that defining the nature of clusters is, again, a subjective decision, and what first may appear as the defining feature may be, as mentioned, merely a **correlate to an even more definitive feature.** Always allow for this possibility when seeking an explanation for cluster membership. Just because you found a clustering solution does not mean it is **"the"** solution. **All models are wrong, some are useful. All scientific conclusions are tentative, some are useful.**

Next, we generate a bivariate plot of petal length and petal width to visualize the separation. We do so through `plt.scatter()`:

```
plt.scatter(iris['petal_length'], iris['petal_width'], c = groups.
labels_, cmap='cool')

Out[262]: <matplotlib.collections.PathCollection at 0x20f05a90>
```

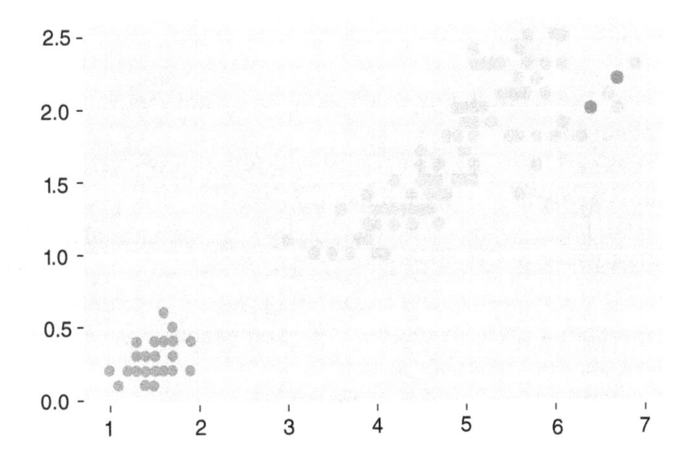

As shown, the group setosa is separating itself quite well from the other two species of flower. Again, in this case, we know there are three clusters, or at least we can presume there are because of the species variable. In other cases, however, the decision may be less clear. There are criteria that can be used to identify the number of clusters in data if results are ambiguous or you have virtually no theory guiding you, but these are beyond the scope of this book. These involve, in part, plotting the total error of the cluster solution by the number of clusters for varying values of k. For details of how this is done in Python, see Grus (2019). Typically, however, an applied scientist will determine the number of clusters based on substantive insight into the problem at hand, and hence in most applied cases, "blind statistical determination" of how many clusters "exist" in the data is much too "abstract" of a solution and ignores the material that is being clustered. And as we have discussed, the "existence" of clusters is laden with philosophical difficulty, but still, such statistical metrics can aid in helping one determine a tentative solution, especially if the data are quite large and you have absolutely no theory guiding you. As we have seen with most of the techniques surveyed in this book, such decisions have to usually be merged with the science you are conducting, and not independent from it. That is, if a three-cluster solution makes more sense scientifically than a four-cluster solution, it should be the three-cluster solution that is typically adopted, at least until new insights emerge as to why the four-cluster solution may be feasible. Otherwise, as we have alluded to throughout this book, you are simply a **servant to the algorithm**, and not merging it with the material you are actually studying. The algorithm, like the computer in the Boeing 737 max (recall those two airline crashes in 2018–2019), cannot guarantee for certain the plane is flying right. The pilot needs to be able to "override" the system when required. Good scientists use statistical modeling to help them learn more about what is being modeled. Though you must be very familiar with the algorithm, your job is not to "appease" it at the cost of your science. Too often, otherwise good scientists become immersed in tweaking their data to fit the model, rather than focusing on fitting the model to the data.

Review Exercises

1. Discuss the purpose of **cluster analysis**. What might motivate a researcher to want to conduct one?
2. How is **cluster analysis** similar in some regards to **factor analysis** or **principal components analysis**? How is it different?
3. In the chapter, it was hypothesized that county might be a useful metric for measuring clusters in Montana during the COVID-19 pandemic. While it may be convenient to use county as a potential clustering variable, why might it not be the best one to explain COVID-19 clusters? What other clustering variables might be better for the state of Montana?
4. Why is the question of the **causes of cluster membership** potentially elusive? In other words, why is it such a difficult question to answer?
5. Related to exercise 4, suppose scientists found a COVID-19 cluster in the United States. Why is determining the "cause" of that particular cluster such a difficult if not impossible question to answer? Why is establishing correlates to cluster membership insufficient?
6. Differentiate between **cluster analysis** and **discriminant analysis**. How are they different? Similar?
7. Why is determining case **"proximity"** a potentially elusive pursuit?
8. Why is cluster analysis not able to answer the question **"What is a cluster?"**
9. Briefly describe the **k-means** approach to clustering. That is, briefly unpack the algorithm. How does it work, in general?
10. Give a rationale for **standardizing** one's data before performing a cluster analysis. Is there anything wrong with comparing cluster solutions across unstandardized and standardized solutions?
11. Consider the **HCV data** from the **Machine Learning Repository**, featuring laboratory values of blood donors and Hepatitis C patients. Perform a **k-means** cluster analysis on variables ALB through to PROT. Try out solutions for two to five clusters, and comment on any findings.
12. Perform a **hierarchical** cluster analysis for the data in #11 and compare results.

References

Bakan, D. (1966). The test of significance in psychological research. *Psychological Bulletin*, 66, 423–437.

Baron, R. M. & Kenny, D. A. (1986). The moderator–mediator variable distinction in social psychological research: Conceptual, strategic, and statistical considerations. *Journal of Personality and Social Psychology*, 51, 1173–1182.

Bartle, R. G. & Sherbert, D. R. (2011). *Introduction to Real Analysis*. Hoboken, NJ: Wiley.

Berkson, J. (1938). Some difficulties of interpretation encountered in the application of the chi-square test. *Journal of the American Statistical Association*, 33, 526–536.

Bishop, C. M. (2006). *Pattern Recognition and Machine Learning*. New York: Springer.

Cohen, J. (1988). *Statistical Power Analysis for the Behavioral Sciences*. New York: Routledge.

Cohen, J. (1990). Things I have learned (so far). *American Psychologist*, 45, 1304–1312.

Cohen, J., Cohen, P., West, S. G., & Aiken, L. S. (2002). *Applied Multiple Regression/ Correlation Analysis for the Behavioral Sciences*. Mahwah, NJ: Lawrence Erlbaum Associates.

Comrey, A. L. (1962). The minimum residual method of factor analysis. *Psychological Reports*, 11, 15–18.

DeCarlo, L. T. (1997). On the meaning and use of kurtosis. *Psychological Methods*, 2, 292–307.

Denis, D. (2004). The modern hypothesis testing hybrid: R.A. Fisher's fading influence. *Journal de la Société Française de Statistique*, 145, 5–26.

Denis, D. & Docherty, K. (2007). Late nineteenth century Britain: A social, political, and methodological context for the rise of multivariate statistics. *Le Journal Electronique d'Histoire des Probabilités et de la Statistique*, 3, 1–41.

Denis, D. (2020). *Univariate, Bivariate, and Multivariate Statistics Using R*. Hoboken, NJ: Wiley.

Denis, D. (2021). *Applied Univariate, Bivariate, and Multivariate Statistics: Understanding Statistics for Social and Natural Scientists, with Applications in SPSS and R (2021)*. Hoboken, NJ: Wiley.

Draper, N. R. & Smith, H. (1998). *Applied Regression Analysis*. Hoboken, NJ: Wiley.

Applied Univariate, Bivariate, and Multivariate Statistics Using Python: A Beginner's Guide to Advanced Data Analysis, First Edition. Daniel J. Denis.

Everitt, B. S., Landau, S., & Leese, M. (2001). *Cluster Analysis*. New York: Oxford University Press.

Fiedler, K., Schott, M., & Meiser, T. (2011). What mediation analysis can (not) do. *Journal of Experimental Social Psychology*, 47(6), 1231–1236.

Fox, J. (2016). *Applied Regression Analysis & Generalized Linear Models*. New York: Sage.

Green, C. D. (2005). Was Babbage's analytical engine intended to be a mechanical model of the mind? *History of Psychology*, 8, 35–45.

Grus, J. (2019). *Data Science from Scratch: First Principles with Python*. New York: O'Reilly Media.

Guillaume, D. A. & Ravetti, L. (2018). Evaluation of chemical and physical changes in different commercial oils during heating. *Acta Scientific Nutritional Health*, 2, 2–11.

Guttag, J. V. (2013). *Introduction to Computation and Programming Using Python*. Cambridge, MA: MIT Press.

Hastie, T., Tibshirani, R., & Friedman, J. (2009). *The Elements of Statistical Learning: Data Mining, Inference, and Prediction*. New York: Springer.

Hays, W. L. (1994). *Statistics*. Fort Worth, TX: Harcourt College Publishers.

Hennig, C. (2015). What are the true clusters? *Pattern Recognition Letters*, 64, 53–62.

Howell, D. C. (2002). *Statistical Methods for Psychology*. Pacific Grove, CA: Duxbury Press.

Izenman, A. J. (2008). *Modern Multivariate Statistical Techniques: Regression, Classification, and Manifold Learning*. New York: Springer.

James, G., Witten, D., Hastie, T., & Tibshirani, R. (2013). *An Introduction to Statistical Learning with Applications in R*. New York: Springer.

Johnson, R. A. & Wichern, D. W. (2007). *Applied Multivariate Statistical Analysis*. Upper Saddle River, NJ: Pearson Prentice Hall.

Jolliffe, I. T. (2002). *Principal Component Analysis*. New York: Springer.

Kirk, R. E. (1995). *Experimental Design: Procedures for the Behavioral Sciences*. Pacific Grove, CA: Brooks/Cole Publishing Company.

Kirk, R. E. (2007). *Statistics: An Introduction*. New York: Cengage Learning.

Kirk, R. E. (2012). *Experimental Design: Procedures for the Behavioral Sciences*. Pacific Grove, CA: Brooks/Cole Publishing Company.

Larose, C. D. & Larose, D. T. (2019). *Data Science: Using Python and R*. Hoboken, NJ: Wiley.

MacKinnon, D. P. (2008). *Introduction to Statistical Mediation Analysis*. New York: Lawrence Erlbaum Associates.

MacQueen, J. (1967). Some methods for classification and analysis of multivariate observations. In: L. LeCam & J. Neyman (Eds.), *Proceedings of the Fifth Berkeley Symposium on Mathematical Statistics and Probability*, Vol. 1. Berkeley, CA: University of California Press, pp. 281–297.

Mair, P. (2018). *Modern Psychometrics with R*. New York: Springer.

McCullagh, P. & Nelder, J. A. (1990). *Generalized Linear Models*. New York: Chapman & Hall.

Montgomery, D. C. (2005). *Design and Analysis of Experiments*. Hoboken, NJ: Wiley.

Mulaik, S. A. (2009). *The Foundations of Factor Analysis*. New York: McGraw-Hill.

Olson, C. L. (1976). On choosing a test statistic in multivariate analysis of variance. *Psychological Bulletin*, 83, 579–586.

Rencher, A. C. & Christensen, W. F. (2012). *Methods of Multivariate Analysis*. Hoboken, NJ: Wiley.

Savage, L. J. (1972). *The Foundations of Statistics*. New York: Dover Publications.

Scheffé, H. (1999). *The Analysis of Variance*. Hoboken, NJ: Wiley.

Siegel, S. & Castellan, J. (1988). *Nonparametric Statistics for the Behavioral Sciences*. New York: McGraw-Hill.

Spearman, C. (1904). The proof and measurement of association between two things. *The American Journal of Psychology*, 15, 72–101.

Stevens, S. S. (1946). On the theory of scales of measurement. *Science*, 103, 677–680.

Stigler, S. M. (1986). *The History of Statistics: The Measurement of Uncertainty before 1900*. London: Belknap Press.

Tabachnick, B. G. & Fidell, L. S. (2007). *Using Multivariate Statistics*. New York: Pearson Education.

Tatsuoka, M. M. (1971). *Multivariate Analysis: Techniques for Educational and Psychological Research*. Hoboken, NJ: Wiley.

Tufte, E. R. (2011). *The Visual Display of Quantitative Information*. New York: Graphics Press.

VanderPlas, J. (2017). *Python Data Science Handbook: Essential Tools for Working with Data*. New York: O'Reilly Media.

Wickham, H. & Grolemund, G. (2017). *R for Data Science*. Boston, MA: O'Reilly Media.

Index

Applied Univariate, Bivariate, and Multivariate Statistics Using Python: A Beginner's Guide to Advanced Data Analysis, First Edition. Daniel J. Denis.

Printed in the USA
CPSIA information can be obtained
at www.ICGtesting.com
JSHW050623221223
53634JS00004B/242